Optical Aspects
of Oceanography

Optical Aspects of Oceanography

Edited by

N. G. JERLOV

Institute of Physical Oceanography,
University of Copenhagen, Denmark

AND

E. STEEMANN NIELSEN

Freshwater Biological Laboratory,
University of Copenhagen, Denmark

ACADEMIC PRESS
LONDON AND NEW YORK
A Subsidiary of Harcourt Brace Jovanovich, Publishers

ACADEMIC PRESS INC. (LONDON) LTD.
24/28 Oval Road,
London NW1

United States Edition published by
ACADEMIC PRESS INC.
111 Fifth Avenue
New York, New York 10003

Library of Congress Catalog Card Number: 73–7038
ISBN: 0–12–384950–0

Printed in Great Britain by
THE ABERDEEN UNIVERSITY PRESS LIMITED
ABERDEEN, SCOTLAND

List of Contributors

ROSEWELL W. AUSTIN, *Scripps Institution of Oceanography, University of California, San Diego, La Jolla, California 92037, U.S.A.*

BRIAN P. BODEN, *Scripps Institution of Oceanography, University of California, San Diego, La Jolla, California 92037, U.S.A.*

GEORGE L. CLARKE, *Woods Hole Oceanographic Institution, Woods Hole, Massachusetts 12543, U.S.A.*

C. S. COX, *Scripps Institution of Oceanography, University of California, San Diego, La Jolla, California 92037, U.S.A.*

SEIBERT Q. DUNTLEY, *Visibility Laboratory, Scripps Institution of Oceanography, University of California, San Diego, La Jolla, California 92037, U.S.A.*

EGBERT KLAAS DUURSMA, *International Laboratory of Marine Radioactivity, IAEA Musée Océanographique, Principality of Monaco.*

GIFFORD C. EWING, *Woods Hole Oceanographic Institution, Woods Hole, Massachusetts 12543, U.S.A.*

PER HALLDAL, *Botanical Laboratory, University of Oslo, Blindern, Oslo, Norway.*

ALEXANDRE IVANOFF, *Laboratoire d'Océanographique Physique de l'Université de Paris VI, Équipe de recherche associée au C.R.N.S., 11 Quai St. Bernard, Tour 24, Paris 5ᵉ, France.*

N. G. JERLOV, *Institute of Physical Oceanography, University of Copenhagen, Haraldsgade 6, 2200 Copenhagen N, Denmark.*

ELIZABETH M. KAMPA, *Scripps Institution of Oceanography, University of California, San Diego, La Jolla, California 92307, U.S.A.*

G. KULLENBERG, *Institute of Physical Oceanography, University of Copenhagen, Haraldsgade 6, 2200 Copenhagen N, Denmark.*

PAUL E. LA VIOLETTE, *U.S. Naval Oceanographic Office, Washington D.C. 20390, U.S.A.*

A. MOREL, *Laboratoire d'Océanographique Physique, 06230 Villefranche-sur-mer, France.*

E. STEEMANN NIELSEN, *Freshwater-Biological Laboratory, University of Copenhagen, 51 Helsingørsgade, 3400 Hillerød, Denmark.*

v

HASONG PAK, *School of Oceanography, Oregon State University, Corvallis, Oregon 97331, U.S.A.*

WILLIAM S. PLANK, *School of Oceanography, Oregon State University, Corvallis, Oregon 97331, U.S.A.*

RAYMOND C. SMITH, *Visibility Laboratory, Scripps Institution of Oceanography, University of California, San Diego, La Jolla, California 92037, U.S.A.*

V. A. TIMOFEEVA, *Black Sea Branch of Marine Hydrophysical Institute, Ukrainian Academy of Sciences, Katzively, Simeiz, Crimea, USSR.*

TALBOT H. WATERMAN, *Department of Biology, Yale University, New Haven, Connecticut 06520, U.S.A.*

H. WEIDEMANN, *Deutsches Hydrographisches Institut, 2000 Hamburg 4, West Germany.*

J. RONALD V. ZANEVELD, *School of Oceanography, Oregon State University, Corvallis, Oregon 97331, U.S.A.*

Preface

This volume is a collection of the papers presented at the Symposium on Optical Aspects of Oceanography. We are greatly indebted to the late John Cruise, Academic Press, for his constant help during the planning of the Symposium.

The Symposium was held at the Institute of Physical Oceanography in Copenhagen, 19th–23rd June 1972. The invited papers were all presented at the sessions. The participants numbering 42 represented laboratories in 15 different countries.

The purpose and the spirit of the meeting was to describe the recent development in ocean optics, laying emphasis on optical application in physical oceanography and marine biology and to provide a forum for workers in optical oceanography.

Biologists have contributed largely to our knowledge of marine optics; on the other hand physicists have often furnished instruments for the biological research. A direct collaboration has taken place in many cases. It is highly desirable that the communication between the two groups is maintained and even improved by discussions but also by presenting current trends in optics in a book where problems are viewed by professionals from the two disciplines.

The reader of this book will find that several articles present a complete review—including recent advances—of the subject under consideration, which is delineated in terms of the basic concepts involved. Therefore the book may be useful not only for specialists but also for undergraduate and postgraduate students in various disciplines.

Copenhagen,   NILS JERLOV
October, 1973   E. STEEMANN NIELSEN

vii

Contents

Chapter 1

Optical Properties of Pure Water and Pure Sea Water

A. MOREL

Chapter 2

Observed and Computed Scattering Functions

G. KULLENBERG

1*

Chapter 3

Refraction and Reflection of Light at the Sea Surface

C. S. COX

Chapter 4

Significant Relationships between Optical Properties of the Sea

N. G. JERLOV

Chapter 5

Structure of Solar Radiation in the Upper Layers of the Sea

RAYMOND C. SMITH

Chapter 6

New Developments of the Theory of Radiative Transfer in the Oceans

J. RONALD V. ZANEVELD

Chapter 7

Underwater Visibility and Photography

SEIBERT Q. DUNTLEY

Chapter 8

Polarization Measurements in the Sea

ALEXANDRE IVANOFF

Chapter 9

Optics of Turbid Waters (Results of Laboratory Studies)

V. A. TIMOFEEVA

Chapter 10

Some Applications of the Optical Tracer Method

HASONG PAK and WILLIAM S. PLANK

Chapter 11

The Fluorescence of Dissolved Organic Matter in the Sea

EGBERT KLAAS DUURSMA

Chapter 12

The Use of Fluorescent Dyes for Turbulence Studies in the Sea

H. WEIDEMANN

Chapter 13

Remote Optical Sensing in Oceanography Utilizing Satellite Sensors

PAUL E. LA VIOLETTE

Chapter 14

The Remote Sensing of Spectral Radiance from below the Ocean Surface

ROSWELL W. AUSTIN

Chapter 15

Light and Photosynthesis of Different Marine Algal Groups

PER HALLDAL

Chapter 16

Light and Primary Production

E. STEEMANN NIELSEN

Chapter 17

Remote Spectroscopy of the Sea for Biological Production Studies

GEORGE L. CLARKE AND GIFFORD C. EWING

Chapter 18

Underwater Light and the Orientation of Animals

TALBOT H. WATERMAN

Chapter 19

Bioluminescence

BRIAN P. BODEN and ELIZABETH M. KAMPA

Chapter I

Optical Properties of Pure Water and Pure Sea Water

A. MOREL

Université de Paris, Laboratoire d'Océanographie Physique,
06230 Villefranche-sur-mer, France

I. Introduction

An optically pure medium is defined as a medium which is totally exempt from any suspended particles. This definition does not imply that the medium is a chemically pure compound, it can be a mixture or a solution as well. This definition only implies that the optical properties, especially scattering and absorption, are only determined by molecules or ions.

Though the problems of scattering and absorption of pure water belongs to physical chemistry, it is also of interest in optical oceanography. As is well known, the waters of the open ocean, particularly the deep waters, are of great purity. Consequently the water itself plays an important part in the observed scattering process. Moreover hypothetically pure sea water forms the "blank" for various optical measurements. Scattering by pure water must be subtracted from the observed scattering to estimate the role played by the particles. A similar subtraction yields the absorption by dissolved matter.

The Committee on Radiant Energy in the Sea (IAPO) set forth definitions for attenuation, absorption and scattering. These concepts are also applicable for an optically pure medium. The subsequent definition, as recalled by Jerlov (1968) are used, with the following relations:

$$c = a + b \tag{1}$$

$$b = \iint_{4\pi} \beta(\theta)\, \mathrm{d}\Omega = 2\pi \int_0^\pi \beta(\theta) \sin \theta \, \mathrm{d}\theta \tag{2}$$

between the attenuation coefficient c, the absorption coefficient a, the total scattering coefficient b, and the volume scattering function $\beta(\theta)$. Nevertheless, in the theoretical discussion, in accordance with the physical chemical literature, the Rayleigh ratio R will be used instead of the volume scattering function at right angle $\beta(90°)$. These quantities have the same definition.

Since scattering is a part of attenuation, it will be examined first. Absorption should be considered in the same manner but, this property lends itself with difficulty to experimentation and, in general, is found from eq. (1).

For this reason, it is not examined separately.

II. Scattering

A. THEORY

The theory of the scattering dipole, developed by Lord Rayleigh (1871), was historically, the first interpretation of the phenomenon of light scattering. In the case of dust free gas, Rayleigh assumed (1899) that the dipoles must be the molecules themselves; in other words, that the optically pure medium scattered because of the discontinuous structure of matter. This theory is successfully applied to isotropic particles small in comparison to the wavelength (the colloidal particles in silica sols, for example). The theory was modified by Rayleigh (1920) and Cabannes (1920) to take into account the anisotropy of molecules. This form is applicable to gases, however, it is not satisfactory for dense media such as liquids.

Paradoxically, scattering by liquids, although more intense than that of gases (having equal volume, but not equal mass), was clearly demonstrated some years later (Martin, 1913). The difficulty of preparing optically pure liquids exempt from fluorescence rendered previous observations questionable.

Smoluchowski (1908) and Einstein (1910) formulated a completely different theoretical approach from statistical thermodynamics. Initially this work was initiated to explain the phenomenon of critical opalescence. This theory also applies to density fluctuations of smaller amplitudes such as those present in a fluid in the ordinary state. Critical opalescence and scattering would thus be phenomena of the same nature differing only in their intensity.

Although the Rayleigh theory is not applicable to liquids, a number of the results obtained from it remain valid in the theory of fluctuations. For this reason it is useful to come back to Rayleigh theory.

(1) *Rayleigh theory*

A particle of any form whatever placed in an electrical field E behaves like a dipole whose induced moment P is given by the electrostatic formula: $P = pE$, where p is the polarizability of the particle. The particle should be small compared to the wavelength so that the applied field E can be considered to be homogeneous. The scattering is then assumed to result from the oscillation of this dipole at the frequency imposed by the exciting radiation.

Given that I_o is the intensity of the incident beam (parallel monochromatic and unpolarized light), d the distance between the observation point and the particle, θ the angle between the direction of propagation of the incident beam and the direction of observation, the scattered intensity $I(\theta)$ in that direction is expressed by:

$$I(\theta) = \frac{I_o}{2\,\mathrm{d}^2} k^4 p^2 (1 + \cos^2 \theta) \qquad (3)$$

where k is the wave number defined by $k = 2\pi/\lambda$, λ being the wavelength. This formula corresponds to the case in which the particle is isotropic, i.e. the polarizability is a scalar. The first well known result of this theory is the wavelength dependence of the scattering according to a λ^{-4} law. The second result is the symmetrical shape of the scattering diagram with respect to the direction perpendicular to the incident beam ($\theta = 90°$).

The incident light being natural, the scattered light is polarized and the polarization depends on θ. The dimensionless functions of intensity, i_1 and i_2, correspond to the two polarized components respectively perpendicular and parallel to the plane defined by the directions of propagation and of observation (i_1 and i_2 are also called, respectively, vertical and horizontal components). i_1 and i_2 are related to the total intensity I by:

$$I(\theta) = \frac{I_o}{2k^2 d^2} [i_1(\theta) + i_2(\theta)]$$

and we have:

$$\begin{vmatrix} i_1 \\ i_2 \end{vmatrix} = k^6 p^2 \begin{vmatrix} 1 \\ \cos^2 \theta \end{vmatrix}$$

Component i_1 is constant, while component i_2 varies with $\cos^2\theta$. At right angle the scattered light is totally polarized ($i_2 = 0$), and totally depolarized ($i_1 = i_2$) at $0°$ and $180°$.

If we consider a unit of volume containing N particles and if the intensities scattered by the particles are considered to be additive, the Rayleigh ratio R (or the volume scattering function at $90°$, β_{90}, defined in the same manner) is:

$$R(\equiv \beta_{90}) = N\frac{I_{90}}{I_o}\,\mathrm{d}^2 \tag{4}$$

$$= \tfrac{1}{2}Nk^4p^2 = N\frac{8\pi^4}{\lambda^4}\,p^2 \tag{5}$$

p has the dimensions of L^3, N those of L^{-3}. If $I(\theta)$ replaces I_{90} in eq. (4), we obtain $\beta(\theta)$ which is expressed by combining (3) and (4) by:

$$\beta(\theta) = \beta_{90}(1+\cos^2\theta) \tag{6}$$

Eq. (3) corresponds to isotropic particles. If they are spherical, the polarizability is given by the Lorentz–Lorenz formula:

$$p = \frac{n^2-1}{n^2+2}r^3$$

where r is the radius of the sphere and n the refractive index. The eq. (5) becomes:

$$R = N\frac{8\pi^4}{\lambda^4}r^6\left(\frac{n^2-1}{n^2+2}\right)^2 \tag{7}$$

The integral over all directions which yields the total scattering coefficient b, according to eq. (2), has, when $\beta(\theta)$ is expressed by eq. (6), the following value:

$$b = \frac{16\pi}{3}\beta_{90} \tag{8}$$

which combined with (7) gives:

$$b = \frac{16\pi}{3}8N\frac{\pi^4}{\lambda^4}r^6\left(\frac{n^2-1}{n^2+2}\right)^2$$

Experiment has shown that even in the case of gases (Strutt, 1918), polarization is not total at right angle. Lord Rayleigh (1920) explained depolarization by the anisotropy of molecules and related the depolarization ratio $\delta = i_2\,(90)/i_1\,(90)$ to the three components of the polarizability vector. Cabannes (1920) showed in addition that anisotropy brought about an increase of scattering which he expressed in function of δ. Given that the isotropic part of the Rayleigh ratio, R_{iso}, has the value expressed by eq. (7), the total Rayleigh ratio, R_{tot}, is:

$$R_{\text{tot}} = R_{\text{iso}} \frac{6+6\delta}{6-7\delta} \tag{9}$$

$6+6\delta/6-7\delta$ is the so called Cabannes factor.

Eq. (6) and (8) are also modified:

$$\beta(\theta) = \beta(90)\left(1 + \frac{1-\delta}{1+\delta}\cos^2\theta\right) \tag{10}$$

$$b = \frac{8\pi}{3}\beta(90)\frac{2+\delta}{1+\delta} \tag{11}$$

β_{90} takes the value of R_{tot} (and no longer R_{iso})

$\frac{1-\delta}{1+\delta} = \frac{i_1-i_2}{i_1+i_2}$ is the degree of polarization.

A more complete description of the polarization state is given by the Krishnan relations which correspond experimentally to the different combinations of orientation of the polarizer and the analyzer. The capital letters refer to the components analyzed in the scattered beam, the subscripts designate the state of polarization of the incident beam (v for vertically, h horizontally, u unpolarized). Whatever the angle θ may be, we have:

$$V_u + H_u = U_v + U_h = \tfrac{1}{2}(V_v + V_h + H_v + H_h)$$
$$= R_{\text{tot}} \quad \text{if } \theta = 90°$$

The angular dependence for each term is expressed by:

$$V_v(\theta) = \text{const.}$$
$$V_h(\theta) = H_v(\theta) = \text{const.} \quad (= 0 \text{ for isotropic particles}) \tag{12A}$$
$$H_h(\theta) = H_v \sin^2\theta + V_v \cos^2\theta$$

hence, at 90°: $H_h = H_v$.

The depolarization ratio δ can be measured with a polarizer in the incident beam or an analyzer in the scattered beam (or with a combination of both) according to:

$$\delta = \frac{H_v + H_h}{V_v + V_h} = \frac{V_h + H_h}{V_v + H_v} = \frac{H_u}{V_u} = \frac{U_h}{U_v} \tag{12B}$$

If the polarizer is put in the incident beam, the components of the scattered beam are expressed by:

$$U_v(\theta) = U_v(90)$$
$$U_h(\theta) = U_v[\delta + (1-\delta)\cos^2\theta] \tag{12C}$$

(2) *Fluctuation theory*

Experience has shown that a given mass of a fluid scatters much more in a gaseous state than in a liquid state which is not consistent with Rayleigh's "molecular" theory. The latter applies to independently scattering particles but cannot apply to liquid because of the strong interaction effects between molecules. However, for the wavelength dependence, the symmetry of the scattering diagram, and the polarization the results so obtained continue to exist in the theory of fluctuations. Mainly this theory gives a new expression for intensity (more precisely, for the isotropic part R_{iso}) which can apply to dense media.

In the Einstein–Smoluchowski theory, scattering is considered to be caused by the random motion of molecules which in a sufficiently small volume causes fluctuations of density and, therefore, of the dielectric constant. The fluctuations to be considered are those whose frequencies are optical frequencies. In this theory, the isotropic part of the Rayleigh ratio is given by:

$$R_{iso} = \frac{\pi^2}{2\lambda_0^4} \Delta V \langle \overline{\Delta \epsilon} \rangle^2 \tag{13}$$

$\langle \overline{\Delta \epsilon} \rangle^2$ is the mean square of fluctuations in the dielectric constant in a small volume element ΔV of the medium, and λ_0 is the wavelength *in vacuo*. The fluctuations of ϵ are assumed to be the result of the density fluctuations, so that

$$\langle \overline{\Delta \epsilon} \rangle^2 = \left(\frac{d\epsilon}{d\rho} \right)^2 \langle \overline{\Delta \rho} \rangle^2$$

where $\langle \overline{\Delta \rho} \rangle^2$ is the mean square of the density fluctuation. These fluctuations are related to the probability of the occurrence of change in the average number of molecules in the volume ΔV. ΔV should be small in comparison to the wavelength, but large enough to obey the laws of statistical thermodynamics. From this theory, an expression of $\Delta V \langle \overline{\Delta \rho} \rangle^2$ is obtained which leads to:

$$R_{iso} = \frac{\pi^2}{2\lambda_0^4} KT\beta_T \rho^2 \left(\frac{d\epsilon}{d\rho} \right)^2$$

or

$$R_{iso} = \frac{2\pi^2}{\lambda_0^4} KT\beta_T \left(\rho n \frac{dn}{d\rho} \right)^2 \tag{14}$$

here K is the Boltzmann constant, T the absolute temperature, β_T the isothermal compressibility, and ϵ is replaced by n^2.

The first way of expressing the density derivative of ϵ is to use a direct relationship between ϵ (or n, the refractive index, with $\epsilon = n^2$) and ρ. Many overly empirical theoretical equations have been proposed. That of Lorentz–Lorenz, for example:

$$\frac{n^2-1}{n^2+2}\frac{1}{\rho} = \text{const.}$$

the derivative of which gives:

$$\rho\left(\frac{\mathrm{d}\epsilon}{\mathrm{d}\rho}\right) = \frac{I}{3}(n^2-1)(n^2+2)$$

and leads to the expression given by Einstein:

$$R_{\text{iso}} = \frac{\pi^2}{2\lambda_0^4}KT\beta_T\frac{(n^2-1)^2(n^2+2)^2}{9} \tag{15}$$

The use of the Sellmeier formula (Laplace) $(n^2-1)/\rho = \text{const.}$, from which $\rho\dfrac{\mathrm{d}\epsilon}{\mathrm{d}\rho} = n^2-1$, leads to an alternative expression of R_{iso} given by King (1923) and Rocard (1925). The Gladstone–Dale empirical formula $(n-1)/\rho = \text{const.}$, from which $\rho\dfrac{\mathrm{d}\epsilon}{\mathrm{d}\rho} = 2n(n-1)$, was also proposed.

It should be observed that in the case of liquids, by combining the Lorentz–Lorenz equation or the others with the density-temperature relation the values of $\mathrm{d}n/\mathrm{d}T$ obtained do not match the experiments. Furthermore, the relationship between n and ρ, should be independent of the pressure and temperature. This has not been found. It is mainly these reasons that have led to abandoning the initial formula of Einstein, or of Vessot–King, in favour of formulas which make direct use of the experimental values $(\partial n/\partial T)_P$ or $(\partial n/\partial P)_T$.

The derivative $\mathrm{d}\epsilon/\mathrm{d}\rho$ is replaced by the partial derivatives $\left(\dfrac{\partial\epsilon}{\partial P}\right)_T$ or $\left(\dfrac{\partial\epsilon}{\partial T}\right)_P$ according to the choice which has been made between T or P to describe the thermodynamic state:
in the first case

$$\rho\frac{\mathrm{d}\epsilon}{\mathrm{d}\rho} = \frac{1}{\beta_T}\left(\frac{\partial\epsilon}{\partial P}\right)_T = \frac{2n}{\beta_T}\left(\frac{\partial n}{\partial P}\right)_T$$

and replaced in (13):

$$R_{\text{iso}} = \frac{2\pi^2}{\lambda_0^4}KTn^2\frac{1}{\beta_T}\left(\frac{\partial n}{\partial P}\right)_T^2 \tag{16}$$

In the second case:

$$\rho\frac{d\epsilon}{d\rho} = -\frac{1}{\alpha_P}\left(\frac{\partial\epsilon}{\partial T}\right)_P = -\frac{2n}{\alpha_P}\left(\frac{\partial n}{\partial T}\right)_P$$

α_P being the volume expansion coefficient, which leads to:

$$R_{iso} = \frac{2\pi^2}{\lambda_0^4}KTn^2\frac{\beta_T}{\alpha_P}\left(\frac{\partial n}{\partial T}\right)_P^2 \tag{17}$$

Coumou *et al.* (1964) gave a complete description involving both partial derivatives. However, the conclusion is that formula (16) is the best approximation, since the corrective term introduced by temperature fluctuations is negligible, particularly in the case of water, complete formulas are useless (Kratohvil *et al.*, 1965; Deželić, 1966). With this equation the isotropic part of the Rayleigh ratio can be determined from physical constants and from experimental values of β_T and $(\partial n/\partial P)_T$.

In the theory of fluctuation, the depolarization or the anisotropic scattering is considered as an effect of fluctuation in the orientation of anisotropic molecules. The theory is not as simple as in the case of the modified Rayleigh theory for an individual anisotropic particle. The anisotropic part of the Rayleigh ratio has been related to other physical quantities dependent on optical isotropy (electric and magnetic birefringence). Cabannes (1929), Prinz and Prinz (1956), Benoit and Stockmayer (1956) indicate that eq. (9) remains valid for liquids and Coumou *et al.* (1964) confirm it experimentally.

Finally, combining (9) and (16) the following is obtained:

$$R_{tot} = \frac{2\pi^2}{\lambda_0^4}KTn^2\frac{1}{\beta_T}\left(\frac{\partial n}{\partial P}\right)_T^2\frac{6+6\delta}{6-7\delta} \tag{18}$$

Eq. (12), which express the angular dependence of different polarized components, can be considered exact for dense media according to experiments and discussion by Deželić and Vavra (1966).

(3) *Electrolyte solutions*

In the case of a solution, the theory of fluctuations remains formally unchanged, but the evaluation of $\langle\Delta\epsilon\rangle^2$ requires, in addition to the previous variables, new thermodynamic variables characteristic to the solution. A complete formula, as written by Stockmayer (1950) for a multicomponent system, involves all the partial derivatives related to each compound, such as:

$$\left(\frac{\partial\epsilon}{\delta m_i}\right)_{T,P,m} \text{ and } \left(\frac{\partial m_i}{\partial\mu_j}\right)_{T,P,\mu}$$

where m is the molality, μ the chemical potential, and the subscripts i and j stand for the components i and j. It is possible to imagine the phenomenon as being the result of two different effects:

(a) Addition of a new term due to concentration fluctuation in the volume element.

(b) Modification of the density fluctuation term when pure solvent is replaced by solution.

The concentration fluctuation term:

The complete formula is considerably simplified for a two-component system (Debye, 1944; Oster, 1948). The additional term, due to the concentration fluctuation R_{cf}, is expressed in the case of an electrolyte solution by:

$$R_{cf} = H \frac{M}{v} C \frac{1}{-\partial (\ln a_0)/\partial C_{P,T}} \tag{19}$$

M is the molecular weight of the electrolyte, v the number of ions, C the concentration of the solute (in g/g), and a_0 the activity of the solvent. The factor H is given by:

$$H = \frac{2\pi^2}{\lambda_0^4} \frac{n_0^2}{N_A} \left(\frac{\partial n}{\partial C} \right)_{P,T}^2 \tag{20}$$

n_0 is the refractive index of the pure solvent and N_A is the Avogadro number. H can be regarded as a constant as long as $\partial n/\partial C$ does not depend on C, which is correct in the case of diluted solutions. In addition, if the solution is ideal, eq. (19) becomes simpler:

$$R_{cf} = H \frac{M}{v} C \tag{21}$$

The density fluctuation term:

Modification of the density fluctuation term can be anticipated with eq. (14) giving the isotropic part of the Rayleigh ratio. The introduction of a solute changes the terms n, β_T, and ρ, and the modified ratio can be written:

$$(R_{iso})_{mod} = R_{iso} \times F$$

with

$$F = \frac{\beta}{\beta_0} \left(\frac{\rho n (\partial n/\partial \rho)}{\rho_0 n_0 (\partial n_0/\partial \rho)} \right)^2$$

the subscript o stands for pure solvent.

As pointed out by Lochet (1953), using the Einstein eq. (15), F remains very close to 1 in the case of aqueous electrolyte solutions.

The reason lies in the contradictory influence of the electrolyte on β_T and n. In general, for water, n increases and β_T decreases when a solute is added (for the small ions acting weakly on n, F is slightly smaller than 1). Numerically, it seems that modification of the density fluctuation term is very small and practically negligible compared to the additional term due to concentration fluctuation.

The depolarization factors:

It seems plausible that if the electrolyte gives small isotropic ions (as Cl^-, Na^+) the concentration fluctuation term is an isotropic contribution. Thus, the depolarization factor should decrease in the case of a solution. Pethica and Smart (1966) confirmed experimentally this decrease. But the effect is less important than theoretically foreseen because the increase of anistropy of water molecules attracted by the ions increases the anistropic part of the Rayleigh Ratio (Rousset–Lochet, 1955).

B. RESULTS AND DISCUSSION

(1) *Experimental*

Experimental determination of scattering constants for optically pure water raises some very difficult problems that can be separated into three categories:

(a) The problem of the absolute calibration of the scattering meter (which exists for all liquids). In essence it is a question of accurately measuring a ratio in the order of 10^{-6} between the scattered flux and the incident flux.

(b) The presence of stray light (difficult to eliminate or even to evaluate). This problem is particularly difficult since the water scatters very weakly (approximately 15 times less than benzene).

(c) Purification, which is more difficult to accomplish for water than for other liquids, as noticed at the time of the first experiments.

Without going into details on problem (a) it must be said that in the case of water, measurements have been made with reference to a standdard, either a standard opal diffusor (see for example Brice *et al.*, 1950) or a standard formed by an optically pure liquid such as benzene or carbon tetrachloride (problems (b) and (c) are less critical for these liquids). The absolute calibration of the instrument can be checked by measuring scattering by an almost monodisperse suspension (polystyrene latexes) and by comparing with the computed Mie intensity functions, or by determining the molecular weight (by eq. (21)) of a known compound (Kratohvil *et al.*, 1965). The use of benzene as a standard has created several problems because of the controversy over the absolute values to be attributed to this liquid. This controversy was raised by Carr and Zimm (1950) but seems to be closed now (Deželič, 1966) with the confirmation of "high values". When the measurement is made relative to a standard liquid, a geometrical-optical correction should be introduced to take into account the change of refractive index (known as the "n^2 correction").

Stray light is particularly due to reflection and to scattering by the glass of the cell. Practical solutions to eliminate it include the use of semioctagonal black painted cells with narrow apertures. It seems that a better solution is to place the (cylindrical) cell in a tank of benzene, the equality of indexes practically eliminates any reflection.

To prepare dust free water different methods have been proposed: ultracentrifuging, envelopment by various precipitates (Sweitzer, 1927), distillation *in vacuo* without ebullition (Martin, 1920), ultrafiltration and filtration through millipore filters (pore sizes 0·1 or 0·22 μm). The latter is simple, efficient, and is used now. All the results to be presented have been obtained by this latter method, except those of Morel (1966) who used the old method of distillation *in vacuo*, and upon comparison, the millipore filtration was found to be slightly less efficient (scattering was 3% greater on the average). It has often been noticed that adding a very small amount of electrolyte, drastically lowering resistivity, makes purification by filtration more efficient. Finally, we must remember that filtration or distillation should be repeated in order to ensure proper cleaning of both the water and cell.

Optical purity may be tested with an intense light beam by viewing at small scattering angles, but it is more reliable to check if the intensities scattered at two symmetrical angles (30° and 150° for example) are equal, as the theory anticipates. However, this criterion is revealed to be insufficient because the very small particles remaining do not exhibit a very dissymetric scattering function. For these particles polarization at 90° is almost total which decreases the measured value of the depolarization factor and modifies the curves derived from eq. (12C) without destroying their symmetry. This is probably the most sensitive criterion.

It must also be pointed out that, for water, light emitted by fluorescence or by Raman effect is not negligible in comparison with scattered light. The precaution must be taken, especially for measurements at short wavelength, of placing monochromatic filters on the incident beam and on the scattered beam (light emitted by fluorescence is not polarized, its presence can be detected by the increase of δ).

TABLE 1. The literature values of R_{tot} ($= \beta_{90}$) for pure water expressed in 10^{-4} m^{-1}, and of the corresponding δ (number in parentheses). The temperature ranges from 18°C to 25°C.

λ(nm)	(1)	(2)*	(3)	(4)	(5)	(6)	(7)	(8)	(9)
366	6·80					4·53			
405	4·05					2·90			
436	2·89 (0·083)	2·95	2·86	2·45 (0·100)	2·32 (0·087)	2·12 (0·09)	2·54 (0·091)		2·82 (0·10)
546	1·05	1·13	1·07	1·08 (0·116)	0·86 (0·076)	0·83 (0·09)	1·08 (0·109)	1·10 (0·05)	1·16 (0·15)
578						0·66			

(1) Kraut–Dandliker, 1955; (2) Mysels–Princen, 1959; (3) Huisman, 1964; (4) Kratohvil et al., 1965; (5) Cohen–Eisenberg, 1965; (6) Morel, 1966–1968; (7) Pethica–Smart, 1966; (8) Lanshina–Shakhparonov, 1966; (9) Parfitt–Wood, 1968.
Data reported as total coefficient b: R_{tot} is computed using $\delta = 0·09$ in eq. (11).

(1) *Pure water*

In Table 1, most of the data for R_{tot} and δ, obtained since 1954 are presented. The older determinations are not recalled for various reasons, such as inadequate knowledge of the geometrical optical correction factor and the wavelength.

TABLE 2. Calculated values of R_{iso} and R_{tot} ($= \beta_{90}$) for water, expressed in 10^{-4} m^{-1}, $T = 20°$C.

		R_{tot}		
λ(nm)	R_{iso}	$\delta = 0.06$	$\delta = 0.09$	$\delta = 0.11$
366	4·36	4·97	5·32	5·55
405	2·80	3·19	3·42	3·57
436	2·04	2·33	2·49	2·59
546	0·78	0·88$_4$	0·94$_7$	0·99
578	0·60	0·68$_4$	0·73$_2$	0·76

Presented in Table 2 are the theoretical values of R_{iso} computed according to eq. (16) with the values chosen from the literature by Kratohvil *et al.* (1965) for the physical quantities β_T, n, and $(\partial n/\partial P)_T$ and for $T = 20°$C. These quantities are sufficiently accurate (except perhaps $\partial n/\partial P$) and the different computations (Mysels, 1964), (Parfitt–Wood, 1968) give very similar results (within 2–3%). They are, on the other hand, different from the previously proposed theoretical values (Dawson–Hulburt, 1937) (Le Grand, 1939), calculated with the Einstein or Vessot–King equation using a different value for β_T. R_{tot} is then computed with eq. (9) for three different values of δ occurring in the Cabannes factor. We see that the influence of this experimental term on the "theoretical" value is great.

As written by Mysels (1964), "because of the problems of stray light and of contaminating dust . . . all the normally expected errors are positive". This remark tends to confirm the lowest values. But the experimental values form two groups: one of low values (columns 5 and 6, Table 1) obtained with the same apparatus (SOFICA) and another of high values generally obtained with the Brice–Phoenix apparatus. Consequently, the preceding remark should be interpreted with care. The differences of purification cannot account for this discrepancy. If, according to the scattering diagrams, water can be considered as optically pure in the experiments by Cohen–Eisenberg and Morel (low values), it can also be considered so in the case of certain high values.

Kratohvil *et al.*, Pethica and Smart observe an apparent 45°/135° dissymmetry very close to 1 or even equal to 1·0. Most probably the differences between the calibration methods should be put forward to explain the differences in the data.

The experimental values of δ are also quite variable. Nevertheless, it seems that the 0·09 value, which represents an average, can be adopted. This value inserted in the calculation leads to the values of R_{tot} in column 4 (Table 2). In relation to the values of R_{tot}, the "high" experimental values are 10% to 20% higher*, Morel's values systematically ranging from 10% to 15% (at 366 nm) lower. The Cohen–Eisenberg values are also lower, but closer to the theoretical values (especially if they are computed with the observed values of δ).

In conclusion, the experimental values are in reasonable agreement with the values prescribed by the theory (it can be noted that agreement is better for liquids other than water). Since there is no decisive argument for choosing between the experimental values, the theoretical values can thus be considered reliable.

The other aspects of the theory were well checked experimentally (particularly by experiments based on relative measurements). Thus, after purification by distillation *in vacuo*, Morel observed scattering functions which agree well with the theoretical curves (Fig. 1). Wavelength selectivity of scattering was demonstrated experimentally by

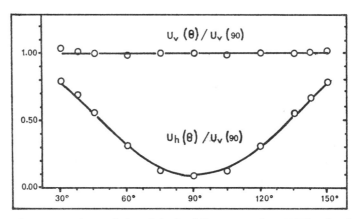

Fig. 1. Curves are theoretical and derived from equations (12C) using δ = 0·09. Dots are experimental (Morel, 1966) and concern water purified by distillation without ebullition *in vacuo*.

* Mysels (1964) had attributed this excess to fluctuations of the degree of association between the water molecules. This excess scattering has been estimated by Litan (1968) and found within the range of experimental error in accordance with the opinions of Kratohvil *et al.* or Cohen–Eisenberg.

Hulburt (1934) who showed that it roughly obeyed a λ^{-4} law. In fact, there is dispersion of n and $\partial n/\partial P$ (in addition to that of the term λ^{-4}) which reinforces selectivity. Table 3, compiled from Tables 1 and 2, shows a good agreement between the computed and observed values. If the wavelength dependence is expressed in terms of a power law, the best exponent is -4.32.

TABLE 3. Wavelength selectivity of scattering.

λ (nm)	Computed from table 2 $\dfrac{R_{\text{iso}}(\lambda)}{R_{\text{iso}}(546)}$	Experimental $R_{\text{tot}}(\lambda)/R_{\text{tot}}(546)$					
		Pure water					Pure sea water
		(1)*	(4)*	(5)*	(6)*	(7)*	(6)*
366	5·60	6·49			5·43		5·43
405	3·59	3·96			3·48		3·56
436	2·62	2·75	2·27	2·70	2·54	2·36	2·57
546	1·0	1·0	1·0	1·0	1·0	1·0	1·0
578	0·77				0·79		0·79

* The numbers in parentheses correspond to the references listed in Table 1.

Influences of pressure and temperature on scattering can be predicted from eq. (14). The values of the partial derivatives at high pressure are questionable. However opposite variations of compressibility (which decreases when pressure increases) and of density (which increases when pressure increases) lead us to believe that the scattering value will remain almost unchanged. On the other hand, temperature dependence can be evaluated with greater certainty. Cohen and Eisenberg (1965) showed theoretically and experimentally that the variation was small. The Rayleigh ratio theoretically has a minimum at about 22°C (explained by the multiplication of β_T which has a minimum at about 45°C, by the absolute temperature). At 5°C and 45°C, the increase is only 2% of the minimum value. This variation is hardly detectable within experimental error.

(3) Pure solutions and pure sea water

The experimental study of scattering by electrolyte solutions was first made by Sweitzer (1927) then by Lochet (1953). Increase in scattering is measurable. Since modification of the density fluctuation term can be neglected, the increase, due only to the concentration fluctuation term, is proportional to the concentration and molecular weight of the electrolyte (only absolutely true if eq. (21) can be applied, i.e. at infinite dilution). Lochet experimentally confirmed this conclusion. By extrapolating his results to zero concentration, he found molecular weights which were in agreement with the formula weights of the electrolytes. Sweitzer's results concerning sodium chloride solutions are presented in Fig. 2 with those of Morel (1966) who used these measurements as a first step in his study of sea water. The results of Pethica and Smart (1966) dealing with potassium chloride solutions are also presented. The Rayleigh ratios of purified solutions are divided by the Rayleigh ratio of similarly purified water. Thus the relative increase is plotted as a function of concentration. Eqs. (20) and (21) enable the calculation of R_{cf}/c which, divided by the value of R_{tot} at the same wavelength (Table 2, 4th column), gives the slope of the theoretical lines. These linear relationships are only valid when approaching zero concentration. It must be noted that the calculated slope is not affected by the wavelength as R_{cf}, like R_{tot}, varies with $n_0^2\lambda^{-4}$.

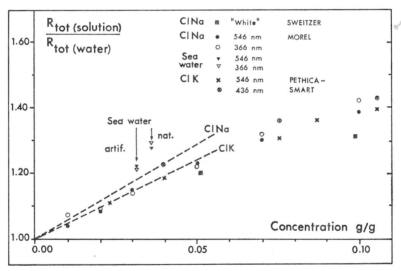

FIG. 2. Relative increase of the Rayleigh ratio for solutions as a function of concentration.

The experimental results of different authors are good verification of the theoretical predictions. At a greater concentration, the evaluation of R_{cf} through eq. (19) rather than through (21) leads to lower values. (Furthermore; $\partial n/\partial c$ is not constant, but decreases as concentration increases). We find that a NaCl solution of 0·035 g/g, which has approximately the same concentration in Cl⁻ ions as sea water of 38‰ salinity,

TABLE 4. Volume scattering function at $90°$ and total scattering coefficient for pure water and sea water as a function of the wavelength.

λ(nm)	350	375	400	425	450	475	500	525	550	575	600	
$\beta_{90}(10^{-4}\,m^{-1})$	6·47	4·80	3·63	2·80	2·18	1·73	1·38	1·12	0·93	0·78	0·68	Pure water
$b^*(10^{-4}\,m^{-1})$	103·5	76·8	58·1	44·7	34·9	27·6	22·2	17·9	14·9	12·5	10·9	
$\beta_{90}(10^{-4}\,m^{-1})$	8·41	6·24	4·72	3·63	2·84	2·25	1·80	1·46	1·21	1·01	0·88	Pure sea water ($S = 35$–$39‰$)
$b^*(10^{-4}\,m^{-1})$	134·5	99·8	75·5	58·1	45·4	35·9	28·8	23·3	19·3	16·2	14·1	

* Computed according to eq. (11) with $\delta = 0.09$ which leads to $b = 16.0 \times \beta\,(90)$.

2

scatters 1·18–1·20 times as much light as pure water. In the same figure
are results (Morel, 1966) for artificial ($S = 34\cdot3\%_{oo}$) and natural
($S = 38\cdot4\%_{oo}$) sea water purified in the same manner (millipore filtration). They, though having the same Cl⁻ ion concentration as the NaCl
solutions, scatter more than said solutions. Without making use of the
theory for a multicomponent system, this result can nevertheless be
explained by the diversity of the cations present and, in addition, by
the fact that anions, other than Cl⁻, are also present. A very approximate calculation can be made: assuming that sea water is an ideal
solution of a hypothetically unique salt having a molecular weight of
70 and by taking the sea water value, 0·20 for $\partial n/\partial C$, the increase of
scattering is found to be on the order of 34% for a salinity of 35‰.

Concluding we can reasonably admit that pure sea water of 35–38‰
salinity scatters 1·30 times more than pure water. This value is used in
Table 4 to compute theoretical scattering of pure sea water according

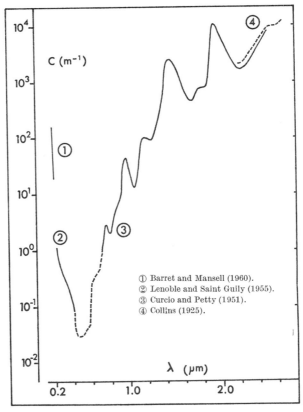

Fig. 3. Attenuation curve for water between 0·2 and 2·8 μm.

to the theoretical values concerning pure water given in Table 2 (for $\delta = 0\cdot09$).

III. ATTENUATION

Jerlov (1968) reviewed the work concerning attenuation and observed that "progress in the investigation of this factor has been relatively slow". After this review there are only a few details to add for it does not seem that the problem has been fundamentally reinvestigated.

As Fig. 3 shows, a very acute minimum for attenuation lies in the visible part of the spectrum. On both sides of this transmission "window", i.e. in ultraviolet, below 200 nm, and in the infrared, above

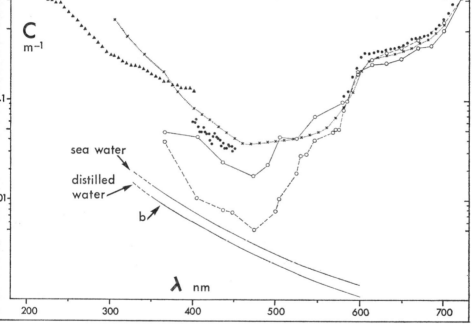

Fig. 4. Attenuation curves in the near ultraviolet and in the visible part of the spectrum.

▲ Lenoble-Saint Guily (1955), path length: 400 cm;
× · · Hulburt (1934) (1945), path length: 364 cm;
● Sullivan (1963), path length: 132 cm;
○— Clarke-James (1939), path length: 97 cm (Ceresin lined tube);
○ - - - James-Birge (1938), path length: 97 cm (Silver lined tube).

Total scattering coefficient for pure water and pure sea water as a function of wavelength, according to Table 4.

700 nm, absorption increases strongly. No theory for remnant absorption between these two limits (200–700 nm) is postulated. Moreover, it must be noted that this residual absorption is 10^4 or 10^6 times weaker than absorption outside the limits.

Regions of high absorption have been studied extensively and the existence of bands explained. Absorption at high frequency (in the far ultraviolet) is related to electron transitions, the bands at lower frequency (in the infrared and distant infrared) are connected with different intramolecular and intermolecular motions. These motions depend on the structure of water, which remains temperature dependent. Consequently, the influence of temperature on certain absorption bands in the infrared (or on the Raman bands which correspond to the same vibrational motions) has been studied extensively (for example, see Walrafen, 1967).

The relatively high transparency of water in the visible and near ultraviolet part of the spectrum varies with wavelength. It is well known that the attenuation coefficient is at a minimum in the blue region (450–500 nm). The curve showing the spectral dependence of c, between 200 and 700 nm, is roughly symmetrical with respect to this minimum (Fig. 4). Rapid increase of c is noted between 570 and 600 nm, followed by further increase from 700 to 760 nm. This wavelength corresponds to the first absorption band in the infrared (Curcio–Petty, 1951).

A. RESULTS AND DISCUSSION

(1) *Experimental*

The problem of experimentally determining attenuation coefficients of pure water and of pure sea water has much in common with the problem of scattering measurements. There is also:

—a radiometrical problem due here to the fact that the fluxes to be compared (the transmitted flux and the incident flux) are almost the same and that the coefficient c to be measured is very small.

—a geometrical problem. The measurement should be made under conditions specified by the definition of the coefficient itself. Practically it becomes a problem of stray light. The incident beam must be freed of divergent rays, the detection beam must be freed of stray light reflected by the cell (tube) and of scattered light travelling around the beam.

—a purification problem which is *a priori* more critical than in the case of scattering measurements, firstly, the necessary volumes are greater and, secondly, because the non eliminated particles can, according to their size, change the total scattering coefficient b much more than the coefficient β (90).

For the most penetrating radiations, the decrease of flux by attenuation is only a few percents for an optical path of one meter, a minimal length allowing the measurement to be made with sufficient precision.

Greater lengths have been used (see Fig. 4, legend) but the solution that proposes increasing the length of the path has the disadvantage of making the second and third problems more difficult to solve. It is also possible to make the beam cross the tube several times by reflection (4 times 488 cm for the Drummeter and Knestrick measurements, 1967).

The problem of obtaining a high collimation, whatever the wavelength, for both the incident beam and the perfectly centered detection beam, can only be partially resolved in practice. The errors due to not respecting the theoretical conditions have been discussed (see references in Jerlov, 1968, page 48).

Purification has always been obtained by distillation often followed by filtration through a fritted glass filter. Millipore filtration apparently was not used, possibly because it may cause an organic contamination that would modify the values in the ultraviolet part of spectrum. To check optical purity, it would be advisable to obtain simultaneous scattering measurements, since the scattering values are better known.

(2) Pure water

The results of James and Birge (1938) and the results of subsequent work are presented in Fig. 4. Findings prior to 1938 have been collected and discussed by James and Birge.

Agreement between the different measurements presented is reasonable for wavelengths greater than 550 nm, but is considerably less adequate for the region of the attenuation minimum where, of course, the measurements are the most difficult. Furthermore, in the violet and ultraviolet part of the spectrum (350–400 nm) the junction of the curves is questionable. The lowest values for c (with a minimum at 473 nm) were obtained with the same apparatus: (James–Birge, 1938) (Clarke–James, 1939). The only difference between the two series of measurements is the cell: a silver lined tube was used in 1938, leading to the lowest values (low curve), while in 1939 a ceresin lined tube was used.

We might think that the sources of discrepancy in these measurements are more often located in the experimental devices than in the purity of water. Moreover, the lowest values are not necessarily the best because of the possibility that the forward scattered light entered the detector. There is no sound basis for choosing between these measurements. Those of Clarke and James are used most frequently.

The search for fine structure (narrow light transmittance region or absorption bands which should be higher harmonics of infrared bands) is the basis of works done by Sullivan (1963) and, more recently (1967) by Drummeter and Knestrick. These two authors approached the problem differently by limiting themselves to relative measurements performed by photographic densitometry. The value of c is unknown but very slight variations of c (1×10^{-3} m^{-1}) can be detected for adjacent wavelengths. Three very weak absorption bands were detected at 470 nm, 515 nm and 550 nm. They are 5–10 nm wide.

(3) Pure sea water

The authors who studied distilled water and filtered sea water with the same apparatus did not find any difference exceeding the precision of the measurements. While studying filtered sea water with a Berkefeld filter, Clarke and James (1939) found slightly lower values than those they obtained with distilled water (in the region of maximum transmission). These measurements tend to confirm the values of James and Birge (1938). The Sullivan (1963) measurements (between 790 and 580 nm) for artificial sea water are perfectly indistinguishable from the measurements concerning distilled water if we plot them on Fig. 4. The ions, at the concentration of sea water, do not have an absorbing action in the visible spectrum, but they do in the ultraviolet, according to Lenoble (1956). Absorption due to said ions would range from 0·05 m^{-1} (at 360 nm) to 0·51 m^{-1} (at 250 nm). Lenoble insists on the difficulty of obtaining reproducible measurements with the salts employed, despite their high purity. Lower values for c were obtained by Copin–Montegut et al. (1971) with natural sea water filtered on a Whatman filter and subsequently irradiated with a u.v. lamp to destroy the organic matter in solution.

Sea water behaves very differently from distilled water at wavelengths smaller than 250 nm. The bromide ion induces strong absorption (Ogura–Hanya, 1966). Nitrate (Armstrong, 1963) and dissolved oxygen (Copin–Montegut et al., 1971) also produce an absorption but slight in comparison to that of bromide. Organic matter dissolved in natural sea water absorbs in the ultraviolet. The practical interest of the preceding studies was renewed by the recent search for quantitative methods for evaluating dissolved organic matter.

The word absorption has often been used instead of attenuation. This is justified because scattering is negligible compared to absorption everywhere else in the spectrum, except in the region 400–500 nm where absorption is minimal. Curves corresponding to the total scattering coefficient b for pure water and for pure sea water from Table 4, have been plotted on Fig. 4. The lack of reliable values for the minimum prevent determining the true role of scattering in the attenuation process, when it is not negligible.

REFERENCES

Armstrong, F. A. J. (1963). *Anal. Chem.*, **35**, 1292.
Barret, J. and Mansell, A. L. (1960). *Nature (London)*, **187**, 138.
Benoit, H. and Stockmayer, W. H. (1956). *J. Phys. Radium*, **17**, 21.
Brice, A., Halwer, M. and Speiser, R. (1950). *J. opt. Soc. Am.*, **40**, 768.
Cabannes, J. (1920). *J. Phys.*, **6**, 129–142.
Cabannes, J. (1929). "La Diffusion Moléculaire de la Lumière." Presses Universitaires de France.
Carr, C. I. and Zimm, B. H. (1950). *J. Chem. Phys.*, **18**, 1616–1626.
Clarke, G. L. and James, H. R. (1939). *J. opt. Soc., Am.*, **29**, 43–55.
Cohen, G. and Eisenberg, H. (1965). *J. Chem. Phys.*, **43**, 3881–3887.
Collins, J. R. (1925). *Phys. Rev.*, **26**, 771.
Copin-Montegut, G., Ivanoff, A. and Saliot, A. (1971). *C.R. Acad. Sci.*, **272**, 1453–1456.
Coumou, D. J., Mackor, E. L. and Hijmans, J. (1964). *Trans. Faraday Soc.*, **60**, 1539–1547.
Curcio, J. A. and Petty, C. C. (1951). *J. opt. Soc. Am.*, **41**, 302–305.
Dawson, L. H. and Hulburt, E. O. (1937). *J. opt. Soc. Am.*, **27**, 199–201.
Debye, P. (1944). *J. Appl. Phys.*, **15**, 338.
Deželić, G. J. (1966). *J. Chem. Phys.*, **45**, 185–191.
Deželić, G. J. and Vavra, J. (1966). *Croat. Chem. Acta*, **38**, 35–47.
Drummeter, L. F. and Knestrick, G. L. (1967). *Appl. opt.*, **6**, 2101–2103.
Einstein, A. (1910). *Ann. Physik.*, **33**, 1275–1298.
Huisman, H. F. (1964). *Proc. Kon. Med. Akad. Wet.*, B, **67**, 367.
Hulburt, E. O. (1934). *J. opt. Soc. Am.*, **24**, 175.
Hulburt, E. O. (1945). *J. opt. Soc. Am.*, **35**, 698–705.
James, H. R. and Birge, E. A. (1938). *Trans. Wis. Acad. Sci.*, **31**, 1–154.
Jerlov, N. G. (1968). "Optical Oceanography." Elsevier, Amsterdam.
King, L. V. (1923). *Proc. Roy. Soc.*, **104**, 333–357.
Kratohvil, J. P., Kerker, M. and Oppenheimer, L. E. (1965). *J. Chem. Phys.*, **43**, 914–921.
Kraut, J. and Dandliker, W. D. (1955). *J. Chem. Phys.*, **23**, 1544–1545.
Lanshina, L. V. and Shakhparonov, M. I. (1966). *Vestn. Mosk. Univ.*, II, 21, 5, 49.
Le Grand, Y. (1939). *Ann. Inst. Océanogr.*, **19**, 393–436.
Lenoble, J. and Saint Guily, B. (1955). *C. R. Acad. Sci.*, **240**, 954–955.
Lenoble, J. (1956). *Rev. opt.*, **35**, 526–531.
Litan, A. (1968). *J. Chem. Phys.*, **48**, 1039–1063.
Lochet, R. (1953). *Ann. Phys.*, **8**, 14–60.
Lord Rayleigh (Strutt, J. W.) (1871). *Phil. Mag.*, **41**, 107–120, 274–279, 447–454.
Lord Rayleigh (1899). *Phil. Mag.*, **47**, 375–384.

Lord Rayleigh (1920). *Proc. Roy. Soc.*, **97**, 435–450; **98**, 57–64.
Martin, W. H. (1913). *Trans. Roy. Soc. Can.*, **7**, 219–229.
Martin, W. H. (1920). *J. Phys. Chem.*, **24**, 478–492.
Morel, A. (1966). *J. Chim. Phys.*, **10**, 1359–1366.
Morel, A. (1968). *Cah. Océanogr.*, **20**, 157–162.
Mysels, K. J. and Princen, L. H. (1959). *J. Phys. Chem.*, **63**, 1696.
Mysels, K. J. (1964). *J. Amer. Chem. Soc.*, **86**, 3503–3505.
Ogura, N. and Hanya, T. (1966). *Nature (London)*, **212**, 758.
Oster, G. (1948). *Chem. Rev.*, **43**, 319.
Parfitt, G. D. and Wood, J. A. (1968). *Trans. Faraday Soc.*, **64**, 805–814.
Pethica, B. A. and Smart, C. (1966). *Trans. Faraday Soc.*, **62**, 1890–1899.
Prinz, N. and Prinz, W. (1956). *Physica*, **22**, 576–578.
Rocard, Y. (1925). *C.R. Acad. Sci.*, **180**, 212.
Rousset, A. and Lochet, R. (1955). *C.R. Acad. Sci.*, **240**, 70–73.
Smoluchowski, M. (1908). *Ann. Physik*, **25**, 205–226.
Stockmayer, W. H. (1950). *J. Chem. Phys.*, **18**, 58.
Strutt, R. J. (1918). *Proc. Roy. Soc.*, **95**, 155–176.
Sullivan, S. A. (1963). *J. opt. Soc. Am.*, **53**, 962–968.
Sweitzer, C. W. (1927). *J. Phys. Chem.*, **31**, 1150–1191.
Walrafen, G. E. (1967). *J. Chem. Phys.*, **47**, 114–126.

Chapter 2

Observed and Computed Scattering Functions

G. KULLENBERG

Institute of Physical Oceanography, University of Copenhagen, Denmark

I. INTRODUCTION

The aim of this work is to review our present knowledge of the light scattering function, i.e. the angular distribution of the scattered light, for sea water, to inter-compare observations from different areas, to compare the observed functions with theoretically computed functions, and to point out gaps in our understanding of the light scattering in the sea. Related problems concerning the particulate matter in the ocean will also be discussed.

Scattering is one of the fundamental processes determining the propagation of light in the sea. The reasons for our interest in the scattering function are numerous: it is an inherent property of the water which is useful as an optical parameter and plays a central role in the theory of radiative transfer. The scattering properties of various water masses are valuable tools in descriptive oceanography. Indirectly, detailed knowledge of the scattering function can yield information about the particle-size distribution and the particle composition by comparison with theoretical calculations of the scattering.

II. Definitions

The scattering function is defined by the relation

$$\beta(\theta) = \frac{\mathrm{d}I(\theta)}{E\,\mathrm{d}V} \quad (m^{-1},\ str^{-1}) \tag{1}$$

where $\mathrm{d}I(\theta)$ is the intensity of the light scattered in the direction θ from the incident beam by the volume element $\mathrm{d}V$ irradiated by the irradiance E. The volume scattering coefficient is obtained by integration

$$b = \int_{4\pi} \beta(\theta)\,\mathrm{d}\omega = 2\pi \int_0^\pi \beta(\theta) \sin\theta\,\mathrm{d}\theta \tag{2}$$

The scatterance in the sea is caused by the water itself, by dissolved salts, and by suspended matter. It is usually assumed that the effects are additive.

The molecular scatterance, i.e. the scatterance caused by the water itself and by the dissolved salts, can be determined theoretically by the fluctuation theory (Smoluchowski, 1908; Einstein, 1910), attributing the scatterance to density or concentration fluctuations by molecular movements. This type of scatterance is proportional to λ^{-4} and has, as regards intensity distribution and polarization, the same properties as Rayleigh scatterance. In the present work the values given by Le Grand (1939) have been used for the molecular scatterance. These are for pure water and the effect of the dissolved salts is thus not corrected for. This effect is usually neglected and can be of importance only for the clearest water.

Generally the particulate matter in the water is the most important contributor to scattering, and the particle scatterance forms the main part of our problem. Particulate matter is brought into the sea by land drainage, rivers and through the atmosphere, and it is created in the sea by organic production as well as by flocculation and adsorption processes. We shall return to the nature and concentration of the particulate matter later.

The scattering process is made up of three phenomena, namely diffraction and refraction plus reflection. The diffraction is determined by the size and the shape of the particle whereas refraction and reflection depend upon the composition, i.e. the index of refraction. This is given relative to the water, and consists of a real and an imaginary part. The latter is in the present case very small. Four dimensionless parameters are formed dependent upon the size, the relative index of refraction m, and the wavelength λ of the incident light. These are the size parameter α, the parameter $m\,.\,\alpha$ giving the wavelength of the light in the particle, the parameter ρ giving the phaseshift of a beam passing through the centre of the particle, and the efficiency factor Q representing the ratio between the scattered intensity and the incident intensity. The definitions are

$$\alpha = \frac{\pi d}{\lambda} ; \qquad \rho = 2\alpha(m-1)$$

$$Q = \frac{1}{\alpha^2} \int_0^\pi (i_1+i_2) \sin \theta \, d\theta$$

where d is the diameter of the particle, assuming spherical particles, i_1 and i_2 are the intensities scattered in the direction θ, with the electric vectors perpendicular to and parallel to the plane of observation, respectively. The incident light is polarized and has unit intensity.

It is noted that the particle scattering coefficient is given by

$$b_p = \frac{\pi}{4} \sum_i Q_i N_i d_i^2 \tag{3}$$

where N_i is the number of particles per unit volume with diameter d_i and efficiency factor Q_i.

Depending upon the size of the parameters α and m, different scattering zones are encountered which can be visualized in the $\alpha - m$ plane (van de Hulst, 1957, p. 131). A third distinguishing parameter is the phaseshift ρ. The limiting cases are demonstrated in Table 1. The case of Rayleigh scattering requires that the particle is much smaller than the wavelength of the light inside the particle. It is noted that in this case the efficiency factor decreases strongly with decreasing size. For $\alpha > 0.4$ the rigorous Mie theory should be used to calculate the scattering intensity and distribution. When the index of refraction $m \to 1$,

TABLE 1. Limiting scattering cases

Rayleigh and fluctuation theory	Rayleigh–Gans theory	Anomalous diffraction
α, ρ small	α large	α, ρ large
$\lvert m \times \alpha \rvert \ll 1$	ρ small	ρ fixed (intermediate)
$\beta \propto \lambda^{-4}$	α fixed	
$Q \propto (m-1)^2 \times \alpha^4$	$m \to 1$	$m \to 1$
	$\lvert m-1 \rvert \ll 1$	$\lvert m-1 \rvert \ll 1$
	$\rho \ll 1$	$Q \sim 2$
	$Q \propto (m-1)^2 \times \alpha^2$	
	$Q \ll 1$ for $\alpha \gg 1$	
	$\leftarrow$ in general: Mie theory $\rightarrow$	

special approximations can, however, be made. When $m \to 1$ for large α-values and small phaseshifts Rayleigh–Gans (Rayleigh, 1871; Gans, 1925) scattering is encountered. In this case the efficiency factor is small. For large particles $(\alpha \to \infty)$ the efficiency factor approaches a constant value of about 2. When $m \to 1$ so-called anomalous diffraction occurs. This is interference between the diffracted light and the light transmitted through the almost transparent particle. Then the scattered intensity becomes concentrated near the direction of the incident light.

The above discussion deals with spheres but the results are in general applicable also for a system of irregular particles. It has been shown that a system of particles as in the sea by and large behaves as a polydisperse system of spherical particles (Beardsley, 1968; Holland and Gagne, 1970). In the sea the forward scatterance is diffraction dominated whereas for angles above 10°–30° refraction+reflection dominate. Hodkinson (1962) demonstrated this by measurements on known suspensions, some of which are representative for the sea. The anomalous diffraction can probably also be significant in the sea when the suspension contains living biological material.

The particle scatterance in sea water is caused by a dilute suspension of irregular particles having various composition and covering a large size-range. Both polarization and dispersion of the scattered light occur to some degree (Jerlov, 1968, p. 39). The problem under consideration is the angular distribution of the scattered light, i.e. the scattering function, and its relation to the size distribution and nature of the particle content in the water.

III. Measuring Techniques

In order to measure the complete scattering function the scatterance must be observed at a number of angles from 0° to 180°. A discussion of the various techniques is given by Jerlov (1968, p. 15) and only certain points will be made here.

It is of central importance that the observations cover both small and large angles, and that they be carried out *in situ*. Most *in vitro* measurements are in the angular interval ca. 10° to ca. 165° and given only in relative units. This prevents (1) a calculation of the scattering coefficients by integration, and (2) a separation of molecular and particle scatterance. In the present review attention has been focused on the observations giving absolute values, and primarily those made *in situ*. The *in vitro* technique is hampered by the risk of contamination and of changes in the particles while sampling and during the time lapse between sampling and measurement.

Unfortunately measurements of the polarization of the scattered light are usually not carried out. They are highly desirable and could be carried out *in vitro*, whereas the scattering function should be measured *in situ*.

The calibration of the instruments presents a special problem. The use of standard scatterers or perfect diffusors is not reliable. The technique of Morel (1966) to use benzene as a known standard gives very accurate results. Kullenberg (1968, 1969) avoids the use of any reference by measuring the components entering eq. (1). This is a reliable technique provided the scattering volume is accurately determined and the geometry of the system is well defined.

IV. Observations of the Light Scattering Function

The existing observations are naturally classified by the measuring technique, *in situ* or *in vitro*, and whether or not the values are given in absolute units. Most observations are limited to the interval $10°-165°$, and only few measurements of the near forward ($\theta \sim 1°$) scatterance are available. Reviews of the historical development have been given by Jerlov (1963; 1968, p. 33) and Spilhaus (1965, p. 43).

The major features borne out by all observations are a smooth curve, a dominating forward scatterance, and a broad minimum around $100°-130°$ with a slight increase in the backward scatterance. There are, however, variations in the details.

A. PARTICLE SCATTERING FUNCTIONS

In Figs. 1 and 2 are shown particle scattering functions observed *in situ* and *in vitro*, respectively. The observations represent lake water (Duntley, 1963), coastal waters (Kullenberg, 1969), Atlantic surface water (Jerlov, 1961), Pacific near-coastal water (Tyler, 1961), Mediterranean (Kullenberg and Berg Olsen, 1972) and Sargasso Sea water (Kullenberg, 1968). Thus the range of very turbid to very clear water is covered. The functions are similar in general form and display the well-known features of the particle scattering function for a polydisperse system such as sea water.

The differences between the functions are most pronounced in the backward region. This reflects both real variations in the scattering particles between the various areas and the inherent experimental difficulties. For scattering angles between $80°$ and $20°$ there are minor individual variations whereas the differences are quite significant for angles below $20°$. The slope of the turbid water functions is steeper than the slope for the clear water functions. This is explained by the difference in particle character between the areas. Most likely anomalous

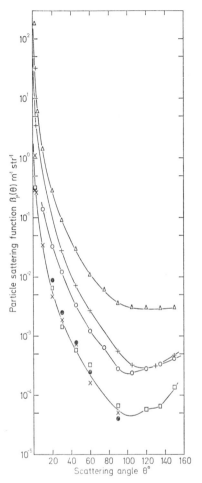

Fig. 1. Particle scattering functions observed *in situ*:

△ Duntley, lake, 1963;
○ Jerlov, Atlantic, 1961;
+ Kullenberg, Baltic, 1969;
× Kullenberg, Sargasso Sea, 1968;
□ Kullenberg and Berg Olsen, Mediterranean, 1972;
● Tyler, Pacific, 1961.

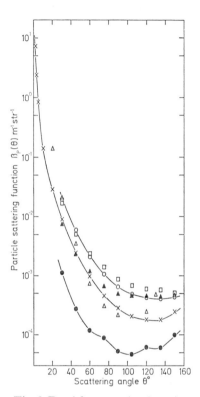

Fig. 2. Particle scattering functions observed *in vitro*:

× Bauer and Morel, Mediterranean, 1967;
△ Beardsley, Atlantic, 1968;
○ Morel, Mediterranean, 50 m, 1966;
● Morel, Mediterranean, 250 m, 1966;
□ Spilhaus, Atlantic, 50 m depth, 1965;
▲ Spilhaus, Atlantic, 700 m depth, 1965.

diffraction is an effective scattering process in areas containing a high amount of living organic matter.

The most striking divergences between individual observations are found in the *in vitro* case as is to be expected. This is to a large extent due to the experimental difficulties inherent in the technique. The observations by Morel (1966) are probably the most reliable with this technique. It should be pointed out that the Bauer–Morel function is similar to the Jerlov (1961) function for angles larger than 15°.

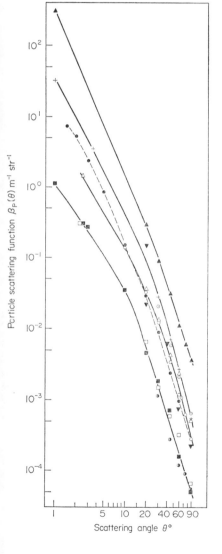

Fig. 3. Particle scattering functions:

● Bauer and Morel, Mediterranean, 1967;

▼ Beardsley, Atlantic, 1968;

▲ Duntley, lake surface, 1963;

○ Jerlov, Atlantic surface, 1961;

■ Kullenberg, Sargasso Sea, 10–75 m depth, 1968;

+ Kullenberg, Baltic surface, 1969;

△ Kullenberg and Berg Olsen, Mediterranean, Gibraltar, 10 m depth, 1972;

□ Kullenberg and Berg Olsen, western Mediterranean, 100 m depth, 1972;

⊙ Morel, Mediterranean, 50 m depth, 1966;

◐ Morel, Mediterranean, 2500 m depth, 1966.

The characteristics of the forward scatterance are borne out by the
representation in Fig. 3. The differences in the slopes are manifest, the
most turbid water has the steepest slope. Morrison (1970) used the same
representation including observations to a scattering angle of 0·2°,
and his general trends are in agreement with the present even though
only two different water types were considered. From 20° angle the
slopes become more similar and seem to converge to a common slope.
The *in vitro* values of Beardsley (1968) differ markedly in shape from
other results. The shape of the Bauer–Morel function, obtained by a
photographic method in the angular interval 1·5°–14·5°, is also different.
An inherent difficulty in every photographic technique is disturbing
stray-light which perhaps can explain the different behaviour of their
function.

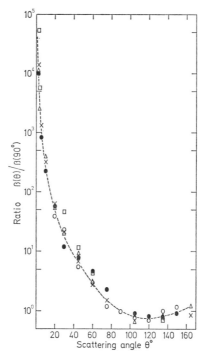

Fig. 4. Ratio of total scatterance to
total scatterance at 90°, $\beta(\theta)/\beta(90°)$:

× Duntley, lake, 1963;

△ Kullenberg, Sargasso Sea, 1968;

□ Kullenberg, Baltic, 1969;

● Ochakovsky, Mediterranean, 1966;

○ Sasaki *et al.*, Pacific, 1960.

B. TOTAL SCATTERANCE

In order to incorporate some of the investigations giving relative values only, examples of normalized functions for the total scatterance are given (Fig. 4). Generally the results conform and deviations occur only in details. Considering these, however, it seems premature to try to define average normalized scattering functions, in particular using other normalization angles than 90°. Obviously, individual scattering functions differ, but it seems possible to define average functions characteristic for certain water masses. More *in situ* observations covering the complete angular interval are needed before reliable characteristic curves can be defined.

An interesting feature shown by the 90°–normalized curves is a consistent decrease of the slope with depth (Kullenberg and Berg Olsen, 1972). This may be a general characteristic which is explained by the change of particles with depth.

The ratio of total scatterance at 45° to total scattering coefficient, $\beta(45°)/b$, is of great interest (Table 2). It has been found to be on an average constant for ocean waters (Jerlov, 1968). To obtain correct values of the ratio an accurate b-value is required. Estimates of b by extrapolation of the scattering function from 30° or even 10° and subsequent integration are not reliable. The scattering should be observed to angles of the order of 1°. This is evident since about 50–60% of the scatterance is found in the interval 0°–5° (Bauer and Morel, 1967; Kullenberg and Berg Olsen, 1972).

TABLE 2. Ratio of total scatterance at 45° to total scattering coefficient, $\beta(45°)/b$

Area	Depth (m)	Wavelength (nm)	Ratio $\times 10^2$	Reference
Atlantic	surface	465; 625	3·4	Jerlov, 1961
East China Sea	surface	~ 450	3·3	Kozlyaninov, 1957
Lake	surface	522	3·3	Duntley, 1963
Mediterranean	surface	545	3·5	Ochakovsky, 1966
Sargasso Sea	10–75	460; 650	3·5	Kullenberg, 1968
Mediterranean	10	633	2·5	Kullenberg and Berg Olsen, 1972
Mediterranean	100	633	3·4	Kullenberg and Berg Olsen, 1972
Baltic Sea	surface	460; 650	2·1	Kullenberg, 1969
Suspension of quartz		—	3·2	Hodkinson, 1962
Suspension of bituminous coal			2·0	Hodkinson, 1962

From laboratory measurements on known suspensions (Hodkinson, 1962) it is found that the ratio depends upon the type of suspension. The oceanic average is close to the value of $3 \cdot 5 \times 10^{-2}$ found for a suspension of quartz particles. This suggests that the average index of refraction for oceanic waters is close to that for quartz. For coastal waters the ratio varies and is significantly lower than in the ocean waters (Kullenberg, 1969). This is in agreement with the different nature and composition of the particles of coastal waters. The average index of refraction is probably lower than the oceanic value.

C. MOLECULAR SCATTERANCE

The scatterance due to the water itself and the dissolved salts form a minor part of the total scatterance. Nevertheless it is significant for large scattering angles, above 45°, in the most clear waters such as the Sargasso Sea and the deep Mediterranean. In the Sargasso Sea the molecular scatterance forms about 60–70% of the scatterance at 90° in the green and the red part of the spectrum whereas it makes only 7% of the total scattering coefficient. In general the influence of the molecular scatterance cannot give rise to any wavelength dependence of the scattering coefficient. The total forward scatterance is virtually independent of wavelength whereas there is an obvious wavelength dependence for larger angles. In more turbid waters the molecular scatterance is negligible.

The scatterance is essentially determined by particles, and the major effect is given by particles larger than the wavelength of the light. This implies that the wavelength dependence is insignificant (van de Hulst, 1957). Since the particulate matter in the sea plays a dominant role in the light scattering problem it seems relevant to discuss the particle content in some detail.

D. PARTICULATE MATTER IN THE SEA

(1) Concentration

The suspended material which is retained by a filter with pore diameter $0 \cdot 45$ μm, sometimes $0 \cdot 25$ μm, is defined as particulate matter. The parameters related to the light scattering problem are the size, form, composition and size distribution of the particles.

The total amount of suspended matter has been the subject of several investigations (Table 3). The most recent results by Jacobs and Ewing (1969) give very low concentrations. Their average value is in good agreement with the value of $0 \cdot 06$ mg l^{-1} given by Kullenberg (1953) calculated by using the scattering observations of Jerlov (1951).

Several studies give higher values, and although one is inclined to accept the lowest values the subject needs further investigation. The coastal waters contain much more particles than the oceanic waters, and the variation in the coastal zones is larger. It is evident that the suspension in sea-water is very dilute, even in coastal areas.

TABLE 3. Amount of suspended particulate matter

Area	Depth (m)	Suspension (mg/l)	Reference
(a) total			
Oceanic	deep water	0·05 (average)	Jacobs and Ewing, 1969
North Atlantic	surface water	0·04–0·15	Folger and Heezen, 1967
Oceanic	—	0·8–2·5 (average)	Lisitsyn, 1959
Pacific, coast	—	1·6	Goldberg et al., 1952
Coastal	—	6·0–18·0	Postma, 1954
(b) organic fraction			
Atlantic	—	0·04–0·17	Riley et al., 1965
North Atlantic	—	0·05–0·2	Gordon, D. C., 1970a
North Atlantic	deep water	0·01–0·02	Gordon, D. C., 1970a
Central Pacific	surface water	0·02	Gordon, D. C., 1971
Central Pacific	deep water	0·005	Gordon, D. C., 1971
(c) inorganic fraction			
Atlantic, offshore	—	0·05–1·0	Armstrong, 1958, 1965
Coastal	—	0·16–1·20	Armstrong, 1958, 1965

The particulate matter contains both organogenic and minerogenic material, and some information is available concerning their relative fractions (Table 3). Highly productive areas contain 60–80% organic matter whereas the oceanic deserts contain much less. Gordon (1970a) found that the organic matter was 25–38% of the total dry weight in the North Atlantic including the northern Sargasso Sea and Copin–Montegut (1972) found the organic fraction to be 64% and 48% in shallow and deep water, respectively in the eastern North Atlantic. As an oceanic average 20–60% organic matter is suggested (Parsons, 1963; Jerlov, 1968).

(2) *Organic matter*

The relative amount of organic matter shows considerable time and space variations. The surface layers can temporarily contain a dominating amount of biological material. The decomposition takes place in the upper few hundred metres. By advective processes organic matter is transported from the productive areas, and recent investigations seem to show that the amount of organic matter varies with depth also below the upper layers (Gordon 1970, 1971). The particulate organic components in the sea are extensively discussed by Riley and Chester (1971, chapter 8). Gordon (1970) made a microscopic study defining four particle categories larger than 5 μm. Aggregates (amorphous particles) were frequent in the surface layers but rare in the deep water. Organic aggregates have been directly observed in the sea, both in surface and deep layers. The size range of these particles is large, and they are of great importance for the light scatterance. The aggregates found in sea water are often almost spherical. Gordon (*loc. cit.*) also reported on flakes, i.e. thin shell-like particles with a circular or elongated form, and fragments formed from decay of detritus. The latter categories showed no significant variation with depth. He also found at all depths amber and clear crystals not associated with the organic matter.

Riley *et al.* (1965) found that the total amount of carbon contained in the aggregates in the northern part of the Sargasso Sea was about 10% of the total dry weight. The seston contained an abundance of mineral particles evident on microscopic examination. This has also been found by others and may be a significant property of the particulate organic matter, suggesting the importance of the adsorption process.

(3) *Inorganic matter*

The inorganic fraction has been studied extensively by Armstrong (1958, 1965) (Table 3). A large portion, 30–70%, of this material is quartz, but also iron and aluminium are present. The relative refractive index of this material is in the range 1·15–1·20. Copin–Montegut and Copin–Montegut (1972) also investigated the mass and composition of the inorganic matter in the eastern North Atlantic finding lower values than Armstrong. Not much is known about the size distribution and shape of the inorganic fraction. According to Sackett and Arrhenius (1962) there are two size fractions in the Pacific abyssal waters, a coarse one for particles larger than 0·5 μm, a fine one for particles in the range 0·01–0·5 μm.

A considerable amount of the minerogenic material is of terrestrial origin, being transported to the oceans by the atmosphere. Rex and Goldberg (1958) found the size distribution of quartz in the sediments to be similar to the size distribution in the atmospheric fall out. This may indicate that the material is not much changed by passage through the water. The size range of the quartz particles is 0·5 to 100 μm. A full discussion of the mineral particles in the sea is given by Riley and Chester (1971, p. 281).

(4) *Vertical distribution of particles*

The vertical distribution of the particulate matter is of great interest, and several investigations have been carried out by optical means (Jerlov, 1968, p. 168). In many areas there are nepheloid layers containing higher particle concentrations than the average. Jacobs and Ewing (1969) discuss the bottom nepheloid layers. In several cases the layers are undoubtedly related to the topography. Betzer and Pilson (1971) initiating a study of the chemical composition of the particles found particulate iron in high concentrations in the nepheloid layers.

The nepheloid layers found in the water column are often related to pycnoclines. A very nice example of scattering structure was found in the Arctic Ocean beneath the ice by Neshyba *et al.* (1968).

Laboratory studies have made it possible to estimate the amount of particulate matter above 1 μm from light scatterance observations (Jerlov and Kullenberg, 1953; Kullenberg, 1953).

(5) *Size distribution*

The size distribution of the particulate matter is the most important parameter for light scatterance. Visual and photographic techniques, microscopic investigations and Coulter counting techniques are at present the means to study the size distribution.

The visual and photographic observations (Nishizawa *et al.*, 1954; Costin, 1970) show that large particles in the range 100–1000 μm are present, but not homogeneously distributed. Clouds of particles are often encountered.

The microscopic investigations are probably the most significant in this connection. In the microscope the shape, area, and size of the particle can be determined. All the microscopic studies show the presence of large particles with a projected area corresponding to a sphere of the size 2 μm or larger. Obviously the cross-sectional area of the large particles dominates over the small particles (Table 4). This is very significant in relation to light scatterance, since it is the projected area which determines the scatterance. The microscopic investigations do not extend below 1–2 μm. For the intermediate sizes the size distributions often conform with an hyperbolic distribution (Fig. 5), but it must be observed that such a distribution greatly underestimates the number of the largest particles. As regards the small particles ($<$ 0·5 μm) nothing can be inferred from the microscopic countings. It is interesting to note that the size distribution given by Kullenberg (1953), determined on theoretical grounds, compares very well with the microscopic distributions.

TABLE 4. Normalized cross-sectional areas for particle distributions in the sea

Reference	Fraction (μm)/relative surface					
Jerlov, 1955	200–100	100–40	40–20	20–10	10–4	$<$ 2
	6·3	50	38	25	11	1
Ochakovsky,						
1966a	$>$ 50	50–25	25–10	10–5	5–2·5	2·5–1
$b = 0{\cdot}10\ \mathrm{m^{-1}}$	8	12	10	1·1	0·5	1
Lisitsyn, 1961	$>$ 100	100–50	50–10	10–5	5–1	$<$ 1
5–7 m	40	—	126	1·0	0·3	1
100 m	22	14	28	1·7	0·3	1
Gordon, D. C., 1970	56	42	28	21	14	7
surface water	1·1	0·6	1·1	0·9	0·9	1
deep water	0·7	2·0	4·4	3·7	1·5	1

During recent years the Coulter counter technique has been adopted for use *in vitro*. An investigation of its relevance has been done by Eckhoff (1969). It is difficult to compare directly the Coulter counter results with the microscopic, since the counter measures particle volume. The results are converted to size by assuming that all particles are spherical which is obviously not true. A lot of the particulate matter in the oceans has the form of flakes, shells or needles. A thin microparticle converts to a small equivalent sphere, but has a large effective scattering area.

Several possible size distributions have been suggested by the Coulter counter results (Bader, 1970; Carder, 1970; Carder *et al.*, 1971; Brun–Cottan, 1971). It seems as if the simple hyperbolic distribution is very representative. Examples of the results obtained by Carder *et al.* (1971) and Brun–Cottan (1971) are given in Fig. 5. Thus the Coulter distributions to a certain extent agree with the microscope results. There is, however, a tendency to get very steep hyperbolic distributions which

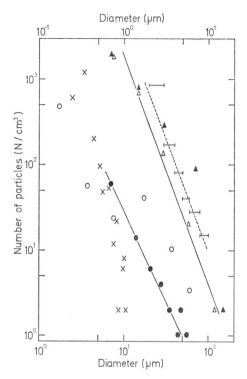

Fig. 5. Examples of particles size distributions:

Upper scale:

△ Kullenberg, Pacific deep, 1953;
|—| Brun–Cottan, Coulter counter, 500 m depth, Mediterranean, 1971.

Lower scale:

● Gordon D.C., microscope, organic matter, surface Atlantic, 1970;
× Carder *et al.*, Coulter counter, Pacific surface, 1971;
▲ Jerlov, microscope, fiord, 1955;
○ Ochakovsky, microscope, Mediterranean, 1966a.

may be ascribed to the conversion from volume to size. In connection with light scatterance this leads to overestimation of the role of the small particles whereas the large particles are in principle excluded. This implies that a large part of the scatterance is excluded. It seems that we have a fair idea about the form and range of the distribution for the intermediate particle sizes, but our knowledge of the ends of the size spectrum is vague. It is clear that an hyperbolic distribution cannot be used for either the largest or the smallest particles. On the contrary one would expect that below a certain size, an asymptotic value of the number per unit volume is reached since there must be a balance between settling, adsorption processes, and dissolution. The adsorption implies that the particle will not decrease in size indefinitely. It has been pointed out by Correns (1937) that the assumption of a Stokes settling velocity for small particles leads to absurd results. The conclusion is that the small particles are removed by other processes, such as flocculation and adsorption.

The problem of the size distribution is important, and it is not satisfactorily solved as yet. We can, however, infer from the above discussion that the large particles, above 1–2 μm, dominate the light scatterance. The role of the small particles can be important for certain angles of the scattering function in special areas.

V. Theoretical Computations of Scattering functions

A. APPLICATION OF THE MIE THEORY

It seems clear from the above discussion that it is particles larger than the wavelength of the scattered light which determine the scattering function. The main influence comes from particles in the range 1 μm to 200 μm. The suspension is dilute and independent scatterance can be assumed. The particle shape is irregular, the needle form is probably frequent. The particle absorption is low compared to the scattering efficiency, and it is a good approximation to use a real refractive index in the range 1·02–1·20.

Several investigations (e.g. Beardsley, 1968; Holland and Gagne, 1970) have shown that the Mie theory for spherical particles is applicable also for a suspension of irregularly formed particles, at least for scattering angles up to about 100°, provided that the total cross-section is the same in both suspensions. Holland and Gagne found excellent agreement between observed and computed scattering functions in the range 10°–120° for a suspension of flat, plate-like particles. The Mie theory was applied, using the true size distribution. For unpolarized light the agreement was not as good as for polarized light, and the backward scatterance differed considerably.

The goal of such computations for sea water is to obtain indirect information about particle size distributions, refractive index and composition. Several attempts have been made, and some of the results will be discussed.

B. EARLIER WORK

Burt (1956) constructed a very useful scattering diagram showing the efficiency factor for particles in the range 0·02–10 μm, as a function of wavelength, size and relative index. The rapid decrease of the efficiency factor with decreasing particle size below $\frac{1}{3}$ μm is clearly demonstrated.

Jerlov (1961) demonstrated the similarity between observed and computed functions for a monodisperse system. Only the general form compares, since the oscillatory trend of the monodisperse curve is completely wiped out for sea water.

Sasaki et al. (1960, 1962) computed scattering functions for monodisperse systems and compared them with observations in the angular interval 30°–150°. They found that a suspension of particles of the order 0·6 μm with an index value 1·20–1·25 generated a scattering function comparable to the observed. However, this result is mis-leading since the observations do not cover the very important forward range, which is required for a meaningful comparison.

Fukuda (1964) constructed a model taking only refraction plus reflection into account for large polyhedric particles. He demonstrated the importance of the index of refraction for large scattering angles by comparing the calculations with observations by Jerlov (1961). The forward diffraction dominated scatterance is not described by the model.

Spilhaus (1965) concluded that a polydisperse model must be used for describing the scattering function in sea water.

The same result was obtained by Ochakovsky (1966a) who compared observed scattering functions in the Mediterranean with functions generated by a monodisperse system. He arrived at a dominating particle size of 4 μm. Ochakovsky further discussed the importance of diffraction in the forward region, considering the phaseshift parameter ρ. Ochakovsky proceeded to apply a model size distribution originally constructed for the atmosphere. Scattering functions were computed assuming the scatterance to be caused by particles of the size 0·75 μm. The agreement between observations and computations is not convincing, and obviously this is explained by the difference in size distribution in the ocean and the atmosphere. A dominating size of 0·75 μm is also at odds with his earlier results as well as with the results of others.

Beardsley (1968) made complete in vitro observations of the scattering matrix for several waters enabling him to make comparisons with the theoretical Mie scattering matrix. He found an approximate agreement. The observed polarization was consistently less than the Rayleigh polarization at 90°. Beardsley concluded:

(1) that the suspension in sea water by and large scatters as a collection of spherical particles;

(2) that the strong forward scatterance as well as the apparent lack of structure in both the phase function and the inherent polarization indicates that the particle size distribution is fairly broad with a dominant diameter greater than 0·5 μm.

The light scatterance and the particle content in the Pacific were studied by Beardsley et al. (1970). The particle content in the interval 2 μm to 11 μm was counted with a Coulter counter. Equation (3) was used to determine a scattering coefficient from the particle counting, using an efficiency factor of 2. The value found in this way was less than the really observed scattering coefficient. The authors concluded that the unexplained difference was due to particles smaller than 1 μm, since these were not included in the counting. This implied that about 50% of the scattering must be due to particles smaller than 1 μm. This result seems, however, doubtful since the larger particles, of the order 20–30 μm, were not specified by the counting either. Only a very limited number of these particles are needed to account for the unexplained scatterance. It is further noted that the shape of the particles must enter into these considerations, not only the number per unit volume.

Kullenberg (1969, 1970) made attempts to describe observed scattering functions by applying the Mie theory to a polydisperse system using various hypothetical size distributions. The integration was carried out over a limited number of α-values using the tables by Ashley and Cobb (1958) and by Gumprecht and Sliepcewich (1953). Only particles larger than 2 μm were included. For the Sargasso Sea a hyperbolic distribution with a mean relative index of 1·20 gave a computed scattering function in good agreement with observed scatterance. The total scattering coefficient, the total number of particles and the suspended weight were also reasonably consistent with observations.

Pak et al. (1971) investigated the Mie scattering by suspended clay particles with an assumed log-normal distribution and a relative index value of 1·15. The size range covered was 0–30 μm. The authors paid special attention to the range of particles responsible for 90% of the scatterance at various angles. They found that generally particles smaller than 1 μm, as small as 0·2 μm, must be included in the distributions for angles above 5°. This result is, however, dependent upon the choice of size distribution. The log-normal distribution overestimates the small particles and underestimates the large particles which is clear from observations. Thus their conclusion is probably not generally valid.

Extensive calculations with the Mie theory covering a range of α-values from 0·2 to 200, corresponding to 0·04 μm to 40 μm for $\lambda = 600$ nm, determining both the efficiency factor, the scattering intensities, and the polarization have recently been carried out by Morel (1972a,b). He specially considered an hyperbolic distribution with an exponent equal to 3·5 and a mean relative index of 1·05. The results show that in this particular case particles larger than 20 μm have a small influence on the scatterance. This is, however, also dependent upon the size distribution. Clearly such a steep distribution does not hold true for the largest particles in sea water, and probably not for particles less than 0·5 μm.

Gordon and Brown (preprint 1972) compared computed scattering functions with observed functions in the Sargasso Sea. They arrived at the result that the average index is 1·05 and the appropriate size range is 0·08 μm to 10 or 20 μm, with an hyperbolic distribution having an exponent of 3. This implies that 60–70% of the particulate matter is organic, with an index value of 1·01, which does not seem realistic. Again the size distribution determined by the Coulter counter technique is doubtful.

C. PRESENT WORK

Kullenberg and Berg Olsen (1972) compare computed scattering functions with those observed in the Sargasso Sea, the Mediterranean Sea and the Baltic Sea. Hyperbolic size distributions are assumed, using a constant value of the exponent for the interval 1 μm to 40 μm. Different values of the exponent and the relative (real) refractive index are investigated. In the Mediterranean the size distribution over the interval 2 μm to 10 μm was observed,* using a Coulter counter, simultaneously with the scattering observations. From the countings an estimate of the exponent was obtained.

The theoretical scattering functions are computed using Deirmendjian's scheme (1969). The integration over the size range is fairly coarse with $\Delta\alpha \sim 5$, the scattering angle increments are 1° and 5° in the intervals 0°–10° and 10°–180°, respectively. The input data are the size range, the size increment, the exponent, and the observed particle scattering coefficient. The computations are truncated at a radius determined by the condition

$$\alpha \times m = 250$$

in order to secure convergence.

* by W. Plank, Oregon State University, Corvallis.

For the surface layer of the Sargasso Sea the best fit to the observations is found for an exponent of 2 and $m = 1 \cdot 20$ (Fig. 6). The agreement between computations and observations (Kullenberg, 1968) is convincing but some caution is necessary since the computations are not complete. It is quite possible that the oscillatory trend in the backward scatterance is a feature generated by the coarseness of the integration.

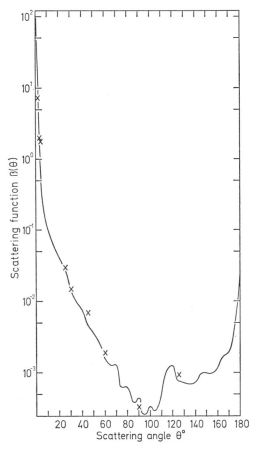

Fig. 6. Computed scattering function and observations for the Sargasso Sea: exponent 2, $m = 1 \cdot 20$, $\lambda = 632 \cdot 8$, $2 \ \mu m \leqslant r \leqslant 19 \cdot 5 \ \mu m$.

Also in the Mediterranean case the computed functions conform with those observed (Figs. 7, 8, 9). The effect of decreasing the index value is to shift the position of the secondary maximum towards 90°. The forward lobe is also altered, the slope of the function becoming steeper. In both the Sargasso Sea and the Mediterranean the scatterance is essentially determined by particles larger than 1–2 μm. The small particles have very little influence. The average index value is also rather high, in the interval 1·15–1·20. This seems quite realistic considering earlier results (Jerlov, 1968, p. 33).

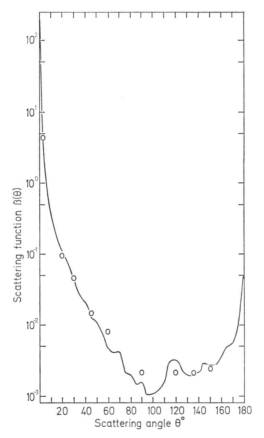

Fig. 7. Computed scattering function and observations in the Mediterranean, Gibraltar, at 100 m depth: exponent 2·2, $m = 1·20$, $\lambda = 632.8$ nm, 1 μm $\leqslant r \leqslant$ 19·5 μm.

For the Baltic Sea, representing coastal water, the problem is rather complicated. There is a large variety of particles, both organic and inorganic, covering a large range of sizes as well as of refractive index values. A reasonable average index seems to be 1·04. The size distribution in such turbid water is poorly known, and the assumed hyperbolic distribution is at best a crude approximation. The exponent is varied between 1·25 and 1·5. In consequence of these difficulties and short-comings, the agreement between computations and observations is less good in this case (Figs. 10, 11). Only the general trends of computed and

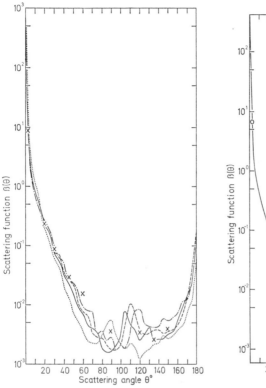 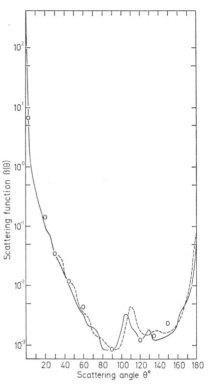

Fig. 8. Computed scattering functions and observations in the Mediterranean, Gibraltar, at 25 m depth: exponent 1·23, $\lambda = 632·8$ nm, $1\,\mu m \leqslant r \leqslant 19·5\,\mu m$, for $m = 1·10, \ldots; m = 1·15$, full drawn; $m = 1·17$, - - -; $m = 1·20$, · · · · ·.

Fig. 9. Computed scattering functions and observations for the western Mediterranean, at 50 m depth: exponent 1·53, $\lambda = 632·8$ nm, $1\,\mu m \leqslant r \leqslant 19·5\,\mu m$, for $m = 1·15$, full drawn; $m = 1·17$, dashed.

observed functions conform. It is evident that (1) both the exponent and the index of refraction are lower than in the ocean, and that (2) more refined size distributions and several index values covering a larger range than so far used should be applied in coastal waters, like the Baltic. More complete observations are also highly desirable.

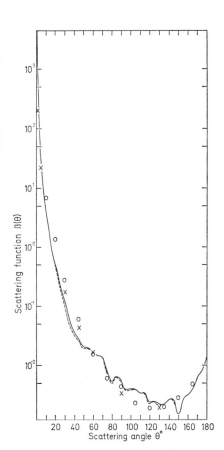

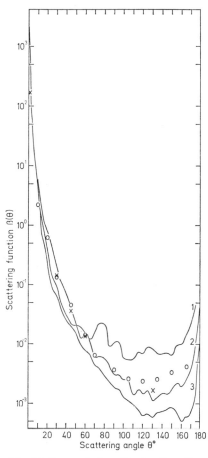

Fig. 10. Computed scattering functions and observations for the Baltic, at 5 m depth: $\lambda = 632 \cdot 8$ nm, $m = 1 \cdot 05$, $1 \mu m \leqslant r \leqslant 19 \cdot 5 \mu m$, for exponent $1 \cdot 25$, dashed; exponent $1 \cdot 5$, full drawn; observations: 655 nm, Station 1 (O); 632.8 nm, Station 3 ($\times$), from Kullenberg (1969).

Fig. 11. Computed scattering functions and observations for the Baltic, at 5 m depth: exponent $1 \cdot 25$, $\lambda = 525$ nm, $1 \mu m \leqslant r \leqslant 19 \mu m$ for $m = 1 \cdot 07$, (1); $m = 1 \cdot 04$, (2); $m = 1 \cdot 03$, (3); observations: 525 nm, Station 2 (O); 632·8 nm, Station 3 ($\times$); from Kullenberg (1969).

REFERENCES

Armstrong, F. A. J. (1958). *J. Mar. Res.*, **17**, 23–34.

Armstrong, F. A. J. (1965). *In* "Chemical Oceanography" (J. P. Riley and G. Skirrow, eds.), vol. 1, p. 409. Academic Press, London.

Ashley, L. E. and Cobb, C. M. (1958). *J. Opt. Soc. Am.*, **48**, 261–268.

Bader, H. (1970). *J. Geophys. Res.*, **75**, 2822–2830.

Bauer, D. and Morel, A. (1967). *Am. Geophys.*, **23**, 109–123.

Beardsley, G. F. (1968). *J. Opt. Soc. Am.*, **58**, 52–57.

Beardsley, G. F., Pak, H., Carder, K. and Lundgren, B. (1970). *J. Geophys. Res.*, **75**, 2837–2845.

Betzer, P. R. and Pilson, M. E. Q. (1971). *Deep-Sea Res.*, **18**, 753–761.

Brun-Cottan, J.-C. (1971). *Cahiers Oceanog.*, **23**, 193–205.

Burt, W. V. (1956). *J. Mar. Res.*, **15**, 76–80.

Carder, L. C. (1970). "Particles in the eastern Pacific Ocean: their distribution and effect upon optical parameters." Ph.D. thesis. 140 pp. Oregon State University.

Carder, K. L., Beardsley, G. F. and Pak, H. (1971). *J. Geophys. Res.*, **76**, 5070–5077.

Copin-Montegut, C. and Copin-Montegut, G. (1972). *Deep-Sea Res.*, **19**, 445–452.

Correns, C. (1937). *Wiss. Erg. d. D. Att. Exp.*, **3**, (135).

Costin, J. M. (1970). *J. Geophys. Res.*, **75**, 4144–4150.

Duntley, S. Q. (1963). *J. Opt. Soc. Am.*, **53**, 214–233.

Eckhoff, R. K. (1969). *J. Sci. Instr. (J. Physics E)*, **2**, 973–977.

Einstein, A. (1910). *Ann. Physik*, **33**, 1275–1298.

Folger, D. W. and Heezen, B. C. (1967). *Abstr. Geol. Soc. Am.*, N. E. Section.

Fukuda, M. (1964). *In* "Physical Aspects of Light in the Sea." (J. Tyler, ed.), pp. 61–64.

Gans, R. (1925). *Ann. Physik*, **76**, (29).

Goldberg, E. D., Baker, M. and Fox, D. L. (1952). *J. Mar. Res.*, **11**, 194–204.

Gordon, D. C., Jr. (1970). *Deep-Sea Res.*, **17**, 175–185.

Gordon, D. C., Jr. (1970a). *Deep-Sea Res.*, **17**, 233–243.

Gordon, D. C., Jr. (1971). *Deep-Sea Res.*, **18**, 1127–1134.

Gordon, H. R. and Brown, O. B. (1972). "A theoretical model of light scattering by Sargasso Sea particulates" (Preprint). University of Miami.

Gumprecht, R. O. and Sliepcevich, C. M. (1953). *J. Opt. Soc. Am.*, **57**, 90–94.

Hodkinson, J. R. (1962). *In* "Electromagnetic Scattering" (M. Kerker, ed.), pp. 87–100. Potsdam, N.Y.

Holland, A. C. and Gagne, G. (1970). *Appl. Optics*, **9**, 1113–1121.

Jacobs, M. B. and Ewing, M. (1969). *Science, N.Y.*, **163**, 380–383.

Jerlov, N. G. (1951). *Rept. Swedish Deep-Sea Expedition*, **3**, 1–59.

Jerlov, N. G. and Kullenberg, B. (1953). *Tellus*, **5**, 306–307.

Jerlov, N. G. (1955). *Tellus*, **7**, 218–225.

Jerlov, N. G. (1961). *Medd. Oceanog. Inst. Göteborg*, **30**, 1–40.

Jerlov, N. G. (1963). *Oceanog. Mar. Biol. Ann. Rev.*, **1**, 89–114.

Jerlov, N. G. (1968). "Optical Oceanography", p. 194. Elsevier Oceanography Series.

Kozlyaninov, M. V. (1957). *Tr. Inst. Okeanol. Akad, Nauk SSSR*, **25**, 134.

Kullenberg, B. (1953). *Tellus*, **5**, 302–305.

Kullenberg, G. (1968). *Deep-Sea Res.*, **15**, 423–432.

Kullenberg, G. (1969). *Rep. Inst. Fys. Oceanog.*, **5**, 16. Københavns Universitet.

Kullenberg, G. (1970). *Rep. Inst. Fys. Oceanog.*, **13**, 22. Københavns Universitet.
Kullenberg, G. and Berg Olsen, N. (1972). *Rep. Inst. Fys. Oceanog.*, **19**, 40. Københavns Universitet.
LeGrand, Y. (1939). *Ann. Inst. Océanog.*, **19**, 393–346.
Lisitsyn, A. P. (1961). Raspredelenie i Sostav Vzvschennogo Materiale v Moryakh i Okeanakh. Sovremennye Osadki Morei i Okeanov (Tr. Soveskehaniya 24–27 maya 1960 g), Moskva, pp. 175–232.
Mie, G. (1908). *Ann. Physik*, **25**, 377.
Morel, A. (1966). *J. Chim. Phys.*, **10**, 1359–1366.
Morel, A. (1972a). "Application de la theorie de Mie au calcul de l'indicatrice de diffusion de la lumiere pour les eaux de mer" (Preprint). Univ. de Paris, Lab. d'Oceanog. Phys.
Morel, A. (1972b). "Au sujet de l'emploi du coefficient total de diffusion pour evaluer la teneur des eaux de mer en particules en suspension" (Preprint). Univ. de Paris, Lab. d'Oceanog. Phys.
Morrison, R. E. (1970). *J. Geophys. Res.*, **75**, 612–628.
Neshyba, S., Beardsley, G. F., Neal, V. T. and Carder, K. (1968). *Science, N.Y.*, **162**, 1267–1268.
Nishizawa, S., Fukuda, M. and Inoue, N. (1954). *Bull. Fac. Fisheries, Hokkaido Univ.*, **5**, 36–40.
Ochakovsky, Yu. E. (1966). U.S. Dept. Comm., Joint Publ. *Res. Ser. Rept.*, **36**, 98–105.
Ochakovsky, Yu. E. (1966a). U.S. Dept. Comm. Joint Publ., *Res. Ser. Rept.*, **36**, 16–24.
Pak, H., Zaneveld, J. R. V. and Beardsley, G. F. (1971). *J. Geophys. Res., Oceans and Atmospheres*, **76**, 5065–5069.
Parsons, T. R. (1963). *In* "Progress in Oceanography" (M. Sears, ed.), Vol. 1, pp. 205–239. Pergamon, New York.
Postma, H. (1954). *Arch. Neerl. Zool.*, **10**, 1–106.
Rayleigh, Lord (1871). *Phil. Mag.*, **41**, 447–454.
Rex, R. W. and Goldberg, E. D. (1958). *Tellus*, **X**, 153–159.
Riley, G. A., Hemert, D. Van and Wangersky, P. J. (1965). *Limn. Oceanogr.*, **10**, 354–363.
Riley, J. P. and Chester, R. (1971). "Introduction to Marine Chemistry", 465 pp. Academic Press, London.
Sackett, W. M. and Arrhenius, G. O. S. (1962). *Geochim. Cosmochim Acta*, **26**, 955.
Sasaki, T., Okami, N., Oshiba, G. and Watanabe, S. (1960). *Rec. Oceanog. Works in Japan*, **5**, 1–10.
Sasaki, T., Okami, N., Oshiba, G. and Watanabe, S. (1962). *Sci. Papers, Inst. Phys. Chem. Res. (Tokyo)*, **56**, 77–83.
Smoluchowski, M. (1908). *Ann. Physik*, **25**, 205–226.
Spilhaus, A. F., Jr. (1965). "Observations of light scattering in sea water." Ph.D. thesis, Dept. of Geol. and Geophys. MJT, 242 pp.
Tyler, J. E. (1961). *Limn. and Oceanog.*, **6**, 451–456.
Van de Hulst, H. C. (1957). "Light Scattering by Small Particles." Wiley, New York, 470 pp.

Chapter 3

Refraction and Reflection of Light at the Sea Surface

C. S. COX

Scripps Institution of Oceanography, University of California, San Diego, La Jolla, California, U.S.A.

I. INTRODUCTION

The sea surface scatters light because it is roughened by action of the wind. Waves and swell of all wavelengths down to the shortest capillary waves combine to refract and reflect light rays into continually changing directions and even to focus small bundles of rays. Bubbles and spume at the surface scatter the light even further.

As a consequence the appearance of the sun reflected at, or refracted through, the sea surface is broken up into myriads of dancing highlights each of which is a distorted image of the sun.

The scattering of light on the sea surface is important for marine organisms for several reasons. The average distribution of directions of propagation of light is broadened by the scattering and even the average intensity is altered. For example when the sun is low in the sky far more sunlight enters the sea on a windy day than on a calm day because under the former condition the rays from the sun enter the sea through steeply tilted surface facets which have many times more transmissivity than facets entered at grazing incidence on the smooth sea. (For the same reason the albedo of the sea depends on its roughness.) Organisms' living very close to the sea surface are exposed to enormous fluctuations of light intensity on sunny days, because of the focusing effect of surface ripples. On windy days when the typical curvature of the sea surface is large the flickers are most pronounced within centimeters of the water surface. On calm days the flickers are apparent to several meters.

The scattering of light by the sea is also important for the visibility of objects on the sea or seen through the sea surface. The horizon on a calm day is almost invisible because the reflection coefficient at grazing incidence approaches unity. Consequently at the horizon the smooth sea takes on nearly the same color and brightness as the sky just above the horizon. On windy days however the rays which have come from the sea horizon have been reflected by steeply tilted wavelets. Hence they must have originated high in the sky and suffered considerable loss of intensity on reflection.

The scattering of light is also important because it allows us to infer the structure of the sea surface and to make such inferences without any physical contact with the sea. Even observations from above the earth's atmosphere have been able to detect and measure the scattering, and associate it with the wind field at the sea surface.

II. Reflection and Refraction of Natural Water Surfaces

When the sea surface is clean the reflection coefficient is given by Fresnel's Law, for parallel and perpendicular polarization,

$$\Gamma_{\parallel} = \sin^2 (i-j)/\sin^2 (i+j)$$
$$\Gamma_{\perp} = \tan^2 (i-j)/\tan^2 (i+j)$$
$$\sin i = n \sin j$$

where n is the index of refraction and i is the angle of incidence in air (Fig. 1). The reflection coefficients are about 2% for normal incidence, and reach 100% for grazing incidence. For perpendicular polarization the reflection coefficient vanishes at Brewster's angle, $i \simeq 53°$ (Fig. 1).

Natural water surfaces are often covered by slicks which make their presence known by their damping action on capillary waves and by a reduction of surface tension. At sea, slicks are often hardly more than monomolecular layers, much thinner than a wavelength of light. They have a negligible effect on the reflectivity of the surface. On the other hand, other slicks, perhaps from oil spills, are thick enough to cause visible interference colors and therefore appreciably influence the reflectivity. Neither type of slick effects the angles of refraction or reflection.

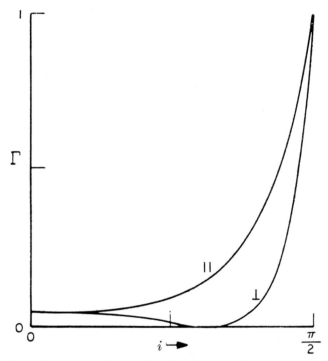

Fig. 1. The reflection coefficient for the air-water interface as a function of zenith angle, i. $\|$ = parallel polarization, $\perp$ = perpendicular.

III. Structure of the Sea Surface

The directional properties and intensities of light beams reflected and refracted at the sea surface depend on the slope of the particular element of sea surface where the beam intersects the sea. The focusing properties of the sea surface depend on the curvature as well. The slope and curvature of the sea surface are mainly controlled by the shorter wavelength waves and ripples on the sea except where bubbles, spume and detritus introduce distortions of the surface. The properties of short waves on wind-blown water surfaces are known very imperfectly; what *is* known has mostly been measured by optical techniques.

Various properties have been measured. In decreasing certainty, the list includes:

(a) Mean square slope components of the water surface.

(b) Probability distribution of surface slopes.

(c) Frequency spectrum of one component of water surface slope.

(d) Mean square curvature.

The mean square and probability distribution of slopes have been measured and correlated with wind in the open sea (Schooley, 1954; Cox and Munk, 1954a,b, 1955). They have also been measured under controlled laboratory conditions (Cox, 1958; Wu, 1971) and compared with the oceanic data (Cox, 1958; Wu, 1972). The comparison is surprisingly good when one realizes that the spectra of turbulence in the air flows are probably very different. In the laboratory experiments, long gravity waves are absent. For this reason there is not the opportunity to model large turbulent eddies near the water surface, a circumstance which probably alters the ripple structure, particularly in the lee of sharp wave crests.

The dispersion relation $\omega(k)$ for infinitesimal ripples becomes greatly distorted by dynamical effects of wind blowing over the waves (Miles, 1959b) and by finite amplitude effects. The dispersion relation becomes smeared by superposition of ripples on the orbital velocity of a broad spectrum of gravity waves and by the shear flow in the oceanic surface boundary layer. For these reasons it is difficult to reduce the frequency spectrum of ripples measured in the laboratory (Cox, 1958) to the needed information on the spatial spectrum.

A useful summary of theoretical ideas by Phillips (1966) calls attention to two modes by which ripples are generated. At low or moderate wind speeds, fleeting trains of ripples are generated at the sharp crests formed by the chance concatenation of short gravity waves. These ripples are clearly seen in the photograph of Fig. 2.

At higher winds, ripples are apparently generated directly by the wind. Phillips associates this with Miles' (1957, 1959a, 1959b, 1960, 1962) shear instability but one must recognize that the wind boundary layer which affects ripples is extremely variable and that the phase velocity of ripples is essentially controlled by the orbital velocity of the large gravity waves on which they are riding. Conditions appropriate to Miles' steady state calculation are far from satisfied.

In a re-examination of the mean square slope data, Wu (1972) adopts the spectral forms proposed by Phillips and evaluates the coefficients from laboratory and open sea measurements of mean square slope. The form chosen for the wavenumber spectrum of slopes (taken regardless of direction) is

$$\chi(k) = B_0/k \text{ for } k_0 < k \leqslant k_T \text{ at all wind speeds and in addition}$$

$$\chi(k) = B_1/k \text{ for } k_T < k \leqslant k_1 \text{ for sufficiently strong winds.}$$

Here k is the wavenumber, regardless of direction, k_0 and k_1 are the limiting wavenumbers, for long waves and short waves respectively,

Fig. 2. Negative of the sea surface photographed from Scripps Institution of Oceanography pier. Wind is blowing onshore (white arrow). Wind speed of 6·5 m/s at 12 m is insufficient to generate ripples directly, but intermittent ripple trains are generated by sharp crests of gravity waves.

beyond which the spectrum is assumed to vanish. The change of spectral form occurs at the capillary-gravity boundary,

$$k_T = \sqrt{g/T}$$

where T the surface tension divided by density is about 73 cm³ s⁻². According to well known data k_0 decreases with increasing wind speed U roughly as

$$k_0 = g/U^2$$

It appears that k_1 increases with increasing wind speed. For winds decreasing to 7 m s^{-1}, k_1 approaches k_T and for low winds the capillary range vanishes altogether. These features are illustrated in a general way by the behavior of the mean square slope (Fig. 3).

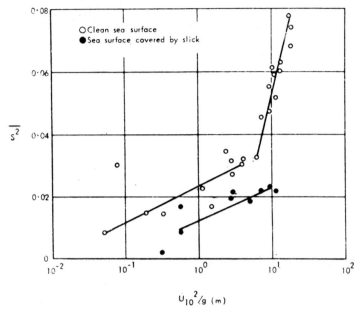

Fig. 3. Mean square slope of the ocean surface as a function of windspeed. Data from Cox and Munk. According to interpretations by Cox (1958), Phillips (1966) and Wu (1972) there are two regimes: below 7 m/s where capillaries are weakly developed and above 7 m/s where capillaries are increasingly developed. Black circles indicate measurements over oil slicks where capillaries and short gravity waves are absent.

$$\sigma^2 = \int_{k_0}^{k_1} k^2 \chi(k)\, \mathrm{d}k$$
$$= B_0 \log (k_T/k_0) \text{ at low winds}$$
$$= B_0 \log (k_T/k_0) + B_1 \log (k_1/k_T) \text{ for higher winds.}$$

The spectral form in the gravity wave region is in agreement with the observations of mean square slope of oil-slick covered sea surfaces where capillary and short gravity waves are removed (black dots) while the clean sea surface shows a rapid growth of capillaries at high winds. In a qualitative way the increase of k_1 with increasing winds is verified by laboratory measurements (Cox, 1958, Fig. 4) which show capillaries extending into higher frequency at higher winds. (In a general way the wavenumber increases with frequency according to the usual dispersion relation.)

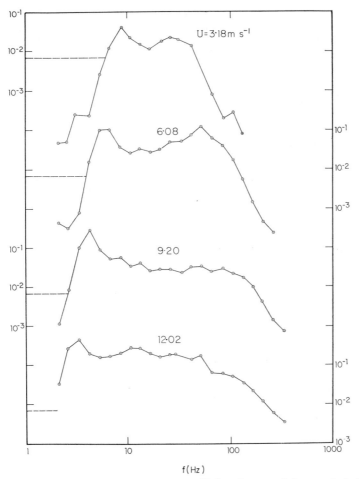

Fig. 4. Slope spectrum times frequency, $fS(f)$, of up- and down-wind slopes as a function of frequency, f, for four wind speeds U as shown. Measurements by Cox (1958) were made in a laboratory wind/water channel.

The areas of specular images formed by reflection of a small light source in the water surface have been measured by Wu (1971) in a laboratory wind-wave tunnel. Wu reports his results in terms of the mean radius of curvature of the water surface, Fig. 5. The radius of curvature is sensitive to the short wavelength part of the wave spectrum, and therefore these measurements provide a test of the form

3*

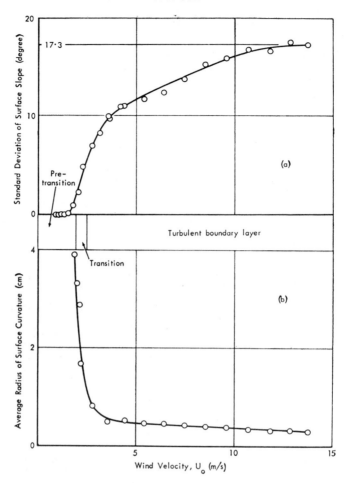

Fig. 5. Laboratory measurements of slope and radius of curvature by Wu (1971). The windspeed is measured close to the water surface.

of the capillary wave spectrum. According to the work of Longuet-Higgins reviewed in Section VI below, the mean area of specular images is inversely proportional to the mean square curvature. According to Phillips' form of the spectrum the mean square curvature

$$-\tfrac{1}{2}B_0 k_0^2 + \tfrac{1}{2}(B_0 - B_1)k_T^2 + \tfrac{1}{2}B_1 k_1^2$$

should increase rapidly with windspeed as soon as capillaries are formed in profusion. Wu's observations showed, however, only the expected rapid decrease of mean radius of curvature associated with the first formation of capillary waves. Beyond a windspeed of $2 \cdot 2$ m s^{-1} the mean decreases only 30% to the highest recorded winds (the laboratory wind of $2 \cdot 2$ m s^{-1} corresponds to $6 \cdot 5$ m s^{-1} for usual anemometer heights in oceanic observations), whereas the decrease expected from Wu's proposal

$$k_1 \chi (\text{wind speed})^2$$

should be several fold. We are forced to conclude that the capillary wave spectrum remains uncertain.

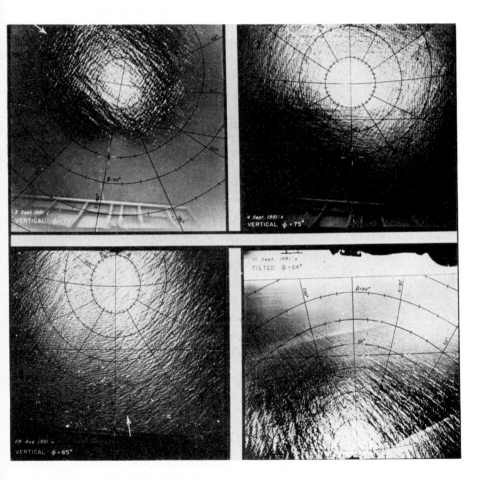

Fig. 6. Upper left and right, and lower left: Sun glitter pattern at wind speeds of $0 \cdot 7$, $3 \cdot 9$ and 14 m s^{-1} respectively. Lower Right: A rectangular artificial slick, with near boundary almost through specular point. Brightness of slick sea surface is reduced for large slopes and increased for small slopes.

IV. Slicks

The effect of oil or natural slicks on the scattering of light at the sea surface arises from two causes, the change of reflection coefficient and the change of sea surface roughness.

The reflection coefficient change is not large. For a multimolecular layered oil slick Cox and Munk (1956) estimated that the reflection coefficient increased by 50% at normal incidence and less at other angles. Many natural slicks are monomolecular. Their influence on the reflection coefficient is negligible, yet the changes in scattering properties are dramatic (Fig. 6). The cause appears to be the complete inhibition of capillary and very short gravity waves. This reduces the r.m.s. slopes and greatly reduces the r.m.s. curvature.

Under light wind conditions, slicks are often irregular in form and may cover very large parts of the sea surface. Possibly the large areas of sea surface with anomalously small scattering visible in photographs from space (Figs. 14 to 21) are slicked areas. Such areas have characteristically sharp edges where capillary waves suddenly appear and scattering changes dramatically.

Slicks are sometimes associated with convergent zones of horizontal water flow, for example, associated with oceanic fronts and internal waves.

At high winds slicks are drawn into narrow bands a few meters wide and hundreds of meters long, parallel to the wind direction (Fig. 12).

V. Usefulness of Ray Approximation

Spindrift, bubbles and detritus on the sea surface cause scattering of an intractable nature. Often such complications can be avoided when winds are less than gale force, by excluding whitecaps and foam patches from analysis. Then the only scatterers are waves with three simplifying aspects: (1) The slopes are usually smaller than 30° hence multiple reflections and shadowing can be ignored, at least for steeply dipping rays. The ratios of visible light wavelength to (2) the radius of curvature and (3) to the wavelength of water waves are both small compared to typical slopes of the waves. Aspects (2) and (3) permit one to use ray optics with adequate accuracy for studying the scattering of visible light. The magnitude of radii of curvature of the sea surface influence the scattering in the following way. According to the ray approximation, there are only separate and discrete points on the ruffled sea surface where light originating at a point source (far above the sea) will be reflected or refracted into a given direction. These specular points form the glittering sparkles apparent on a sunny day. On the other hand, the wave nature of light requires that an adjoining region around each of these points must enter cooperatively into the reflection process. Provided these adjoining regions do not overlap we can conclude that the ray approximation of distinct specular points is appropriate.

We estimate the size of the region by means of Rayleigh's criterion that the optical path length of rays cooperating in the reflection process should not differ by more than $\lambda/4$, where λ is the light wavelength. If a characteristic radius of curvature is R and i is the angle of incidence the size of the adjoining region is roughly

$$2R\sqrt{(\lambda/R)}\cos i$$

The separation of specular points is governed by the wavelength L of the shortest water waves; for example specular images of vertically back reflected light would occur on successive crests and troughs of the waves. Thus we require that

$$2R\sqrt{(\lambda/R)}\cos i \ll L/2$$

or equivalently that $\sqrt{\lambda/R}\cos i$ be small compared to the maximum slope of the wavelets.

Appropriate values are $R \geq 5$ mm, r.m.s. slope = 0·1 to 0·3. Hence for $\lambda = 500$ nm (green light), the inequality is barely satisfied. On occasion (see below) the specular points coalesce and are annihilated on contact. When this occurs we must expect that dramatic interference effects would be formed by reflections of monochromatic light source.

The requirement that the wavelength of water waves be large compared to the light wavelength is necessary so that the angles of Bragg scattering be small. In fact if these angles are small compared to the r.m.s. slope of the ocean waves then the overall intensity distribution of scattering will not be much modified by the wave nature of light. According to the studies outlined in Section III, the cutoff wavelength corresponding to k_1 is greater than a millimeter or so even when winds of 15 m s^{-1} blow over the sea, while r.m.s. slopes are of order 0·1–0·3. Hence at least the gross aspects of scattering are adequately treated by the ray approximation when radiation with wavelengths short compared to 100 μm is incident.

Clearly criterion (2) is more stringent than (3). We expect the ray approximation to be barely adequate to specify the separation between sparkles when visible light is used but the overall scattering pattern will be adequately represented when much longer waves are used.

VI. Distribution of Scattering in Time and Space

When the statistical properties of waves are known, it is possible to calculate by ray optics the statistical properties of scattering. One needs to consider a hierarchy of statistical properties of the waves and of the scattering; the probability distribution of water slopes at an instant is related to the overall smoothed distribution in direction of reflected and refracted beams at that instant (Cox and Munk, 1954b),

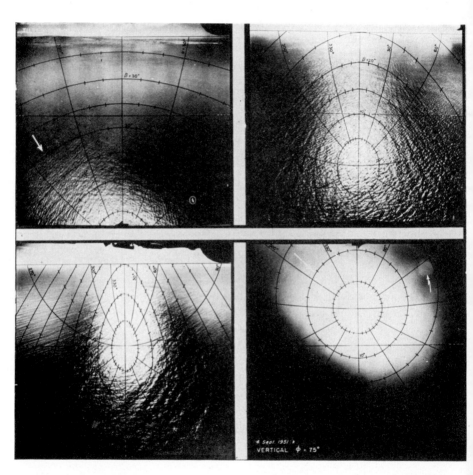

Fig. 7. Upper left and right and lower left: Sun glitter pattern at solar elevations of $\phi = 75°$, $50°$ and $10°$, respectively. The superimposed grids consist of lines of constant slope azimuth (radial) drawn for every $5°$. The white arrow indicates wind direction. Lower right: Photometric photograph obtained by optical smoothing (a lensless camera).

(Figs. 7 and 8), but is insufficient knowledge to calculate the variation of scattered light corresponding to the statistical distribution of individual specular points. For the latter, the joint probability distribution of water slopes at two separated points is required. To find the time variation of scattering one also needs to know how this joint probability unfolds in time.

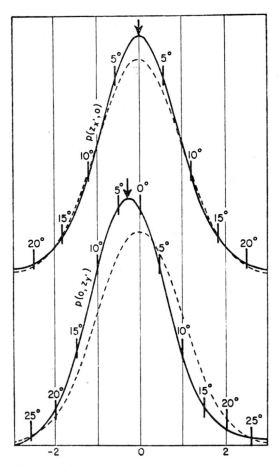

Fig. 8. Principal sections through the probability distribution surface. The upper curves are along the crosswind axis, the lower ones along the up/downwind axis (positive upwind). The solid curves refer to the observed distribution, the dashed curves to a Gaussian distribution of equal mean square slope components. The thin vertical lines show the scale for slope components normalized by the rms component of slopes. The two scales for tilt refer to a wind speed of 10 m s^{-1}; skewness shown in the lower curve is computed for this wind speed. The modes are marked by arrows.

The required joint probability distributions are closely related to the space-time spectrum of water waves. In fact, if the probability distribution can be assumed to be Gaussian then knowledge of the spectrum is sufficient information.

In a remarkable series of papers Longuet-Higgins (1956, 1957a,b, 1960a,b,c) has computed many of the statistical properties of specular points. In order to make detailed calculations he has adopted the hypothesis of Gaussian probability distributions and, for the time variation, assumed that the waves have a dispersion relation.

Among the calculated statistical quantities we illustrate two: the distribution of areas of images of a small light source and the creation and annihilation of specular points. The probability distribution of areas depends essentially on the solid angle subtended by the light source ω and an integral over the wave spectrum closely related to the mean square curvature. This is to be expected because the area of a specular image projected on to a horizontal plane is $\omega/(4 \times \text{total curvature})$. Suppose the spectrum of elevation of the waves $E(k)$ can be represented as a product of a wavenumber factor $F(k)$ and a normalized directional factor $G(\theta)$, $E = FG$. The distribution of image areas in units of the area $\omega/(2M_4\sqrt{6\Theta})$ is shown in Fig. 9. Here M_4 is the fourth moment of the wave spectrum

$$M_4 = \int_{-\infty}^{\infty}\int k^4 E(k) \, \mathrm{d}k_x \, \mathrm{d}k_y$$

and Θ is an angular factor

$$\Theta = \int_{-\pi}^{\pi}\int_{-\pi}^{\pi} G(\theta_1)G(\theta_2) \sin^4 (\theta_1 - \theta_2) \, \mathrm{d}\theta_1 \, \mathrm{d}\theta_2.$$

It is remarkable that the shape of the distribution function is only slightly dependent on the shape of the wave spectrum (the two lines are almost extreme limits for any spectral shape) and that there is a

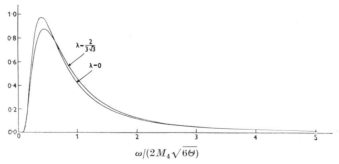

Fig. 9. The probability distribution of areas of images.

sharply defined lower limit to the areas of specular images. The latter point has nothing to do with the shape of the spectrum—we only require that the fourth moment of the spectrum exist and that the Gaussian hypothesis be correct.

Longuet-Higgins shows that the location of specular points is associated with maxima, minima and saddle points of a certain function, which for specular points vertically below and a light source vertically above the observer, coincides with the shape of the water surface. The maxima and minima therefore are points where the total curvature is positive, the saddle point where the total curvature is negative.

Specular points are generally annihilated in pairs: A maximum or minimum of the function moves toward a saddle point. The process is illustrated in Fig. 10. The specular points are identified by a cross and a circle in the left diagram. In the center diagram the two points have just coalesced and on the right we see no possible specular images.

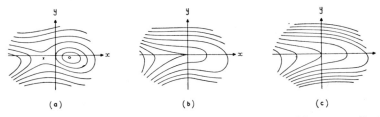

Fig. 10. The annihilation of a pair of specular images. Lines are effectively contours of the water surface. The creation of specular points is similar but reversed in time.

The foregoing studies relate to the statistics of individual specular points. For some purposes it is valuable to examine the spatial spectrum of brightness associated with a field of specular images. It is plausible that the long wavelength parts of this "brightness spectrum" are related to the spectrum of long waves for the following reason. In the absence of long waves and swell and if the sea is uniformly ruffled by wind, the average pattern of brightness produced by an overhead light source is a bright core of scattered light below the observer which gradually diminishes in intensity at increasing distances from the core. The effect of long waves superimposed on the ruffled pattern is to tilt the ruffled sea to and fro. Therefore if one examines the scattered light in a region where the distribution (due to the ruffling elements only) is decreasing linearly in brightness with increasing distance from the core, one expects to find the smoothed brightness to vary with the slope of the long waves. We shall put these remarks into quantitative form. To keep the discussion simple we take a one dimensional example.

Suppose the light source has an infinitesimal angular size $2\delta S_0$. According to ray theory there will be a reflection of light to the observer whenever the slope of the water surface $S(X) = \partial z(x)/\partial x$ has a certain slope, S_0, within the tolerance δS_0 as illustrated in the middle part of Fig. 11. The critical slope S_0 will vary from place to place over the sea

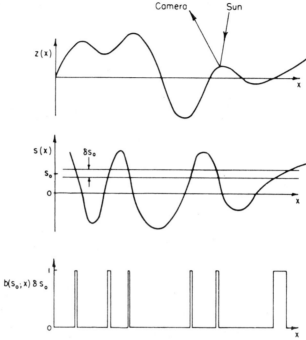

Fig. 11. Upper diagram shows wave elevations $z(x)$. Middle shows the slope $\delta z/\delta x$. Reflections of a small source occur when $\delta z/\delta x = S_0$. Lower panel shows the resultant distribution of brightness as a function of position along the sea surface.

surface, but in a sufficiently small region, if the source and observer are high over the sea, this variation may be neglected, as assumed in the figure. The distribution of brightness across the sea surface will vary as shown in the lower part of the figure. We now consider some statistical properties of the spatial brightness function:

(1) It is clear that the mean brightness is directly proportional to the probability of finding a sea surface element with slope S_0 to $S_0 + \delta S_0$.

(2) The correlation of brightness at two separated points may be shown to depend only on the joint probability distribution for slopes at the two separated points. If this joint distribution is

Gaussian it may be completely derived from a knowledge of the spatial spectrum of water waves by the methods of Rice (1944). In this way one can relate unambiguously the spectrum of the spatial distribution of brightness with the spectrum of the spatial distribution of waves. Calculations by Cox (1956) show that for low spatial frequencies the brightness spectrum depends on the long wavelength part of the wave spectrum and on the mean square slope of all waves. At high spatial frequencies the spatial spectrum of brightness depends on the mean square slope and mean square curvature. This development provides a rational explanation of the visibility of ocean wind waves when they are within a sun glitter pattern (see Fig. 12).

Fig. 12. The tradewind sea photographed from an elevation of 3600 m. The waves are visible because they tilt the roughened sea surface alternately towards and away from the observer thus modulating the intensity of back scattered sunlight. The area photographed was at the edge of the sun glitter area where the mean brightness was decreasing towards lower left. Note scattering from white caps independent of position. Barely visible are streaks associated with wind blown slicks.

VII. Optical Observations of Sea Surface Roughness

In the laboratory two techniques have been used. Cox (1958) describes use of a method which produces changes of brightness at the water surface in proportion to one component of slope of the water surface. By recording the brightness changes in space and time one finds the space-time history of surface slopes. The method (Fig. 13) requires use of an extended light source of graded brightness below the water surface and one or more directional photometers above.

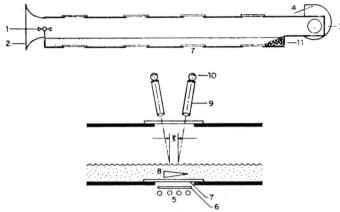

Fig. 13. Wind and water tunnel for measurements of slopes of waves generated by wind. The stippled area in the lower third of the tank indicates water. The cross-sectional area of the air passage is 26·3 by 26·3 cm. The dimensions of the water channel are: 14 cm (depth), 26·3 cm (breadth), 6·1 m (length). Numerals refer to the following details: 1, cup anemometer; 2, entrance nozzle; 3, suction centrifugal fan; 4, damper for controlling wind speed; 11, gravel beach for absorbing waves. On an enlarged scale are shown: 5, light source composed of four cylindrical incandescent light bulbs operated on direct current; 6, diffusing glass; 7, plate glass windows (top and bottom of tank); 8, hollow wedge filled with inky water (placed directly upon or beneath lower plate glass window but here shown raised for clarity); 9, telescope tube which focuses an image of water surface on a pinhole directly in front of photocell; 10, photocell.

Wu (1971) describes use of a device which projects a narrow beam of light at the sea surface at a controllable angle of incidence. He then measures the intensity of back reflected specular images from a small patch of water surface and makes counts as a function of the angle of incidence (hence tilt of the water surface). This method gives him a measure of the curvature of the water surface and the frequency of occurrence of selected water slope components. (The frequency of occurrence of slopes is not the same as the probability distribution of slopes unless the slope distribution in space-time is Gaussian. Wu's experiments are in agreement with the oceanic measurements by Cox and Munk that the slopes are only approximately Gaussian.)

Oceanic measurements have been made to measure the probability distribution of slopes by associating the smoothed spatial distribution of back reflections of sunlight in terms of the probability of occurrence at the particular slope required to reflect light from the sun toward the observer. The method requires that the observer be well above the ocean. Observations can be made with the sun at any elevation (Figs. 6 and 7) but interpretation of the results is simplified if the sun is high so that multiple reflections, shadowing, and hiding can be neglected.

The method has been used by Schooley (1954) and Cox and Munk (1956). The probability distribution of slopes according to the latter observations is shown in Fig. 8. It is roughly Gaussian but has significant skewness associated with the direction of the wind.

In another experiment Schooley has estimated the mean square curvature by photographing reflections of a flash bulb in a river.

These methods depend on a concentrated light source, the sun or a flash bulb. Stilwell (1969) has developed a method of measuring the spatial spectrum of wave slopes using an extended source for example the *overcast* or *clear sky away* from the sun. His method refines one originally used by Barber (1949). In this method the slope of waves is deduced from the distributions of brightness over the sea surface. He chooses conditions in which the reflected brightness is roughly a linear function of the wave slope. The variation of brightness can be the result of the variation of reflectance with angle of incidence (hence with tilt of the waves), or of the variation of brightness over the sky dome, or both. Stilwell's method of analysis relies on spectrum analysis of the brightness distribution as recorded on photographs, the spectrum analysis being carried out optically by diffraction of a monochromatic light beam.

From the early days of space exploration, it has been evident that optical study of the sea surface could be carried out by satellites in earth orbit. By this means it should be possible to get information on the roughness and from this deduce the surface windspeed or wind stress.

As a tool of oceanography and meteorology this would have the great value of synoptic observation over largely inaccessible ocean areas. It suffers from the obvious difficulties that the earth is a cloudy planet, that it can be used only on the daylit hemisphere, and there only when the sun is at a high elevation, therefore only at low latitudes.

Some interesting analyses have nevertheless been carried out by Levanon (1971), Kornfeld (1972) and Strong (1970) using data from the synchronous ATS 1 satellite which is "moored" over the central

Pacific. Kornfeld reports that he has been able to correlate his roughness estimates with island observations of wind. Using Cox and Munk's relation of mean square slope to windspeed the correlation indicates that wind can be estimated to 1 or 2 m s^{-1}. It is even possible to estimate wind direction roughly.

Analyses of this type are of course limited to equatorial waters because the maximum excursion of the center of the sun glitter pattern is only 10° north and south of the equator. Fortunately the sky is often clear near the equator. The principal problem is to deal with back scattering due to clouds, some of which may not be resolved. The method used by Levanon and Kornfeld is to observe the changes of brightness which occur as the earth rotates and the glitter pattern moves through the region of interest. Under the assumptions that the light scattering from clouds is relatively insensitive to sun elevation, and that the wind field is stationary with respect to the earth, it is possible to separate cloud and sea scattered signals and even derive a separate slope probability distribution for each point of interest on the sea surface.

Fig. 14. View of Pacific Ocean from ATS 1 satellite, 23rd March 1970 at 1818. The time shown for Figs. 1 to 14 is Universal Time.

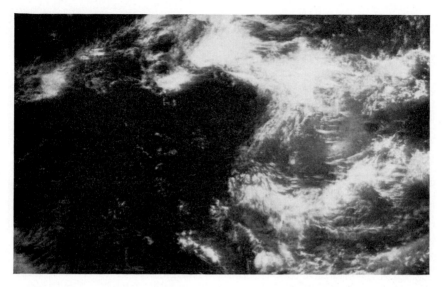

Fig. 15. View of Pacific Ocean from ATS 1 satellite, 23rd March 1970 at 1842.
Note sun glitter between two cloud banks at right.

Fig. 16. View of Pacific Ocean from ATS 1 satellite, 23rd March 1970 at 1906.

Fig. 17. View of Pacific Ocean from ATS 1 satellite, 23rd March 1970 at 1930.

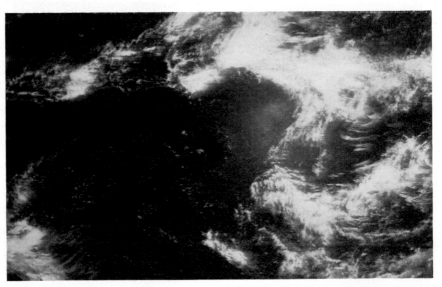

Fig. 18. View of Pacific Ocean from ATS 1 satellite, 23rd March 1970 at 1953.

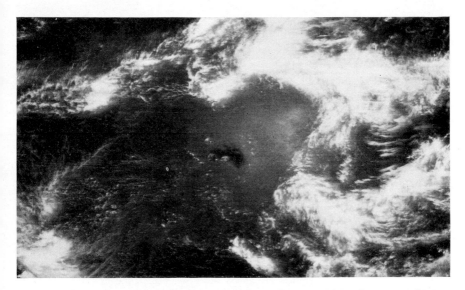

Fig. 19. View of Pacific Ocean from ATS 1 satellite, 23rd March 1970 at 2017.
Calm area or slick is visible as dark patch on left fringe of sun glitter.

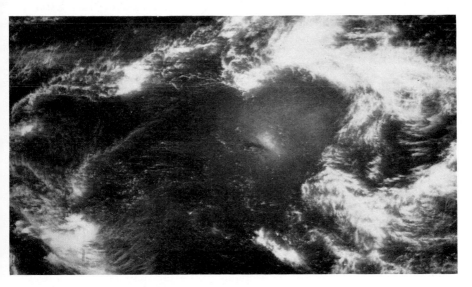

Fig. 20. View of Pacific Ocean from ATS 1 satellite, 23rd March 1970 at 2041.

Fig. 21. View of Pacific Ocean from ATS 1 satellite, 23rd March 1970 at 2105.
Sun reflection is enhanced in calm or slick patch.

Images from ATS 1 are transmitted in digital form and reconstructed
and photographed on the ground. An example, kindly provided by
Dr. A. Strong of NASA, is shown in Fig. 14 which shows the Pacific
Ocean extending from the coast of California (upper right) to the morning terminator cutting across the central Pacific. The sun glitter is
visible as a hazy patch in right center. As the earth and satellite rotate
(Figs. 15 to 21) the sun glitter patch moves westward.

Interesting features are slick or calm areas which are visible as
dark areas at the edge of the glitter pattern but become brilliantly
illuminated when the center of the glitter pattern moves over them.

VIII. DISCUSSION AND CONCLUSION

The processes of light scattering at the sea surface are important
biologically, thermodynamically and as a tool to investigate the roughness of the ocean surface. It appears that much more needs to be learned
in all these directions. For example, it appears that little is yet known
of the response of living organisms to the flickering light near the surface of the sunlit sea. The causes of the flickering—the spectrum of
short waves—has not been measured yet with precision under oceanic
conditions. It appears that methods using light scattering will be
applicable to further studies of this sort, and that some kind of scattering measurements, possibly of radar radiation will provide a useful
method for spacecraft measurement of the sea surface roughness.

References

Barber, N. F. (1949). *Nature (London)*, **164**, 485.
Cox, Charles (1956). Proc. 1st Conf. on Coastal Engineering Instruments, 1–17.
Cox, Charles (1958). *J. Mar. Res.*, **16**, 199–225.
Cox, Charles and Munk, Walter (1954a). *J. Mar. Res.*, **13**, 198–227.
Cox, Charles and Munk, Walter (1955). *J. Mar. Res.*, **14**, 63–78.
Cox, Charles and Munk, Walter (1956). *J. Opt. Soc. Amer.*, **44**, 838–850.
Kornfield, J. (1972). Ph.D. Thesis, University of Wisconsin.
Levanon, N. (1971). *J. Phys. Oceanog.*, **1**, 214–220.
Longuet-Higgins, M. S. (1956). *Proc. Cambridge Phil. Soc.*, **52**, Part 2, 234 245.
Longuet-Higgins, M. S. (1957a). *Proc. Cambridge Phil. Soc.*, **54**, Part 4, 439–453.
Longuet-Higgins, M. S. (1957b). *Phil. Trans. Roy. Soc. (London)*, **249**, No. 966, 321–387.
Longuet-Higgins, M. S. (1960a). *J. Opt. Soc. Amer.*, **50**, 838–844.
Longuet-Higgins, M. S. (1960b). *J. Opt. Soc. Amer.*, **50**, 845–850.
Longuet-Higgins, M. S. (1960c). *J. Opt. Soc. Amer.*, **50**, 851–856.
Miles, J. W. (1957). *J. Fluid. Mech.*, **3**, 185–204.
Miles, J. W. (1959a). *J. Fluid Mech.*, **6**, 568–582.
Miles, J. W. (1959b). *J. Fluid Mech.*, **6**, 583–598.
Miles, J. W. (1960). *J. Fluid Mech.*, **7**, 469–478.
Miles, J. W. (1962). *J. Fluid Mech.*, **13**, 433–448.
Phillips, O. M. (1966). "The Dynamics of the Upper Ocean", Cambridge University Press, London.
Rice, S. O. (1944). *Bell System Tech. J.*, **23**, 282–332.
Schooley, A. H. (1954). *J. Opt. Soc. Amer.*, **44**, 37–40.
Stilwell, Jr., Denzil (1969). *J. Geophys. Res.*, **74**, 1974–1986.
Strong, Alan E. and Ruff, Irwin S. (1970). *Remote Sensing Environ.*, **1**, 181–185.
Wu, Jin (1971). *J. Opt. Soc. Amer.*, **61**, No. 7, 852–858.
Wu, Jin (1972). *Phys. of Fluids*, **15**, No. 5, 741–747.

Chapter 4

Significant Relationships between Optical Properties of the Sea

N. G. JERLOV

Institute of Physical Oceanography, University of Copenhagen, Denmark

I. INTRODUCTION

The optical parameters provide many elements in the mosaic of problems presented by the properties of the sea. It is documented that optical data may be utilized in various ways to gain information about oceanographic conditions. An essential incitement for further progress regarding optical applications to oceanography would be to reveal existing relationships between the observed parameters. The possibilities of establishing such connections on the basis of available experimental data will be discussed in this chapter with a view to characterizing constituents of the sea.

Of these, the basic effect due to the water itself is nearly constant; it is only slightly dependent on temperature and pressure. Likewise the sea salts exert a weak influence on optical properties limited to the ultraviolet range. Our interest will exclusively be focused on the optical role of the variety of particulate matter and dissolved substances, chiefly yellow substance.

Several approaches to exploring the nature of such matter in different water masses are forthcoming:

(1) Comparison between theoretical and experimental parameters, for instance scattering.

(2) Investigation of spectral variance of a specific property for determining selective attenuation.

(3) Establishing for constant wavelength (λ) a system of functional relations between the separate properties.

II. Inherent Properties

A. DEFINITIONS

Using the following notations:

c = total attenuation coefficient

c_w = total attenuation coefficient for pure water

a = total absorption coefficient

a_p = absorption coefficient for particles

a_y = absorption coefficient for yellow substance (380 nm)

b = total scattering coefficient

b_p = scattering coefficient for particles

β = scattering function

we have by definition:

$$c = a+b \tag{1}$$

$$c = a_p+a_y+b_p+c_w \tag{2}$$

$$b = \int_{4\pi} \beta \, dw \tag{3}$$

B. METERS

A few words should be said about the meters for recording inherent properties with a view to assessing the accuracy normally attainable in such measurements.

By means of modern scattering meters the scattering function can be obtained over all angles which permits integration for evaluating the total scattering coefficient. It is generally thought that scattering is a sensitive indicator of the particle component. On the other hand, comparison between data published by different workers reveals that the absolute values of scatterance do not always compare well, obviously because of difficulties involved in the calibration of the meters.

A beam transmittance meter (c-meter) with a light beam of only 1–2 m length is not a suitable device for measuring the small variations in the attenuance of clear ocean water. In this case the calibration of the instrument is also a crucial factor. In turbid water, however, the c-meter serves as an excellent tool, especially if adapted to record spectral attenuance. It is highly desirable to include the far ultraviolet and the infrared in the observations in order to detect substances with characteristic absorption patterns.

What has been said about c-meters also holds true for such a-meters which directly measure absorptance in the sea. The *in vitro* method of determining absorptance on filtered samples may be useful for turbid water but is impaired by considerable errors for clear ocean water.

C. SPECTRAL ATTENUATION

For many years we have made comparative measurements *in situ* of the spectral attenuance in the red (655 nm) and in the ultraviolet (380 nm) for the purpose of determining the wavelength selectivity for particles and yellow substance. These data are shown in Table 1 together with other results obtained by means of laboratory monochromators. It should be noted that Burt and Visser compared the attenuance of sea samples with that of distilled water. In other cases the difference $(c-c_w)$ has been calculated by using the following values for pure waters:

$$c_w(655 \text{ nm}) = 0\cdot30 \text{ m}^{-1}$$
$$c_w(380 \text{ nm}) = 0\cdot04 \quad \text{(Clarke and James, 1939)}$$

A diagram in Fig. 1 is prepared on the basis of these two groups of $(c-c_w)$ coefficients. The majority of plots are well represented by a straight line. This seems rather remarkable considering the different meters used and the accuracy generally attainable in attenuation measurements. There is evidence for concluding that the line

$$(c-c_w)_{380} = 1\cdot8(c-c_w)_{655} \tag{4}$$

is a good approximation for describing the wavelength selectivity of particles and yellow substance in ocean waters, possibly with the exception of the clearest waters, for which some uncertainty remains on account of the low accuracy of these coefficients. Since scattering is generally fairly independent of wavelength, absorption by these components should be responsible for the whole selectivity. This point will be reconsidered later.

Table 1. Observations of the attenuation coefficient (c) and of the coefficient relative to pure water ($c-c_w$).

c-meter Type	Path-length m	Region	c m^{-1} 380 nm	c m^{-1} 655 nm	$c-c_w$ m^{-1} 380 nm	$c-c_w$ m^{-1} 655 nm	Reference
Monochromator	1	Continental Slope Atlantic Ocean	0·14	0·36			Clarke and James (1939)
Beckman DU	1	Off Peru			0·14	0·06	Burt (1958)
					0·22	0·13	
					0·35	0·21	
					0·60	0·34	
Beckman DU	1	Bermuda waters			0·21	0·12	Visser (1968)
					0·82	0·44	
Transmittance meter	2	Caribbean Sea	0·14	0·36			Jerlov (1951)
		Galapagos	0·27	0·43			
	1	Kattegat	0·58	0·53			Jerlov (1955)
	1	The Sound	0·95	0·56			
	1	Baltic Sea	1·19	0·57			
	1	Bermuda waters	0·24	0·40			Ivanoff et al. (1961)
			0·38	0·48			
			0·45	0·52			
			0·64	0·61			
	1·5	Sargasso Sea	0·06	0·32			Kullenberg et al. (1970)
	1·5	Mediterranean	0·12	0·35			"Helland-Hansen" exp.
			0·21	0·38			
			0·26	0·42			

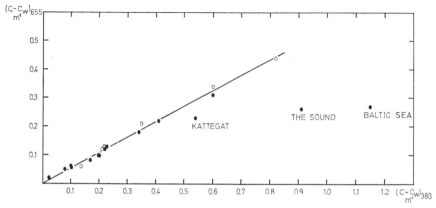

Fig. 1. Relationship between attenuation coefficients $(c-c_w)$ for the red (655 nm) and for the ultraviolet (380 nm). Open circles refer to direct measurements of $(c-c_w)$.

Three plots do not fit in the scheme, namely those for the Kattegat, the Sound and the Baltic which is attributed to the abundance of yellow substance brought to the Baltic by rivers. For these waters which are abnormally rich in yellow substance eq. (4) is not valid. In this case effective application has been made of an empirical relation which yields directly the absorption coefficient for yellow substance:

$$(c-c_w)_{380} - K(c-c_w)_{655} = a_y \tag{5}$$

This rests on the assumption that for particles the selective absorptance is proportional to the attenuance in the red. Large areas in the Baltic are characterized by a K-value of 1·6.

D. RELATION BETWEEN ABSORPTION AND SCATTERING BY PARTICLES

Great interest is focused on the optical behaviour of particles. The question arises if at present there is sufficient experimental material for establishing any general relation between the absorptance and the scatterance by particles.

For this purpose available data of $(c-c_w)$ and the particle scattering coefficient b_p for 655 nm have been brought together in Table 2. The very small scattering by pure water, b_w, is neglected for this wavelength. Furthermore, Table 2 contains a few similar results for these two parameters obtained by Kullenberg at three Mediterranean stations for 632 nm. In this case the attenuation coefficient for pure water $c_w = 0·26$ (Clarke and James, 1939). No correction for the λ shift is applied to the compared properties of $(c-c_w)$ and b_p which are found to be virtually independent of wavelength in the red (Burt, 1958).

4

TABLE 2. Observations of the attenuation coefficient relative to pure water $(c-c_w)$ and of the scattering coefficient for particles (b_p).

Region	$c-c_w$ m^{-1}		b_p m^{-1}	
	380 nm	655 nm	380 nm	655 nm
Sargasso Sea	0·05	0·04	0·03	0·02
	0·06	0·03	0·04	0·02
Caribbean Sea	0·11	0·06	0·06	0·06
Eq. Central Pacific	0·11		0·05	
Romanche Deep	0·14		0·07	
Mediterranean	0·11	0·04	0·04	0·03
	0·16	0·11	0·06	0·07
		0·15		0·11
Galapagos	0·25	0·13	0·09	0·08
Bermuda waters	0·20	0·10	0·10	0·11
	0·25	0·16	0·11	0·12
		0·32		0·23
Kattegat	(0·54)	0·23	0·16	0·15
Baltic Sea	(1·15)	0·27	0·21	0·20
Bothnian Sea	(1·72)	0·38	0·31	0·28

The diagram in Fig. 2, based on the tabulated values, includes highly different waters from the Sargasso Sea to the northern Baltic. There is no absorption by yellow substance in the red and thus the particle absorptance $a_p = c - c_w - b_p$. It follows from the trend of the straight line fitted to the plots that

$$(c-c_w) = 1\cdot43 \times b_p(655 \text{ nm}) \qquad (6)$$

Hence the particle absorptance

$$a_p = 0\cdot43 b_p(655 \text{ nm}) \qquad (7)$$

i.e. particles scatter about twice as much as they absorb.

These simple relations are presented with some reservations for the accuracy of the absolute values of the properties and do not claim to be of universal character considering the fact that the balance between scattering and absorption is dependent on particle size. The significant point is that the ratio of absorptance to scatterance shows *small* variations from the clearest ocean to turbid coastal areas, which suggests that particles on an average possess similar optical properties in the red.

With some reluctance an analogous diagram (Fig. 2) is presented for the ultraviolet (380 nm) which is founded solely on the author's own measurements in the oceans. In this spectral range, absorption is due

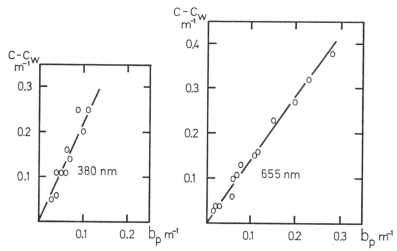

Fig. 2. Beam attenuation coefficients $(c - c_w)$ as a function of particle scattering coefficient (b_p) for the ultraviolet (380 nm) and for the red (655 nm).

to particles, a_p, as well as to yellow substance, a_y, $(c - c_w = a_p + a_y + b_p)$ and the scarcity of observations does not admit of any reliable separation between these two factors. Waters with an abnormally high value of a_y such as the Baltic cannot be incorporated in the diagram. The group of plots are again represented by a straight line indicating that

$$c - c_w = 2 \cdot 1 b_p \ (380 \ \text{nm}) \tag{8}$$

and

$$a_p + a_y = 1 \cdot 1 b_p \ (380 \ \text{nm}) \tag{9}$$

A comparison between eqs. (6) and (8) yields, if particle scattering is non-selective, $(b_p(380 \ \text{nm}) = b_p(655 \ \text{nm}))$

$$(c - c_w)_{380} = 1 \cdot 5 (c - c_w)_{655} \tag{10}$$

and

$$(a_p - a_y)_{380} = 2 \cdot 5 (a_p)_{655} \tag{11}$$

The small disagreement between eqs. (4) and (10) may well be ascribed to observational errors.

III. Relations between Inherent and Apparent Properties

The attenuation of radiance (L) in the sea is described by the classical equation:

$$\frac{\mathrm{d}L}{\mathrm{d}z} \cos \theta = -cL + \int_{4\pi} \beta L \, \mathrm{d}w \tag{12}$$

When dealing with this equation an apparent property K can be introduced

$$K = -\frac{1}{L}\frac{dL}{dz} \tag{13}$$

This coefficient becomes dependent on a series of parameters (λ, c, β, θ, and z). It is recalled that for the surface layer K is even negative for certain directions, and that it approaches a constant value with increasing depth. In spite of its depth dependence, significant connections between K and inherent properties exist.

It follows that the analogous irradiance coefficient K_d (E_d = downward irradiance)

$$K_d = -\frac{1}{E_d}\frac{dE_d}{dz} \tag{14}$$

is also dependent on depth as well as on solar elevation.

By integrating eq. (12) over the sphere, Gershun's formula is obtained (E_u = upward irradiance, E_0 = scalar irradiance)

$$\frac{d(E_d - E_u)}{dz} = -aE_0 \tag{15}$$

With

$$K_E = -\frac{1}{E_d - E_u}\frac{d(E_d - E_u)}{dz} \tag{16}$$

it is found that

$$\frac{a}{K_E} = \frac{E_d}{E_0}\left(1 - \frac{E_u}{E_d}\right) \tag{17}$$

The irradiance ratio E_u/E_d is approximately proportional to the ratio of the back-scattering coefficient to the attenuation coefficient. It attains a maximum value of 10% for blue light in the clearest water (Jerlov, 1951; Lundgren and Højerslev, 1971).

Equation (17) brings out that a/K slightly corrected may be identified with the significant ratio of cosine collection to equal collection (E_d/E_0). Because of the existence of upward irradiance, E_u, the coefficient a is always less than K. It is gathered from Table 3 which presents a compilation of a/K data that a maximum value of 0·93 occurs for the highly directed light in the red (633 nm) whereas a minimum value of 0·62 is found in the violet (427 nm).

For asymptotic radiance distribution, eq. (12) reduces to the form

$$L = \frac{\int_{4\pi} \beta L \, d\omega}{c - k \cos \theta} \tag{18}$$

where K is a constant irrespective of depth and solar elevation. Using eq. (18) the attenuation coefficient c can be derived from measured values of L, β and k. Some workers have employed this method to determine the inherent properties of a water mass (Table 4). For the Sargasso Sea attention is drawn to the small variation of the ratio k/c in the spectral range 375–525 nm.

TABLE 3. Values of a/K derived from irradiance measurements between 10 and 50 m.

Region	Solar elevation (degrees)	Wavelength nm	a/K	Reference
Baltic Sea	52	535	0·75	Jerlov and Liljeqvist (1938)
Mediterranean Off Gibraltar	68	372	0·88	Højerslev (1972)
	74	427	0·69	
	68	477	0·69	
	70	533	0·73	
	71	572	0·67	
	75	635	0·79	
South of Sardinia	72	372	0·78	
	64	427	0·62	
	60	477	0·73	
	73	533	0·84	
	68	572	0·85	
	74	633	0·93	

The relationship between the factors k/c and b/c has been the subject of extensive studies on artificial suspensions by Timofeeva and Gorobetz (1967). These results conform well with the thorough theoretical interpretation by Prieur and Morel (1971) for waters of different particle content.

TABLE 4. Optical properties derived from near-asymptotic radiance distributions in the sea.

Region	Depth	Wavelength nm	Observed		Calculated		k/c	Reference
			k m^{-1}	b m^{-1}	c m^{-1}	a m^{-1}		
Baltic Sea	100	535	0·12	0·25	0·34	0·09	0·35	Jerlov and Nygård (1968)
Sargasso Sea	150	375	0·042	0·030	0·053	0·023	0·79	Lundgren and Højerslev (1971)
	250	425	0·41	0·24	0·48	0·24	0·85	
	400	475	0·27	0·16	0·32	0·16	0·85	
	150	525	0·46	0·23	0·55	0·32	0·84	

IV. Colour Index

A. SIGNIFICANCE OF OCEANIC COLOUR OBSERVATIONS

Quantitative measurements of ocean colour from aircraft yield valuable information for locating distinct water masses and for the assessment of primary production. With a remote detection system allowance must be made for atmospheric effects as "air light" and light reflected at the sea surface. On the other hand, the oceanographic concept of colour refers adequately to colour *in situ* produced only by back-scattering from water molecules and suspended particles. The subsurface upwelling light and thus *in situ* colour is dependent exclusively on the spectral composition of incident light penetrating the sea surface and on the inherent optical properties of the water.

An attempt is made here to explore the potential of *in situ* colour observations for the optical characterization of water masses.

B. DEFINITION OF COLOUR INDEX

Conventionally, an objective numerical value of the colour of the sea can be derived from spectral distributions of upwelling light which are subjected to a colour analysis in terms of the C.I.E. chromaticity coordinates. This system considers any colour as synthesized by a mixture of three components which may be described as red, green and blue. The evaluation of colour in this way is a fairly complicated procedure.

During a conference on "The Colour of the Sea" held in Woods Hole Oceanographic Institution on 5–6 August 1969, the possibility was discussed of recording only in the blue and in the green when studying spectra of back-scattered light from the sea, obtained from aircraft as a measure of chlorophyll concentration (Clarke *et al.*, 1970). In view of the fact that chlorophyll exhibits maximum absorption in the blue whereas it absorbs very little in the green, the variations in the blue/green radiance ratio would be partly due to chlorophyll (Ramsey, 1968) which is associated with primary production.

A new model of colour representation should essentially be based on experimental evidence. The spectral distribution of upward irradiance typical for clear ocean water (Fig. 3) bears witness to the following features:

(1) The curves at all depths are peaked at 450 nm, i.e. at a wavelength smaller than the 460 nm at which maximum transmittance for downward irradiance occurs. This characteristic shift is attributed to highly wavelength-selective scattering by the water itself.

(2) The red irradiance is relatively very low.

According to the general principle in the attenuation mechanism in normal ocean water, overall decreased transmittance reduces the shortwave part of the spectrum more than the longwave part, and shifts the maximum of transmittance towards longer wavelengths because of selective absorption by particles and yellow substance. This leads to a colour change from blue via green to brown. Only in very turbid waters such as the Baltic, which is abundant in yellow substance, does the red component in the upwelling light become comparable with the blue.

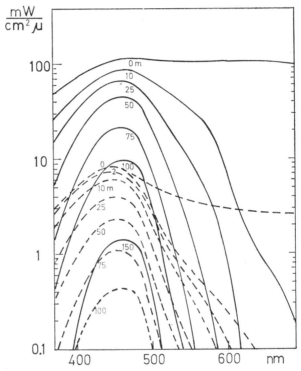

Fig. 3. Comparison between spectral distribution of downward (——) and upward (- - -) irradiance for a solar elevation of 55–60° in the Sargasso Sea. (Lundgren and Højerslev, 1971).

It seems permissible to employ a simplified model for arriving at a characteristic colour parameter. In view of the spectral composition of upwelling light, one could substitute for the conventional three-stimulus colour system, a colour system with only two components, i.e. nadir radiance in the blue (450 nm) and in the green (520 nm).

Avoiding a complicated numerical representation of colour a colour index is defined as the ratio

$$F = \frac{L(180°)_{450\,nm}}{L(180°)_{520\,nm}}$$

C. COLOUR METER

The meter was built in 1969 but has not been used in a systematic way until the summer of 1971 during a cruise in the Mediterranean. The housing of the meter contains two photovoltaic cells facing downwards in order to record nadir radiance. The two cells are mounted in short Gershun tubes limiting the field of view to $\pm 15°$. The cells are provided with interference filters of 447 and 521 nm respectively. The two amplified signals have been recorded separately for the purpose of showing the mutual relationship for different experimental conditions. By using logarithmic amplifiers the colour index can be obtained directly on a pointer instrument. The meter is calibrated in energy units.

The colour meter is suitable for routine work and needs no special arrangements since it is lowered manually from shipboard.

D. OBSERVATIONS

(1) Depth dependence

It is obvious that the index will be a function of depth except for the trivial case when the attenuation of nadir radiance is the same for 447 nm as for 521 nm. The general influence of depth on the index is brought out in Fig. 4 which summarizes data from different water masses in the Mediterranean.

The high index measured southwest of Sardinia increases with depth, about 10% per 8 m, since the 447 radiance is less attenuated than is the 521 radiance. It should be added that Lundgren and Højerslev (1971) found an even higher increase of 25% per 8 m for the extremely clear Sargasso water.

On the contrary, the index is considerably reduced with increasing depth in the very turbid water mass encountered near Gibraltar on 12 July 1971 due to stronger attenuation in the blue than in the green.

Next day the situation is changed and resembles the trivial case mentioned as the index is little dependent on depth.

(2) Effect of solar elevation

The solar elevation is a complicating factor in underwater light measurements (Jerlov and Nygård, 1969). Since the index is a ratio its change with increasing depth is not expected to be much effected by variations in the elevation. The spectral changes imposed by the water

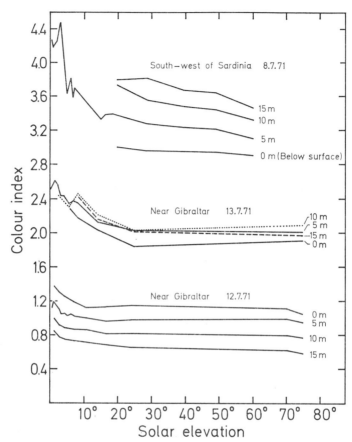

FIG. 4. Colour index as a function of depth and of solar elevation at three stations in the Mediterranean.

on upwelling radiance are largely decisive for *in situ* colour. But due allowance should be made for the spectral character of the incident light. For instance, with the sun near the horizon the blue skylight becomes dominant in the incident light which results in a high index.

In the three water masses considered in Fig. 4 the index was recorded from sunrise to noon at least at one depth. The trends of the curves indicate that the solar elevation effect is not important for elevations above 15° especially at small depths. At elevations < 15° the effect is small for the most turbid water near Gibraltar where the attenuation by the water chiefly determines the colour, whereas it is considerable for the clearest water off Sardinia mostly due to spectral changes in the incident light.

(3) *Effects of clouds*

The influence of clouds on the index is similar to the solar elevation effect. Screening of the sun by a cloud often leads to a drastic reduction of sunlight compared with skylight, and as a consequence a higher index is obtained. However, the effect is usually small. A few evaluations in Table 5 show that the index increases about 10% when the upwelling radiance (447 nm) is reduced to one half by a cloud obscuring the sun.

TABLE 5. Influence of clouds on colour index (just below the surface).

Station	Date	Radiance 447 nm	Colour index
Off Gibraltar	31.5.71	153	1·42
		101	1·43
		75 clouds	1·60
		46 clouds	1·56
South-west of Sardinia	5.6.71	171	2·61
		124 clouds	2·88
East of Sardinia	12.6.71	181	2·58
		68 clouds	2·71

The colour index with an overcast sky has not been investigated so far. Any appreciable change of the index from clear to overcast sky is not probable, judging from observed spectral distributions of global radiation (Taylor and Kerr, 1941).

(4) *Conclusions regarding the method*

The above discussions of experimental evidence provide the basis for formulating directions of use for the colour meter. The recommendations take the following practical form:

The colour index is preferably recorded (1) at small depth (0–2 m); (2) with an unobscured sun; (3) with a sun at more than 15° elevation.

The information yielded by the colour index measured in this way is limited to the uppermost 10 or 20 m of the sea. It is for the upper layers, however, that optical characterization is of most direct interest.

(5) *Regional distribution in the Mediterranean*

The method has been tested during a cruise in the summer of 1971 with the Norwegian research ship "Helland-Hansen" in the western Mediterranean. The observed colour data plotted on the chart of Fig. 5 show at Gibraltar a colour index between 1·13 and 1·92 indicating that the inflowing Atlantic surface water is characterized by a high and fluctuating turbidity. This water is mixed with clear Mediterranean water during its flow eastward which brings about a gradual increase of the index. It is interesting to note that southwest of Sardinia the index has attained a value of 2·92, close to the maximum value found on the cruise. This is consistent with the general circulation pattern of the Mediterranean which often shows a westward current south of Sardinia (with clear Mediterranean water) whereas the eastward flow (with the admixture of turbid Atlantic water) occurs nearer to the African coast. A significant detail is the unmistakable tendency of lower index close to Sardinia than in offshore water obviously due to contamination from land.

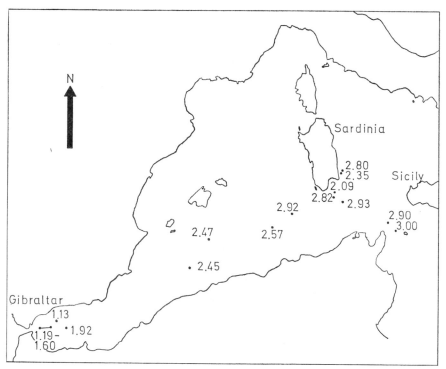

Fig. 5. Regional distribution of colour index in the western Mediterranean.

E. RELATION TO OTHER OPTICAL PROPERTIES

The colour index reacts to the optical properties of the upper layers. Obviously it varies inversely with the beam attenuation coefficient. But it is not realistic to relate the index to inherent optical properties in view of their low accuracy compared with the precision in the determination of the index.

As regards the apparent properties, it is documented that they possess striking regularities. When a new parameter such as the colour index is introduced it is relevant to find out how it is linked to other apparent properties. Only experimental tests of method in various water masses can assess such relationships.

We are not taking any interest in the trivial case of comparing the colour index with the attenuation of one of its radiance components. More significant and also a test of the usefulness of the method is to relate the index to downwelling light. By way of example the attenuation of number of quanta (350–700 nm) as well as that of blue irradiance (465 nm) for the stations, displayed in Fig. 5, are chosen as suitable parameters. In Fig. 6 the colour index is represented as a function of the depth at which (1) the percentage of surface quanta is 10% and (2) the percentage of blue surface light is 30%, in both cases for a solar elevation of 60°–70°.

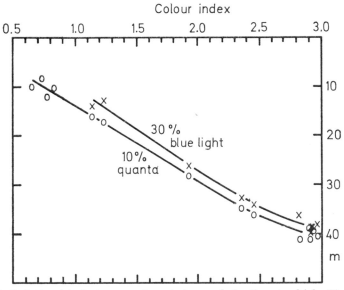

Fig. 6. Relationship between colour index and depths at which (1) the percentage of surface irradiance of quanta (350–700 nm) is 10% and (2) the percentage of blue (465 nm) surface irradiance is 30%.

The relationship between colour index and 10% quanta level is rendered more complete and representative by a few observations (solar elevation 60°) made by Berg Olsen (personal communication) in the turbid upwelling area near Dakar. The diagram thus covers a large range of oceanic water types characterized by colour indexes from 0·7 to 3·0. It follows from the clear associations exhibited in Fig. 6 that the index can readily be fitted into the pattern of apparent properties. It should be observed that the attenuation of downwelling light is defined for a certain elevation whereas the index is independent of elevation if above 15°.

F. CONCLUSION

A colour index defined as the ratio of nadir radiance at 450 nm to that at 520 nm can be determined in a few minutes from shipboard. Judging from data so far collected and from arguments presented above, the index method may be employed to advantage for optical classification of water masses. There is, however, a need for broadening the experimental basis for such a classification before the usefulness of the method can be definitely established.

REFERENCES

Burt, W. V. (1958). *Deep-Sea Res.*, 5, 51–61.
Clarke, G. L., Ewing, G. C. and Lorenzen, C. J. (1970). *Science, N.Y.*, 167, 1119–1121.
Clarke, G. L. and James, H. R. (1939). *J. Opt. Soc. Am.*, 29, 43–55.
Højerslev, N. (1972). Københavns Universitet, *Rep. Inst. Fys. Oceanog.*, 20.
Ivanoff, A., Jerlov, N. and Waterman, T. H. (1961). *Limnol. Oceanog.*, 6, 129–148.
Jerlov, N. G. (1951). *Rep. Swedish Deep-Sea Expedition*, 3.
Jerlov, N. G. (1955). *Medd. Oceanog. Inst. Göteborg*, 25.
Jerlov (Johnson), N. G. and Liljequist, G. (1938). *Svenska Hydrograf. Biol. Komm. Skrifter, Ny Ser. Hydrog.*, 14.
Jerlov, N. G. and Nygård, K. (1968). Københavns Universitet, *Rep. Inst. Fys. Oceanog.*, 1.
Jerlov, N. G. and Nygård, K. (1969). Københavns Universitet, *Rep. Inst. Fys. Oceanog.*, 4.
Kullenberg, G., Lundgren, B., Malmberg, Sv. Aa., Nygård, K. and Højerslev, N. (1970). Københavns Universitet, *Rep. Inst. Fys. Oceanog.*, 11.
Lundgren, B. and Højerslev, N. (1971). Københavns Universitet, *Rep. Inst. Fys. Oceanog.*, 14.
Prieur, L. and Morel, A. (1971). *Cah. Oceanog.*, XXIII, 1.
Ramsey, R. C. (1968). Unpublished.
Taylor, A. H. and Kerr, G. P. (1941). *J. Opt. Soc. Am.*, 31, 3–8.
Timofeeva, V. A. and Gorobetz, F. I. (1967). *Izv. Acad. Nauk SSSR, Ser. Geofiz.*, 3, 291–296.
Visser, M. P. (1968). *Kon. Ned. Meteorol. Inst. Rep.*, W. R. 67–002.

Chapter 5

Structure of Solar Radiation in the Upper Layers of the Sea*

RAYMOND C. SMITH

*Visibility Laboratory, Scripps Institution of Oceanography,
University of California, San Diego, La Jolla, California, U.S.A.*

I. INTRODUCTION

The sun's electromagnetic energy which penetrates into the upper layers of the sea is a basic pre-requisite for life in the oceans, is characterized by fundamental physical processes, and is essential to many activities of man. Solar radiation supplies the energy for ocean ecosystems by conversion to chemical energy through photosynthesis and to heat energy by absorption. The sun's radiant energy is selectively absorbed and scattered as it penetrates the upper layers of the sea, and these basic processes distinctively alter the structure of the radiant energy field. From a description of the underwater radiant energy field, many of the inherent properties of the water and information on basic physical processes can be obtained. This knowledge and information is a requirement for the study and solution of many problems of interest to man.

To describe the time rate of flow of radiant energy, one must specify its magnitude (the square of the electric field vector), its polarization (direction of oscillation of the electric field vector), its wavelength (frequency of oscillation of the electric field vector), and its direction of propagation. The spectral characteristics (Tyler and Smith, 1970) and polarization (Lundgren, 1971) of underwater radiant energy are

* This work was supported by a grant from the National Science Foundation, NSF–GA–19738.

reviewed elsewhere. Radiance, the energy flux per unit solid angle per unit area normal to the direction of propagation incident on a point, specifies the remaining characteristics of radiant energy. A radiance distribution, the totality of radiance values for every direction about the point, gives a complete description of the geometrical structure of the radiant energy field. The significant features of underwater radiance distributions, along with techniques for their experimental measurement and theoretical analysis, are the principal subjects of this review.

The structure of the radiance distribution in the upper layers of the sea is dependent upon factors which modify the solar radiation in the earth's atmosphere, conditions of the air-sea interface and optical properties of the sea water. The radiance distribution above the sea surface depends primarily upon the altitude of the sun, the scattering-absorbing properties of atmospheric molecules and particles, the meteorologic conditions and the radiant energy reflected back from the sea surface. Recent measurements and a review of earlier measurements of the solar constant have been made by Thekaekara *et al.* (1969, 1970) and Arvesen *et al.* (1969). Sky radiance distributions have been reported for a variety of sun angles and meteorological conditions by Gordon *et al.* (1966a,b,c). Monte Carlo methods have been used to study the reflected and transmitted radiation in the atmosphere over a wide range of conditions and this work has been summarized by Plass and Kattawar (1971). The effect of wind stress on the surface roughness of the air-sea interface has been investigated by a number of authors (including Duntley, 1954; Cox and Munk, 1954; Cox, 1958; Wu, 1969) and discussed in a review article by Ursell (1956). The relationship of sea surface slopes to the underwater radiance distribution has been discussed by Cox and Munk (1955) and Gordon (1969). Primary emphasis in this chapter will be on the radiance distribution of solar energy as a function of depth below the sea surface and how this radiance distribution is altered and characterized by the optical properties of sea water.

II. UNDERWATER RADIANCE DISTRIBUTIONS

Underwater radiance distributions have been obtained by means of a radiance distribution camera system (Smith *et al.*, 1970). This instrument contains two cameras placed back to back, each equipped with a 180° field of view ("fisheye") lens, and it is fabricated so that film can be exposed by remote control. The projection geometry of the 180° field of view lens is shown in Fig. 1, where the projection origin of the lens (origin of the $x-y$ coordinates in the figure) is at a

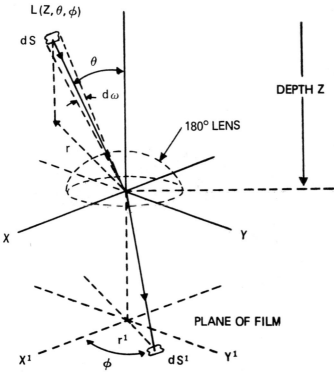

Fig. 1. Schematic diagram showing the projection geometry of the 180° field of view (fisheye) lens for obtaining radiance distributions.

depth z underwater. The radiance, $L(z, \theta, \varphi)$, is the energy per unit time per unit area per unit solid angle incident on the origin from the direction (θ, φ). Referring to Fig. 1, θ is the zenith angle of an incident ray from an infinite object and (r', φ) are the coordinates of the image of this ray on the plane of the film. As described by Miyamoto (1964) and discussed in detail by Smith et al. (1970), this lens has the property of projecting a hemisphere onto a plane such that:

$$r' = f\theta, \tag{1}$$

where f is the lens focal length. In addition, the irradiance received by the film during the time of exposure, $E'(r', \varphi)$, has a one-to-one correspondence to the field radiance given by:

$$E'(r', \varphi) = L(z, \theta, \varphi)\left(\frac{n}{n'}\right)^2\left(\frac{\pi}{4F^2}\right)\left(\frac{\sin\theta}{\theta}\right). \tag{2}$$

Here n and n' are the indices of refraction of object and image space, respectively, and F is the f/number of the optical system. By the

5

methods of photographic photometry, the density of an area on a film negative can be related to the exposure of that area, and this can, in turn, be related to the desired field radiance.

Before considering quantitative data, the significant qualitative features of an underwater radiance distribution will be illustrated by means of a positive print, obtained from a data film negative, taken by the radiance distribution camera system. Figure 2 illustrates the upper

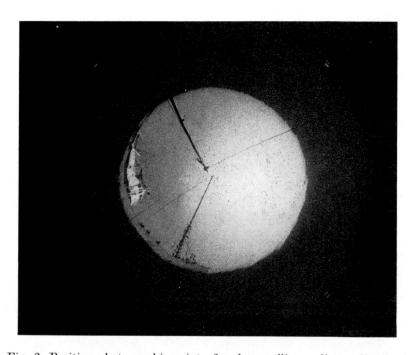

Fig. 2. Positive photographic print of a downwelling radiance distribution, obtained from a data film negative, taken by the radiance distribution camera system. In this photograph the zenith ($\theta = 0°$) is in the centre, the horizontal directions ($\theta = 90°$) form the outside circumference of the circular format, and the intermediate zenith angles are directly proportional to the radial distance from the centre (eqn. 1). The Snell circle is seen here as the brighter, inner circle extending to the critical angle ($\theta = 48\cdot6°$). Dark, radial lines are the ship's boom and the instrument's electrical and suspensions cables. Other information: depth 1·75 m, zenith angle of the sun 29° 30′, refracted angle of the sun 21° 45′, clear skies (less than 0·1 cloud cover), glassy-calm sea surface, Beaufort Wind Scale No. 0, instrument suspended 6 m from side of ship (R/V Helland-Hansen), 60 nm bandwidth (full width at half maximum) spectral response approximately centered at the wavelength of maximum transmittance of the water under investigation (λ max. $\approx$ 473 nm obtained using Wratten No. 48 filter), Mediterranean Sea (Sta. D2) 8 July 1971, 38° 22′ N 07° 12′ E.

hemisphere of a radiance distribution at a depth of 1·75 m. As noted in the caption to Fig. 2, the environmental conditions were ideal when the exposure for this photograph was taken. In this figure the zenith is at the centre, the horizontal directions form the outside circumference of the circular format, and intermediate angles from the zenith are proportional to the radial distance from the centre (eq. 1).

The role of refraction in structuring the distribution of radiance underwater is illustrated in Fig. 2 where many principal features can be explained by means of Snell's law of refraction:

$$m \sin i = n \sin j. \tag{3}$$

Figure 3 schematically depicts the application of this law to the experimental situation that existed when the photograph shown in Fig. 2 was obtained. Consider an observer (or his radiance instrument) at a depth of 1·75 m underwater viewing the upper hemisphere in a plane containing the observer, the zenith and the sun (the vertical plane of

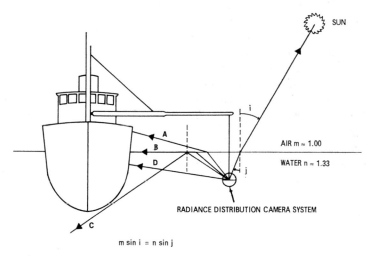

Fig. 3. Schematic diagram depicting the application of Snell's Law to the experimental situation that existed when the exposure for fig. 2 was obtained. See text for details.

the sun). To the right of the zenith (to the right of the centre in Fig. 2) the refracted image of the sun is seen as a bright spot with a diffuse glow surrounding it. To the opposite side of the zenith from the sun the refracted image of the ship above its water line can be seen along a path of sight such as "A". As the angle of view is increased from the zenith, the water line of the ship (and the horizon at other azimuth angles) is seen at the critical angle, j_c, defined as:

$$\sin j_c = \frac{m}{n}\sin 90°. \tag{4}$$

The critical angle is approximately 48·6° for the sea-air interface and is shown as path of sight "B" in Fig. 3. Thus the total 180° sky hemisphere above water is compressed by refraction into a 97·2° cone underwater. This is clearly seen in Fig. 2 as the brighter inner circle containing the refracted images of the sun, sky and ship above its water line.

Total internal reflection from the surface occurs, e.g. path "C", when the direction of sight is increased to a zenith angle greater than 48·6°. As the direction of sight is further increased from the zenith, the path of sight "C" eventually strikes the bottom of the ship. From this angle, until the water line of the ship is viewed directly, the reflected (thus inverted) image of the ship below its water line is seen. Finally, for paths of sight from the water line downward, the ship is seen along a direct underwater path "D". The dark shadow along the lefthand circumference of Fig. 2 is due to the reflected and direct image of the ship below the water line.

A small maximum can appear in near surface radiance distributions at zenith angles of about 70° to 90° which depends upon the relative absorbing and scattering properties of the water. Consider paths of sight for zenith angles of 48·6° to 90° and assume the ship was not present. The radiance as viewed along a path such as "C" is composed of backscattered radiant energy which is internally reflected from the sea-air interface plus radiant energy which is forward scattered between the surface and the observer. Both the upwelling internally reflected and the downwelling component of the radiance depend upon the absorptance and scatterance along the path of sight considered. The internally reflected component will decrease as the zenith angle is increased due to increased absorption along the longer path length to the surface. The forward scattered radiant energy, on the other hand, may increase as the path length to the surface is increased until the gain due to forward scatterance is balanced by increased absorptance. If such an increase occurs, a small maximum will appear at sun zenith angle a few degrees less than 90° (Tyler, 1958).

Quantitative values of the relative radiance versus zenith angle in the vertical plane of the sun are shown in Fig. 4. This downwelling data was obtained from a microdensitometer scan of the same negative used to produce the positive print shown in Fig. 2, where the down-

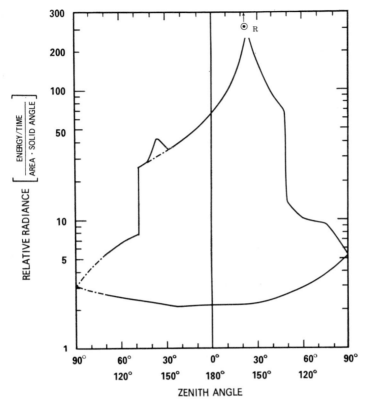

Fig. 4. Relative radiance versus zenith angle for up- and downwelling radiance in the vertical plane of the sun. The data for the upper curve were obtained from the same negative that was used to produce the positive print shown in Fig. 2. The upper curve gives the downwelling radiance from the zenith to the horizontal (90°), the lower curve gives the upwelling radiance from the horizontal to the nadir (180°). Dashed portions of the curves are extrapolations of the data to eliminate perturbations in the radiance distribution due to the ship. $\odot_R$ indicates the refracted angle of the sun. The angular resolution of the radiance data in this figure is 7·2°.

welling radiance is given by the upper curve from the zenith to the horizontal toward and away from the sun. Principal features discussed above can be seen in this curve. The greatest radiance values occur in the direction of the refracted image of the sun, $\odot_R$. Since a relatively coarse scanning resolution of 7·2° was used to obtain this curve, the sharp detail, such as the small sun's image, will be somewhat smoothed by the scanning process. The steep drop in radiance values at 48·6° on both sides of the zenith marks the Snell circle. The small maximum in the curve on the side away from the sun is light reflected from the ship above its water line. The maximum in the radiance on the side of the sun between 70° and 90° is due to the interplay of the internally reflected and forward scattered radiant energy. In this particular example, the perturbation of the radiance distribution due to the ship obscures a possible similar maximum on the side opposite the sun.

Considerable care was taken to minimize perturbations in the underwater radiance distribution caused by the ship and instrument cables. However, as with any underwater radiant energy measurement, such perturbations are very difficult to eliminate completely. It is an advantage of the photographic technique of measuring underwater radiance that one can directly see and identify the source of any perturbation on the film, as well as quantitatively measure its magnitude. Thus, by directly identifying sources of perturbation and by making use of the inherent symmetry about the vertical plane of the sun in underwater radiance distributions, the data can usually be accurately corrected for ship and cable perturbations.

Upwelling radiance, which is composed only of backscattered radiant energy, was obtained with a downward facing camera and 180° field of view lens. Upwelling radiance, obtained simultaneously with the corresponding downwelling radiance, is shown as the lower curve in Fig. 4. This curve goes from the horizontal (90°) on the side away from the sun, through the nadir (180°), to the horizontal on the side of the sun. Figure 4 should not be confused with a polar plot of radiance. There is no "origin" within the curves of this figure. It is merely a plot of relative radiance versus zenith angle with the curve from 90° to 180° folded back underneath the 0° to 90° curve. In this way the radiance distribution is plotted on a logarithmic scale, which allows the full range of radiance values to be displayed accurately, and yet the curves maintain the heuristic attributes of a polar plot.

This technique of plotting radiance allows a subtle feature of the radiance distribution to be readily observed. Note that the minimum radiance value is not at the nadir but at the refracted anti-solar point. This is because, in relatively clear water near the surface, the instrument casts and then "sees" its own shadow.

Figure 5 is a photograph showing this aureole effect (Minnaert, 1954). The various slopes in the water's surface cast a streak of light or shade behind it; all these streaks run parallel to the line from the refracted image of the sun through the projection origin of the observing lens. The instrument then records these lines as meeting perspectively in the refracted anti-solar point, that is, in the shadow image of the instrument. Thus, the radiance as seen by the instrument has a minimum at its own shadow image surrounded by a circle of slightly increased radiance. This effect is of little practical consequence but is nevertheless a pleasing phenomena to observe and is a further example of how our instruments, however subtly, perturb the environment we are attempting to study.

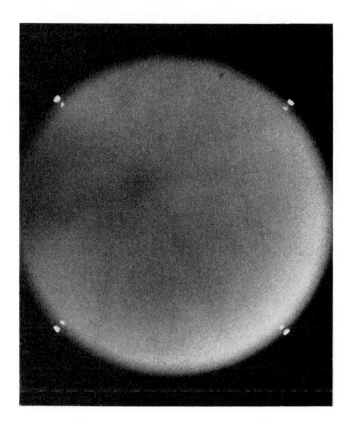

Fig. 5. Positive photographic print of an upwelling radiance distribution, obtained from a data film negative, taken by the radiance distribution camera system. View of upwelling radiance at 1·75 m showing the aureole effect. See text for details.

Figure 6 shows a positive photographic print of, and Fig. 7 shows the quantitative data for, a radiance distribution obtained at a depth of 20·4 m and with a sun zenith angle of 75° 30′. The remaining oceanographic conditions are the same as for Fig. 2. In Figs. 6 and 7 the refracted angle of the sun lies very close to the edge of the Snell circle and the diffuse glow surrounding the sun's image spreads out beyond this circle. This glow is broader than that seen at the shallower depth, because we are viewing the narrow angle forward scatterance of the sun over a longer path length. The Snell circle is still obvious but its edge is more diffuse, again because of the increase scatterance along a longer path at this greater depth. The dark shadow near the zenith is

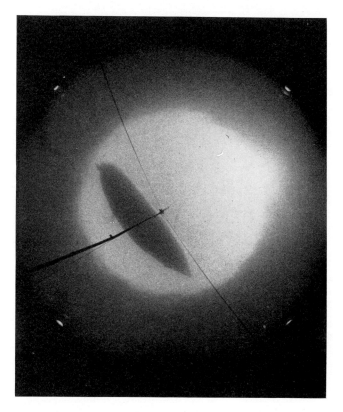

Fig. 6. Positive photographic print of a downwelling radiance distribution. Other information: depth 20·4 m, zenith angle of the sun 75° 30′, refracted angle of the sun 46° 30′, remaining conditions are the same as for Fig. 2.

the image of the bottom of the ship and is seen as a dip in the radiance curve in Fig. 7. This figure indicates a slight increase in the radiance just prior to reaching the Snell circle, and an increase in brightness can also be seen at all azimuth angles near the Snell circle in the photograph shown in Fig. 6. This increased radiance is probably due to the distribution of atmospheric light.

The figures discussed above, obtained under ideal environmental conditions, illustrate that the structure of solar radiant energy just below a calm surface of the sea is primarily dependent upon the radiance distribution above the surface and Snell's law of refraction. A cloudy atmosphere, which obscures the sun, will decrease and diffuse the underwater radiance peak in the direction of the refracted angle of the sun. An increased sea state will produce a glitter pattern about the

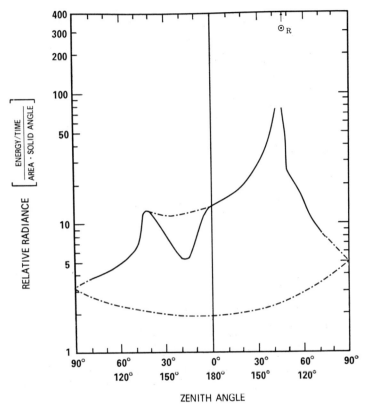

Fig. 7. Relative radiance versus zenith angle for downwelling radiance in the vertical plane of the sun. Data obtained from the same negative used to produce Fig. 6. Upwelling (dashed) curve estimated.

sun's image and diffuse the edge of the Snell circle. Even with clear skies and a calm surface, the radiance distribution will lose its sharp structure as the absorptance, the scatterance, or the depth of observation is increased.

The relative radiance versus zenith angle for three depths is shown in Fig. 8. Moderately rough seas produced a glitter pattern about the sun and diffused the Snell circle, as shown by the 5 m curve. With increasing optical depth this image detail is lost and the shape of the radiance

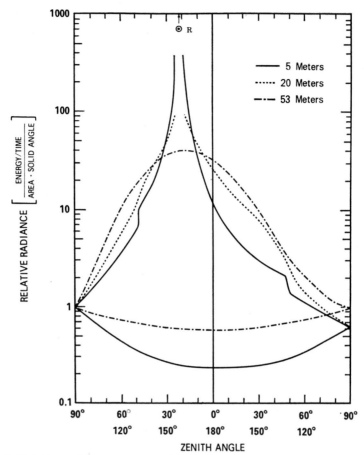

Fig. 8. Relative radiance versus zenith angle for up- and downwelling radiance in the vertical plane of the sun. The curves have been arbitrarily normalized to unity at $\theta = 90°$ so that they may be readily compared. Other information: zenith angle of the sun 28°, refracted angle of the sun 21° 30', clear skies (less than 0·1 cloud cover), moderate breeze (Beaufort No. 4) with gusts to 25 knots, wind chopped moderate sea (1 to 1½ m waves) in lee of island, instrument suspended 4 m from stern of ship (R/V Ellen B. Scripps), 60 nm bandwidth (FWHM) spectral response approximately centered at the wavelength of maximum transmittance of the water under investigation (λ max. $\approx$ 497 nm obtained using Wratten No. 64 filter), Gulf of California (Fresnel II Cruise—Sta. 2) 16 March 1971, 25° 26' N 111° 08' W. The upwelling curve for 20·4 m is very nearly the same as the upwelling curve for 4·6 m and is not shown.

distribution becomes increasingly smooth, as shown by the 20 and 53 m curves. In addition, with increasing optical depth, the point of maximum radiance moves toward the zenith away from the direct refracted angle of the sun. It is postulated (Whitney, 1941) that the radiance distribution at great optical depths is symmetrical about the vertical axis and that its shape is dependent only upon the scatterance and absorptance of the water.

Preisendorfer (1959) has shown that the asymptotic radiance hypothesis is equivalent to the statement that the direction (θ, φ) and depth (z) dependence of the radiance distribution multiplicatively uncouple at great depths. Duntley (1963), extrapolating from Tyler's (1960) Lake Pend Oreille data, estimated that an asymptotic distribution was not reached until the depth was greater than 20 attenuation lengths $(1/c)$. To be precise, the depth at which an asymptotic distribution is reached depends primarily on the scattering length $(1/b)$. The absorption length $(1/a)$ then determines how many, if any, photons remain when asymptotic depths have been reached. For normal ocean water types, in the spectral region of maximum transmittance, the evidence indicates that the incident energy density will be reduced by a factor of at least a million or more before an asymptotic radiance distribution can be obtained. This is particularly true for waters where the absorption is greater than the scattering and for spectral regions where the the ratio of absorption to scattering is greater than unity. Thus, unless the radiance distribution input to the ocean is relatively diffuse due to cloudy skies and a rough sea surface or the water has a low ratio of absorption to scattering, the simplification which the asymptotic radiance hypothesis provides in the theoretical analysis of the underwater radiant energy field is not generally valid in the upper layers of the sea.

The structure of solar radiation in the upper layers of the sea, where an asymptotic radiance distribution does not in general obtain, is at once the most complex, the most interesting and the most difficult to measure accurately. The manner in which the optical properties of the sea distinctively alter the radiance distribution in the upper layers of the sea are of particular interest, and this can best be analyzed by means of the theory of radiative transfer.

III. Equation of Transfer for Radiance

The theoretical study of radiative transfer in the sea has been discussed by many authors (e.g. LeGrand, 1939; Timofeeva, 1957; Jerlov, 1961, 1968; Lenoble, 1961; Preisendorfer, 1961, 1964, 1965, 1968, 1972; Tyler and Preisendorfer, 1962; Tyler, 1968; Feinstein et al., 1970). In the following review, which draws especially from the work of Preisendorfer and Tyler, emphasis is placed on the equation of transfer for radiance and on the optical properties of sea water which are derivable from underwater radiance distributions as a function of depth. This will provide a theoretical framework in which experimental radiometric measurements of solar radiation in the sea can be understood and compared.

The fundamental equation which governs the variation of radiance in a scattering and absorbing medium is the equation of transfer for radiance L:

$$\frac{l}{v}\frac{\delta L(z, \theta, \varphi, t)}{\delta t} + \frac{\mathrm{d}L(z, \theta, \varphi, t)}{r} =$$

$$-c(z)L(z, \theta, \varphi, t) + L^*(z, \theta, \varphi, y) + L_\eta(z, \theta, \varphi, t) \qquad (5)$$

where

$$L^*(z, \theta, \varphi, t) = \int_{4\pi} \beta(z, \theta, \varphi; \theta', \varphi')L(z, \theta, \varphi, t)\,\mathrm{d}\omega(\theta', \varphi'). \qquad (6)$$

The path function, L^*, is the radiance per unit length in the direction of the line of sight, generated by radiant energy scattered into the line of sight from all directions about the point z. $\beta(z, \theta, \varphi; \theta', \varphi')$ is the volume scattering function at point z for radiant energy incident in the direction (θ', φ') and scattered off in the direction (θ, φ). L_η is the source function, v is the velocity of light in the medium, and c is the total volume attenuation coefficient:

$$c(z) = a(z) + b(z). \qquad (7)$$

Here a is the volume absorption coefficient and the total scattering coefficient, b, is given by

$$b(z) = \int_{4\pi} \beta(z, \theta, \varphi; \theta', \varphi')\,\mathrm{d}\omega(\theta', \varphi'). \qquad (8)$$

It is usual to assume that the radiance distribution in the sea is in a steady, or at least quasi-steady, state so that the time dependent term in eq. 5 can be neglected. Before neglecting this term, we note that to do so requires the assumption of a constant input of radiance to maintain the system in a steady state. Thus, we will be dealing with an irreversible dissipative system, analogous to friction, where energy is lost or entropy gained. This is roughly equivalent to saying that the phenomena represented by eq. 5 are inherently statistical events and that we will be viewing the interaction of radiant energy with matter on a macroscopic level.

By assuming that the underwater radiance distribution is in a steady state, supplied by a constant source of radiance on the surface, that the source function is negligible, and that the radiant energy is nearly monochromatic and unpolarized, eq. 5 is reduced to the classical form:

$$\frac{dL(z, \theta, \varphi)}{dr} = -c(z)L(z, \theta, \varphi) + L^*(z, \theta, \varphi), \tag{9}$$

where $-r \cos \theta = z$. The first term on the right gives the space rate of loss of $L(z, \theta, \varphi)$ by attenuation along a direction of travel; the second term gives the space rate of gain of $L(z, \theta, \varphi)$ by rescattering of radiant energy back into the original direction of travel.

Preisendorfer (1968) has discussed in detail the principal methods of formulating and solving the basic equations of radiative transfer theory. Among these methods, which will be briefly outlined below, are iterative procedures and Monte Carlo techniques.

The Monte Carlo method of solving the equation of transfer (Cashwell and Everett, 1959; Meyer, 1954) can be based mathematically on the theory of stochastic processes and, in essence, attempts to follow the probable history of a single photon introduced into a scattering-absorbing medium. This history, by an ergodic argument, can be said to be representative of the instantaneous distribution of an aggregate of particles simultaneously introduced into the system. Plass and Kattawar (1969) and Raschke (1971) have calculated the complete radiation field in the atmosphere-ocean system using a Monte Carlo method. By making a realistic model for each component of scattering and absorption, both in the atmosphere and the ocean, they have calculated radiance values which show many of the essential features of underwater radiance distributions. The full potential of this technique will be realized when experimental measurements of the radiance distribution above and below water, along with the measurement of important optical properties, can be used for comparison with the calculations. For example, if the underwater radiance distribution is sufficiently sensitive to the shape of the volume scattering function, a comparison of trial calculations with experimental radiance distribution data could be used to obtain the optimum shape, including the forward scattering peak, of the volume scattering function.

A first iterative solution of the equation of transfer (eq. 9) has been used by Preisendorfer (1964) to give a model of radiance distribution in natural hydrosols. He develops this model assuming that c and $\beta(\theta)$ are independent of the depth z and that the path function can be approximated, making use of the solution of the two-flow Schuster equations for irradiance by

$$L_*(z, \theta, \varphi) = L_*(0, \theta, \varphi) \exp(-Kz) \tag{10}$$

where K is independent of depth. The radiance at depth z is then shown to be

$$L_r(z, \theta, \varphi) = L_0(z_t, \theta, \varphi) \exp(-cr) +$$
$$\frac{L^*(z, \theta, \varphi)}{c + K \cos \theta}(1 - \exp(-(c + K \cos \theta)r)). \quad (11)$$

Here $L_r(z, \theta, \varphi)$ is the radiance at depth z from the direction (θ, φ) and $L_0(z, \theta, \varphi)$ the radiance from the same direction evaluated at the target point at depth z_t a distance r away. Equation 11 illustrates the dependence of $L(z, \theta, \varphi)$ on the path function, the volume attenuation coefficient, and the attenuation coefficient for irradiance, K.

The above model, as well as other models developed by Jerlov and Fukuda (1960), Lenoble (1958, 1961, 1963) and Schellenberger (1963), have been reviewed by Jerlov (1968). One basic feature of these models is that they strive to relate the principal optical properties of the sea and the underwater distribution of radiance in a systematic manner.

IV. OPTICAL PROPERTIES OF THE OCEAN

Radiance is the most basic radiometric quantity. From radiance distribution data on the natural radiant energy field underwater, many of the important optical properties which relate to radiative transfer processes in the ocean can be calculated. The usefulness of radiance distribution data is outlined in Fig. 9 (Tyler, 1968). In this figure $L(z, \theta, \varphi)$ represents radiance distribution data taken at various depths, z, where θ is the zenith and φ the azimuth angle of a radiance value. From the radiance distribution data, the optical properties shown on the chart in Fig. 9 can be determined.

The irradiance, E, is the radiant flux incident on an infinitesimal element of surface containing the point under consideration, divided by the area of that element. E_d is the downwelling irradiance, i.e. it is the flux per unit area measured by a horizontally oriented cosine collector (Smith, 1969) facing upward. Similarly, E_u is the upwelling irradiance. They are defined in terms of the radiance distribution by the following equations:

$$E_d(z) = \int_{\varphi=0}^{2\pi} \int_{\theta=0}^{\pi/2} L(z, \theta, \varphi) \cos \theta \, d\omega \quad (12)$$

$$E_u(z) = \int_{\varphi=0}^{2\pi} \int_{\theta=\pi/2}^{\pi} L(z, \theta, \varphi) \, |\cos \theta| \, d\omega \quad (13)$$

where $d\omega = \sin \theta \, d\theta \, d\varphi$.

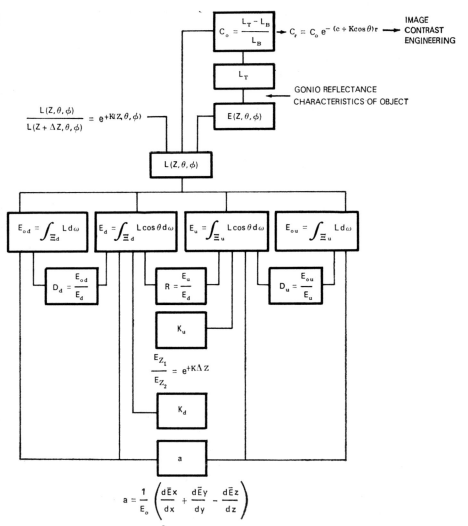

Fig. 9. Chart illustrating optical properties which can be derived from radiance distribution data. These properties are defined and discussed in the text.

The irradiance reflectance, R, at depth z is defined as

$$R(z) = E_u(z)/E_d(z). \qquad (14)$$

R may be thought of as the reflectance of a hypothetical plane surface at depth z in the medium. R depends upon and exhibits information about the scattering properties of the entire medium above and below the level z.

The integral of a radiance distribution, at a point at depth z, over all directions about the point gives the scalar irradiance, E_0:

$$E_0(z) = \int\int_{4\pi} L(z, \theta, \varphi)\, d\omega. \tag{15}$$

The scalar irradiances due to radiant energy received separately from the upper and lower hemispheres can be written:

$$E_{0d}(z) = \int_{\theta=0}^{2\pi}\int_{\theta=0}^{\pi/2} L(z, \theta, \varphi)\, d\omega \tag{16}$$

$$E_{0u}(z) = \int_{\varphi=0}^{2\pi}\int_{\theta=\pi/2}^{\pi} L(z, \theta, \varphi)\, d\omega \tag{17}$$

where

$$E_0(z) = E_{0d}(z) + E_{0u}(z). \tag{18}$$

The scalar irradiance, when divided by the velocity of light in the medium, yields the total amount of radiant energy per unit volume of space at the given point, i.e. the radiant energy density. Smith and Wilson (1972) have summarized the arguments for considering scalar irradiance, suitably filtered to measure total quanta (350–700 nm) (Jerlov and Nygård, 1969), an optimum measure of the energy available for photosynthesis and they have described a new technique to experimentally measure E_0.

The down- and upwelling distribution functions are defined as:

$$D_d(z) = \frac{E_{0d}(z)}{E_d(z)} \tag{19}$$

$$D_u(z) = \frac{E_{0u}(z)}{E_u(z)} \tag{20}$$

These functions are a simple means of characterizing the depth dependence of the shape of the radiance distribution. In addition, these distribution functions play important roles in the equations of applied radiative transfer theory as developed by Preisendorfer.

The attenuation coefficients for irradiance are defined as the logarithmic depth derivatives of the irradiance functions. Thus,

$$K_d(z) = \frac{-1}{E_d}\left(\frac{dE_d}{dz}\right), \tag{21}$$

or alternatively,

$$\frac{E_d(z_2)}{E_d(z_1)} = \exp\left(-K_d(z_2-z_1)\right), \tag{22}$$

where K_d has units of reciprocal length, and z is the depth at which E_d is measured. Similar K-type functions can be defined for upwelling irradiance and for the scalar irradiances. Physically, the K functions are the quantities that specify the individual depth dependence of irradiance functions. Historically, the K functions were derived from the experimental fact that, in general, radiant energy decreases exponentially with depth. These functions have been used in connection with both theoretical and experimental aspects of radiant energy in natural waters.

Preisendorfer (1958) has shown by a method which makes use of the divergence relation of the radiant energy field (Gershun, 1939), that the volume absorption function, a, can be computed without the requirement of previous knowledge of the volume attenuation function, c, or the volume scattering function, b. This may be done by dividing the rate of change with depth of the net upwelling irradiance by the scalar irradiance, that is,

$$a(z) = \frac{-1}{E_0(z)} \frac{\mathrm{d}}{\mathrm{d}z} (E_d(z) - E_u(z)). \qquad (23)$$

This relation holds for an emission-free optical medium that is in the steady state and in which the index of refraction is constant within the medium.

The upper part of Fig. 9 summarizes the application of radiance measurements to the problems of image contrast engineering. For an object to be radiometrically detectable, it must have a different radiance from its background in the direction of view of the sensor. Whatever difference does occur constitutes an optical signal. This optical signal at the object can be described by its inherent contrast (Duntley, 1952, 1962, 1963) defined as:

$$C_0 = \frac{L_T - L_B}{L_B}, \qquad (24)$$

where L_T is the inherent radiance of the object and L_B is the inherent radiance of the background against which the object is seen. Thus the first major step in any detection problem is the assessment of the magnitude of the inherent contrast. In computing the inherent contrast of a submerged object, the background radiance, L_B, can be obtained directly from the radiance distribution. The inherent radiance of the object, L_T, depends on the distribution of the flux incident upon the object and the angular reflectance of the object. Thus knowledge of the gonioreflecting properties of the object and the radiance distribution of the light field in which the object is immersed leads directly to the determination of inherent contrast for any path of sight.

6

Once the inherent contrast has been determined, the equation of contrast reduction (which can be derived from eqs. 24, 11, and 10),

$$C_R = C_0 \exp\left(-(c+K\cos\theta)r\right) \tag{25}$$

can then be applied to problems of the visibility of submerged objects. Here C_R is the apparent contrast seen by a sensor at a distance, r, along a path, c is the volume attenuation function, and K is the attenuation function for radiance, defined as:

$$K(z, \theta, \varphi) = \frac{-1}{L(z, \theta, \varphi)} \frac{dL(z, \theta, \varphi)}{dz}, \tag{26}$$

which can be calculated from the radiance distribution data. The θ, as used in eq. 25 and Fig. 9, is the angle between the vertical and the line of sight. Using eq. 25, the apparent contrast can be calculated along any path of sight once the inherent contrast of the object has been determined. Thus, the key to the problem of determining the contrast and detectability of submerged objects is a knowledge of the natural radiance distribution in the water in question.

It should be clear from the summary provided in Fig. 9 that the spatial distribution of radiance at various depths provides one of the most useful descriptions of the underwater radiant energy field. This usefulness was recognized by early workers, concerned with the optics of the ocean, and has inspired a number of them to construct instruments for the measurement of the spatial distribution of the sun's radiant energy underwater.

V. RADIANCE INSTRUMENTS

Early studies of the angular distribution of submarine daylight using a screened photometer were reported by Pettersson (1938) and by Jerlov and Liljequist (1938). By use of this instrument they were able to measure the radiance in concentric azimuthal zones of the upper hemisphere and to observe that the submarine light field tended to become more vertical in direction with increasing depth. Jerlov and Liljequist (1938) also made use of a radiance or Gershun tube (1939) photometer which directly limited the solid angle of acceptance of the incident radiant energy. Since this early work, radiance tube photometers have been used with ever increasing sensitivity and sophistication by numerous workers (Whitney, 1941a,b; Timofeeva, 1951, 1957, 1962; Duntley et al., 1955; Tyler, 1960; Jerlov and Fukuda, 1960; Jerlov, 1965; Sasaki et al., 1955, 1958, 1960, 1962; Schellenberger, 1967; Lundgren, 1971). Underwater orientation of the radiance tube instruments have been by means of servo systems and/or ingenious gimbaling principles. Reviews of these instruments and the resulting radiance data have been given by Jerlov (1968) and by Tyler (1968).

With the use of underwater radiance tube photometers these workers have discovered fine details in and the general behaviour of the angular structure of underwater radiant energy. In addition, they have shown that the experimental data is consistent with simple theoretical radiative transfer models and have demonstrated that radiance distributions approach an asymptotic state at great depths. However, in general, most of the reported experimental work to date has been limited to radiance measurements in only a few selected azimuthal planes and the horizontal plane. It was not until 1957 that sufficiently complete sets of data, so that all the optical properties summarized in Fig. 9 could be calculated, were obtained by Tyler (1960). Computations from these data were described by Tyler *et al.* (1959) and by Tyler and Shaules (1964). In addition, relatively complete radiance data for the Sargasso Sea and comparison of these data with the theory of radiative transfer have been reported by Lundgren and Højerslev (1971). These workers have confirmed the usefulness of radiance distribution data and stimulated further efforts to devise a more simple method for its measurement.

Underwater radiance tube photometers have proven to be accurate, to have high sensitivity, and to be versatile in the sense that they can be constructed so as to make radiance measurements for various spectral bandwidths or so as to measure the polarization of the incident radiation. The principal disadvantages of these instruments have been the problem of orientation and the time required to accumulate a set of data. These disadvantages arise because of the inherent nature of radiance distribution data.

To specify a radiance value requires giving both a magnitude and a direction. It has been found in practice that the maximum radiance for any fixed zenith angle occurs in the azimuth of the sun, which can then be used as an azimuthal reference direction. Securing an accurate zenith orientation, particularly in the oceanographic environment, has proven more difficult. Compounding the difficulty is the quantity of information in a complete radiance distribution.

A complete radiance distribution contains a great quantity of information. For example, to completely record a radiance distribution incident on a single point from the total hemisphere about that point with a radiance tube of $7 \cdot 2°$ angular resolution would require about one thousand radiance measurements, i.e. taking into account a magnitude and two direction components, about three thousand individual bits of information. Increase the angular resolution and the quantity of information rapidly increases. Radiance tube instruments are slow in obtaining such large quantities of information. Here "slow" is used in its most classical definition, to mean that the sun has moved appreciably during the time interval required to make the desired set of measurements. The full potential of radiance distribution measurements, detailed comparison with radiative transfer theory and completed knowledge of the ocean optical properties, depends upon obtaining accurate radiance distribution as a function of depth in a time small compared to any changes in the input radiance distribution of the sun and sky or to changes in the optical properties of the water itself.

The radiance distribution camera system, discussed briefly at the beginning of this chapter and in detail by Smith *et al.* (1970), was designed to record underwater radiance distributions as a function of depth accurately, rapidly and completely. The information-detecting capacity of photographic emulsions is very high (Zweig *et al.*, 1958; Jones, 1961) and thus the photographic approach is well adapted to the acquisition of radiance data. In addition, a complete depth profile of radiance distribution can be obtained in a time which is short compared to appreciable changes in the sun zenith angle. The accuracy of the radiance distribution camera system is limited by the techniques of photographic photometry (Mees, 1954). Recent preliminary analysis of radiance camera data show that the goal of 25% absolute accuracy and 10% relative precision for the radiance measurements can consistently be attained. The ultimate sensitivity of the instrument depends upon the emulsion speed and latitude of the best available commercial photographic films. Lengthy exposure times, to increase sensitivity of measurement, are of course not possible without reintroducing some of the disadvantages of the radiance tube instruments. To date the camera system has obtained radiance distributions to about seven optical depths ($K_d z \approx 7$) in a wide variety of water types and under a variety of environmental conditions.

The principal advantage of the photographic method of recording radiance, the rapid and complete accumulation of data, dictates that rapid and efficient data processing techniques be developed. The Visibility Laboratory has developed techniques for interfacing a high-speed microdensitometer with a computer for the purpose of processing radiance data stored on photographic film. Computer programs are used for computing radiance distributions and the optical properties outlined in Fig. 9 from the data film. It should be noted that the photographic film is a compact and convenient method of storing raw data. The original data film can be easily rescanned numerous times. Thus, the data can initially be reduced using a relatively coarse resolution and later rescanned with a finer resolution as circumstances dictate. This allows one to maximize the useful information, while minimizing the burden of data quantity.

Radiance distributions as a function of depth are the most basic radiometric quantities for describing the structure of solar radiation in the upper layers of the sea. Radiance, through the equation of transfer and its relation to other optical properties of the ocean, systematically unifies theoretical concepts and experimental results. Through the evolution of radiance instruments we are on the threshold of obtaining even more complete information on the angular structure of underwater radiant energy. This will provide a more complete knowledge of the optical properties of natural waters and should increase our understanding of the processes of radiative transfer in the ocean.

REFERENCES

Arvesen, J. C., Griffin, R. N. and Pearson, B. D. (1969). *Appl. Opt.*, **8**, 2215–2232.
Cashwell, E. D. and Everett, C. J. (1959). "A Practical Manual on the Monte Carlo Method for Random Walk Problems." Pergamon Press, Oxford.
Cox, C. S. and Munk, W. (1954). *J. Opt. Soc. Amer.*, **44**, 838–850.
Cox, C. S. and Munk, W. (1955). *J. Mar. Res.*, **14**, 63–78.
Cox, C. S. (1958). *J. Mar. Res.*, **16**, 199–230.
Duntley, S. Q. (1952). "The Visibility of Submerged Objects." Visibility Lab., Mass. Inst. of Technology, Cambridge, Mass.
Duntley, S. Q. (1954). *J. Opt. Soc. Amer.*, **44**, 574.
Duntley, S. Q., Uhl, R. J., Austin, R. W., Boileau, A. R. and Tyler, J. E. (1955). *J. Opt. Soc. Amer.*, **45**, 904(A).
Duntley, S. Q. (1962). *In* "The Sea" (M. N. Hill, ed.), Vol. 1, pp. 452–455. Wiley-Interscience, New York.
Duntley, S. Q. (1963). *J. Opt. Soc. Amer.*, **53**, 214–233.
Feinstein, P. L., Piech, K. R. and Leonard, A. (1970). *In* "Electromagnetics of the Sea", pp. 38–1–38–10. Agard Conference Proceeding, No. 77, Paris.
Gershun, A. (1939). *J. Math. Phys.*, **18**, 51–151.
Gordon, J. I. and Church, P. V. (1966a). *Appl. Opt.*, **5**, 793–801.
Gordon, J. I. and Boileau, A. R. (1966b). *Appl. Opt.*, **5**, 803–813.
Gordon, J. I. and Church, P. V. (1966c). *Appl. Opt.*, **5**, 919–923.
Gordon, J. I. (1969). *In* "Directional Radiance of the Sea Surface", Ref. 69–20. Scripps Institution of Oceanography.
Ivanoff, A. (1957). *Ann. Geophys.*, **13**, 22–53.
Jerlov (Johnson), N. G. and Liljequist, G. (1938). *Svenska Hydrograf. Biol. Komm. Skrifter, Ny Ser. Hydrog.*, **14**, 1–15.
Jerlov, N. G. and Fukuda, M. (1960). *Tellus*, **12**, 348–355.
Jerlov, N. G. (1961). *Medd. Oceanog. Inst. Göteborg, Ser. B.*, **8**, 1–40.
Jerlov, N. G. (1965). *In* "Progress in Oceanography" (Mary Sears, ed.), Vol. 3, pp. 149–157. Pergamon Press, New York.

Jerlov, N. G. (1968). "Optical Oceanography." Elsevier Publishing Co., Amsterdam.

Jerlov, N. G. and Nygård, K. (1969). Københavns Universitet, *Rep. Inst. Fys. Oceanog.*, **10**, 1–19.

Jones, R. C. (1961). *J. Opt. Soc. Amer.*, **51**, 1159–1171.

LeGrand, Y. (1939). *Ann. Inst. Oceanog.*, **19**, 393–436.

Lenoble, J. (1958). *Ann. Inst. Oceanog.*, **34**, 297–308.

Lenoble, J. (1961). *Compt. Rend.*, **252**, 2087–2089.

Lenoble, J. (1963). *Compt. Rend.*, **256**, 4638–4640.

Lundgren, B. and Højerslev, N. (1971). Københavns Universitet, *Rep. Inst. Fys. Oceanog.*, **14**, 1–33.

Lundgren, B. (1971). Københavns Universitet, *Rep. Inst. Fys. Oceanog.*, **17**, 1–34.

Mees, C. E. K. (1954). "The Theory of Photographic Processes." Macmillan, New York.

Meyer, H. A. (1954). *In* "Monte Carlo Methods", Symposium. Statistical Laboratory, University of Florida.

Minnaert, M. (1954). "The Nature of Light and Color in the Open Air." Dover Publications, New York.

Miyamoto, K. (1964). *J. Opt. Soc. Amer.*, **54**, 1060–1061.

Pettersson, H. (1938). *I.C.E.S. Rapports et Proces-Verbaux des Reunions*, **108**, 9–12.

Plass, G. N. and Kattawar, G. W. (1969). *Appl. Opt.*, **8**, 455–466.

Plass, G. N. and Kattawar, G. W. (1971). *J. Atmos. Sci.*, **28**, 1187–1198.

Preisendorfer, R. W. (1958). *Scripps Inst. Oceanog.*, Ref. Rep., 58–41.

Preisendorfer, R. W. (1959). *J. Mar. Res.*, **18**, 1–9.

Preisendorfer, R. W. (1961). *Union Geod. Geophys. Inst. Mon.* **10**, 11–30.

Preisendorfer, R. W. (1964). *In* "Physical Aspects of Light in the Sea" (J. E. Tyler, ed.), pp. 51–60. University Hawaii Press, Honolulu, Hawaii.

Preisendorfer, R. W. (1965). "Radiative Transfer in Discrete Spaces." Pergamon Press, New York.

Preisendorfer, R. W. (1968). *J. Quant. Spectrosc. Radiat. Transfer*, **8**, 325–338.

Preisendorfer, R. W. (1972). "Hydrologic Optics." Gordon and Breach, New York.

Raschke, E. (1971). *Beitr. Phys. Atmos.*, **45**, 1–19.

Sasaki, T., Okami, N., Watanabe, S. and Oshiba, G. (1955). *J. Sci. Res. Inst.*, **9**, 103–106.

Sasaki, T., Watanabe, S., Oshiba, G. and Okami, N. (1958). *J. Oceanog. Soc. Jap.*, **14**, 1–6.

Sasaki, T., Watanabe, S., Oshiba, G. and Okami, N. (1960). *Rec. Oceanog. Works Jap.*, (Special Number 4), 197–205.

Sasaki, T., Watanabe, S., Oshiba, G. and Kajihara, (1962). *Bull. Jap. Soc. Sci. Fish.*, **28**, 489–496.

Schellenberger, G. (1963). *Gerlands Beitr. Geophys.*, **72**, 315–327.

Schellenberger, G. (1965). *Acta Hydrophys.*, **10**, 79–105.

Schellenberger, G. (1967). *Gerlands Beitr. Geophys.*, **76**, 69–82.

Schellenberger, G. (1967). *Gerlands Beitr. Geophys.*, **76**, 321–333.

Smith, R. C. and Tyler, J. E. (1967). *J. Opt. Soc. Amer.*, **57**, 589–595.

Smith, R. C. (1969). *J. Mar. Res.*, **27**, 341–351.

Smith, R. C., Austin, R. W. and Tyler, J. E. (1970). *Appl. Opt.*, **9**, 2015–2022.

Smith, R. C. and Wilson, W. H. (1972). *Appl. Opt.*, **11**, 934–938.

Thekaehara, M. P., Kruger, R. and Duncan, C. H. (1969). *Appl. Opt.*, **8**, 1713–1732.

Thekaehara, M. P., ed. (1970). *In* "The Solar Constant and the Solar Spectrum Measured from a Research Aircraft", pp. 85. NASA Technical Report NASA TR R-351.

Timofeeva, V. A. (1951). *Dokl. Akad. Nauk SSSR*, **76**, 831–833. (English translation.)

Timofeeva, V. A. (1957). *Dokl. Akad. Nauk SSSR*, **113**, 556–559. (English translation.)

Timofeeva, V. A. (1962). *Izv. Akad. Nauk SSSR, Ser. Geofiz.*, **6**, 1843–1851. (English translation.)

Tyler, J. E. (1958). *J. Mar. Res.*, **16**, 96–99.

Tyler, J. E., Richardson, W. H. and Holmes, R. W. (1959). *J. Geophys. Res.*, **64**, 667–673.

Tyler, J. E. (1960). *Bull. Scripps Inst. Oceanog.*, **7**, 363–412.

Tyler, J. E. and Preisendorfer, R. W. (1962). *In* "The Sea" (M. W. Hill, ed.), Vol. 1, pp. 397–451. Wiley-Interscience, New York.

Tyler, J. E. and Shaules, A. (1964). *Appl. Opt.*, **3**, 105–110.

Tyler, J. E. (1968). *J. Quant. Spectrosc. & Radiat. Transfer*, **8**, 339–354.

Tyler, J. E. and Smith, R. C. (1970). "Measurements of Spectral Irradiance Underwater." Gordon and Breach, New York.

Tyler, J. E., Smith, R. C. and Wilson, W. H. (1972). *J. Opt. Soc. Amer.*, **62**, 83–91.

Ursell, F. (1956). *In* "Surveys of Mechanics" (G. K. Batchelor, ed.). Cambridge University Press, London.

Whitney, L. V. (1941a). *J. Mar. Res.*, **4**, 122–131.

Whitney, L. V. (1941b). *J. Opt. Soc. Amer.*, **31**, 714–722.

Wu, J. (1969). *J. Geophys. Res.*, **74**, 444–455.

Zweig, H. J., Higgins, G. C. and MacAdam, D. L. (1958). *J. Opt. Soc. Amer.*, **48**, 926–933.

Chapter 6

New Developments of the Theory of Radiative Transfer in the Oceans

J. RONALD V. ZANEVELD

School of Oceanography, Oregon State University, Corvallis, Oregon, U.S.A.

I. INTRODUCTION

The theory of radiative transfer is of central significance to the study of optical oceanography. From this theory are derived the various methods for measuring and predicting the behavior of optical parameters in the ocean. The interrelationships of the main areas of study within optical oceanography are best depicted in a diagram such as Fig. 1.

The four main areas of study are the submarine lightfields or radiance distributions due to both natural and man-made light sources, which are characterized by the apparent optical properties; the light scattering and attenuation properties of sea water, called the inherent optical properties of sea water; the suspended and dissolved materials that participate in light scattering and attenuation; and the hydrographic parameters in the ocean such as currents, temperature, dissolved oxygen, etc. Between each of the areas of study exist both experimental and theoretical relations, many of which are little understood.

Radiative transfer in the ocean is the study of the interaction of light with the water and the material suspended and dissolved in it.

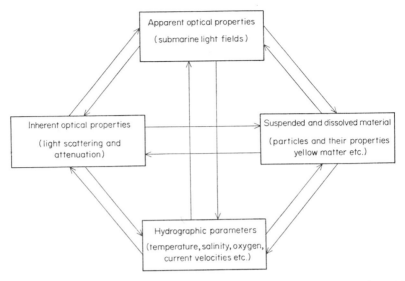

Fig. 1. The interrelationships of the various groups of concepts in optical oceanography.

This interaction is, however, highly dependent upon the scattering properties of the suspended matter so that a consideration of the particles in the ocean cannot be strictly excluded from the theory of radiative transfer.

The purpose of this chapter is to provide a brief review of the main topics of interest during recent years in the theory of radiative transfer in the oceans. An effort was made not to overlap subject matter with other chapters in this volume. The subjects of polarization and the relationship between particle size distributions and volume scattering functions will thus not be discussed. In this chapter all effects due to polarization will be neglected. All light will be considered to be mono-chromatic.

The principles and mathematical techniques of theoretical hydro-logical optics have recently been reviewed by Preisendorfer (1968), Jerlov (1968) and in the AGARD conference proceedings (1970). Only developments more recent than these reviews will be considered.

II. The Classical Problem

The classical radiative transfer problem in the ocean may be defined as follows: to derive an exact analytical expression for the submarine daylight field at all depths when the inherent optical properties of the ocean are known everywhere.

This problem has been solved for the case of a homogeneous plane parallel medium. Solutions may be found in Chandrasekhar (1950), Ambarzumian (1958), Sobolev (1963) and Preisendorfer (1965). Recent reviews may be found in Jerlov (1968) and Timofeyeva and Neuymin (1968).

Due to the large number of methods already available for the solution of the classical problem little attention has been paid to it in recent years. With the availability of large numbers of radiance measurements (Smith *et al.*, 1970) the problem may receive renewed interest as experimental verification of theoretical work becomes feasible.

Sugimori and Hasemi (1971) have extended Jerlov's (1968) method for the calculation of the submarine light field up to third order scattered light. This solution by-passes the equation of radiative transfer but solves for the scattered light from first principles. For thin layers the method is promising, but for greater depths more than third order scattering must be considered.

One of the most powerful theoretical tools for the study of radiative transfer is the Monte Carlo method. Two recent papers (Plass and Kattawar, 1972 and Kattawar and Plass, 1972) have used Monte Carlo calculations to predict the upward and downward flux as well as the upward and downward radiance for a realistic atmosphere-ocean system. The calculations are made for several wavelengths and various scattering models for the ocean. The method takes into account all known processes of absorption and scattering. The authors cite a probable error of the order of 5% for their calculations.

Other methods for the solution of the classical problem that are available have not recently been applied to the oceans but to similar problems in the atmosphere. Some of the methods most readily applicable to the ocean are the modified Fourier transform method (Dave and Gazdag, 1970), and improved methods employing spherical harmonics (Devaux and Herman, 1971).

Feinstein *et al.* (1970) have shown that the singular eigenfunction approach may successfully be applied to oceanic problems. In this case the light field and the volume scattering function are represented as in Chandrasekhar (1950) but the resulting set of differential equations is solved by using singular eigenfunction expansions. The drawback of this approach is the high degree of mathematical complexity and the difficulty of developing the necessary numerics.

Heggestad (1971) has contributed an approximate method for the calculation of light transmission through optically thick clouds that is suitable for oceanic calculations. This model is limited by the consideration of narrow angle scattering events only, but multiple scattering is included.

III. The Narrow Light Beam

The increased use and development of the laser in the last decade has spurred theoretical considerations of the behavior of a narrow light beam in the sea. This problem is of considerable interest not only because of applications to communication and ranging, but solutions may directly be applied to visibility problems. Any radiating object may be considered as a collection of point sources. If the response of the medium to a point source of light is known, the response of the system to any object may be obtained by convolving the object with the point source response. The point source response may also be used to calculate the Modulation Transfer Function of sea water. This function entirely specifies the visibility in an oceanic medium.

Much of the theoretical work on this problem has been carried out by Romanova (1968a, 1968b, 1969, 1970, 1971). The study involves a solution of the equation of radiative transfer in the form (Romanova, 1968a):

$$\cos \theta \frac{\delta L}{\delta z}(x, y, z, \theta, \varphi) + \sin \theta \sin \varphi \frac{\delta L}{\delta x}(x, y, z, \theta, \varphi) +$$

$$\sin \theta \cos \varphi \frac{\delta L}{\delta y}(x, y, z, \theta, \varphi) + L(x, y, z, \theta, \varphi) =$$

$$\frac{b}{4\pi c} \int_0^{2\pi} \int_0^{\pi} \beta(\theta, \theta', \varphi - \varphi') L(x, y, z, \theta', \varphi') \sin \theta' \, d\theta' \, d\varphi'$$

where: θ is the zenith angle and φ the azimuth

x, y, z is a cartesian coordinate system

$L(x, y, z, \theta, \varphi)$ is the radiance

c is the beam attenuation coefficient

b is the total scattering coefficient

$\beta(\theta, \theta', \varphi - \varphi')$ is the volume scattering function

$$\frac{1}{4\pi} \int_{-1}^{+1} \int_0^{2\pi} \beta(\cos \theta) \sin \theta \, d\theta \, d\varphi = 1$$

The singly scattered light is calculated directly. The multiple scattered light is obtained from a determination of the moments I_{nk} of the light field due to multiple scattered light. The moments are given by:

$$I_{nk}(z, \theta, \phi) = \iint_{-\infty}^{+\infty} x^n y^k L_{\text{multiple}}(x, y, z, \theta, \phi) \, dx \, dy$$

The volume scattering function is represented by a series of Legendre functions. Using the small angle approximation, a solution is obtained for the light field adjacent to the axis of the beam at not too great distances from the axis of the beam. The mean deviations $\bar{x}$, $\bar{y}$ of the light field from the axis of the beam were calculated as well as the mean dispersions $\overline{x^2}-\bar{x}^2$ and $\overline{y^2}-\bar{y}^2$ and the correlation coefficients between these dispersions. Comparison of the theoretically obtained mean deviations with experimental values shows reasonable agreement.

Bravo-Zhivotovskiy *et al.* (1969) have also provided a solution for the narrow beam problem. They employed the narrow angle approximation. The distribution of irradiance across a beam is calculated. The volume scattering function is approximated as an exponential function of the angle. Calculated values agreed well with experimental observations.

Romanova (1968b) has obtained approximate theoretical expressions for the light field due to a narrow beam at great depths in a scattering medium. It is shown that the mean deviations at great depths do not depend on the optical depth. At a given depth the dispersion increases as the ratio of the total scattering coefficient and the beam attenuation coefficient increases. In a subsequent paper Romanova (1971) calculates the effective size of the light spot at the boundaries of a scattering layer when the layer is illuminated by a narrow beam. Again the response to the beam is characterized by the mean deviations, the dispersions of the mean deviations and a correlation coefficient. In this case generalized invariance principles as outlined by Chandrasekhar (1950) and Ambarzumian (1968) are employed to obtain a solution.

Granatstein *et al.* (1972) have developed a single scattering theoretical model for the reflectance of a narrow beam in turbid water. The reflectance is related to the concentrations of suspended and dissolved materials in the water. The model shows good correlations with laboratory measurements.

Perhaps the best way of studying the narrow beam in sea water is the Monte Carlo method. This method avoids many of the mathematical complexities involved in analytical solutions of the equation of radiative transfer. The Monte Carlo method has been used by Chilton *et al.* (1969) in a study of the attenuance of a collimated beam as a visibility parameter. Golubitskiy and Tantashev (1969) have also used the Monte Carlo approach and report accuracies of 20–30% compared with experimental observations. Hessel and LaGrone (1970) have used a Monte Carlo method to predict the irradiance distribution on a plane perpendicular to the laser beam at any distance from the source. Laboratory measurements show excellent agreement with computed values.

Ivanov and Ganich (1968), Makarevich *et al.* (1969) and Levin (1969) have reported experimental measurements on the structure of a narrow beam under water for various volume scattering functions and beam attenuation coefficients.

The Modulation Transfer Function carries the same information as the point source response function. The Modulation Transfer Function of sea water has been calculated using a numerical method by Zaneveld and Beardsley (1969). Wells (1969) has obtained a theoretical solution using the small angle approximation. By means of the transformation between the Modulation Transfer Function and the narrow beam response (Zaneveld *et al.*, 1970) it is possible to calculate the point source response from the Modulation Transfer Function.

A great amount of work has been carried out recently on the study of light wave propagation in turbulent media (for instance Consortini and Ronchi, 1970; Furutsu, 1971; Sodha *et al.*, 1971; Lutomirski and Yura, 1972; Torrieri and Taylor, 1972; Berreman, 1972; Brown, 1972). Turbulence has been shown to have considerable influence on submarine light fields (Honey and Sorenson, 1970; Hodgson, 1972). A complete theoretical study of the narrow light beam in sea water must not be limited to considerations of turbulence only or particulate scattering only. Both effects must be included.

In his study of the collimated beam Yura (1971) has included both the scattering due to suspended particles and the effect of turbulence induced index of refraction variations. He also calculated the Modulation Transfer Function for both processes. The study is limited by the assumption of single scattering and a narrow angle approximation.

The study of the narrow beam has been extended to include pulsed sources. The equation of radiative transfer in this case must be written as:

$$\frac{1}{v}\frac{\delta L}{\delta t}(\vec{\Omega})+\mathbf{V} \cdot L(\vec{\Omega}) = -cL(\vec{\Omega})+ \int_{4\pi} \beta(\vec{\Omega} \cdot \vec{\Omega}')L(\Omega')\,d\vec{\Omega}'$$

where v is the speed of light.

Romanova (1969, 1970) has contributed two papers dealing with the non-stationary light field due to a narrow beam. One deals with the light field at great depths (Romanova, 1969). The other deals with the light field in the surface layer (Romanova, 1970). Analytic solutions for this problem are extremely difficult so that no complete solutions are available yet. Kochetkov (1970) has provided a solution for the time dependence of the light from a pulsed source scattered through 180°. The asymptotic solution for large time periods is given. Some of the results obtained for the transmission of light pulses through the atmosphere should be applicable in modified form to the ocean (for instance Kerr *et al.*, 1969).

Dolin and Savel'yev (1971) have derived an expression for the back scattered signal from a pulsed source for a turbid medium. As the solution is for single scattering only and employs the small angle approximation it is of limited usefulness.

Minin and Goncharov (1970) have applied Monte Carlo techniques to the reflection of light pulses by a semi-infinite layer of scattering medium. The relative ease with which solutions are obtained indicates that for time-dependent problems the Monte Carlo method is better suited than purely analytical approaches.

IV. The Asymptotic State

The shape of the radiance distribution at great depths, $L_\infty(\theta)$, and the coefficient of attenuation of the asymptotic distribution, k_∞, continue to pose some interesting problems. No exact analytical relationship between k_∞ and the inherent optical properties has been found yet. Herman and Lenoble (1968) have obtained a method for the calculation of the shape of the asymptotic radiance distribution and the asymptotic attenuation coefficient. Both the volume scattering function and the asymptotic distribution are developed into a series of Legendre functions. A recurrence relation for the coefficients of the radiance in terms of the coefficients of the volume scattering function is then obtained. By limiting the number of coefficients, solutions are then found for the shape of $L_\infty(\theta)$ and the relation between k_∞/c and b/c.

Beardsley and Zaneveld (1969) used a numerical solution of the equation of radiative transfer to obtain the dependence of the near-asymptotic apparent optical properties on the inherent optical properties. Timofeeva (1971) has contributed some experimental results relating the shape of the asymptotic radiance distribution to the ratio of radiance attenuation and beam attenuation, k_∞/c. These studies do not directly include the shape of the volume scattering function. The first complete analysis of the dependence of the shape of the asymptotic distribution on the shape of the volume scattering function was carried out by Prieur and Morel (1971). The equation of radiative transfer at great depths is solved by a numerical iterative approach. The scattering functions are constructed by a linear combination of the scattering function for pure water, b_m, and the scattering function for particles, b_p. Calculations were made for several ratios of b_m/b_p. Results are summarized in a series of figures giving the shape of $L_\infty(\theta)$ for various values of b_m/b_p and k_∞/c. Furthermore the relationship between k_∞/c, b/c and the shape of the volume scattering function is summarized in Fig. 2. It should be noted that $1-\dfrac{b}{c} \leqslant \dfrac{k_\infty}{c} \leqslant 1$ so that all values relating k_∞/c and b/c must lie in the region enclosed by $\dfrac{k_\infty}{c}=1-\dfrac{b}{c}$, $\dfrac{k_\infty}{c}=1$ and $\dfrac{b}{c}=1$. For constant shape of the volume scattering function lines as drawn on the figure can be constructed. The shape of the scattering function thus has a pronounced effect on the relation between k_∞/c and b/c.

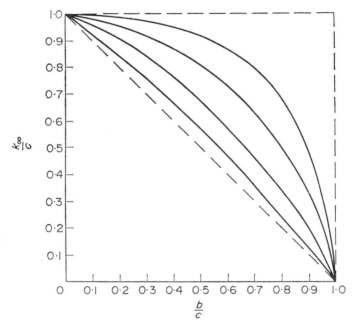

Fig. 2. The relation of k_∞/c and b/c for different shapes of the volume scattering function.

Similar results were obtained by Zege (1971). In this paper expansions for $L_\infty(\theta)$ and $\beta(\theta)$ similar to those of Herman and Lenoble (1968) are used, resulting in recurrence relations between the expansion coefficients. In the case of weak absorption or strong absorption approximate expressions for the dependence of k_∞/c on b/c and the expansion coefficients of $\beta(\theta)$ can be given. Finally a numerical technique is used confirming the results of Prieur and Morel (1971).

In the work of Lundgren and Højerslev (1971) an alternate proof of the asymptotic light field hypothesis is given. It is shown that as depth increases indefinitely the attenuation coefficients

$$k(\theta, \phi, z) = -\frac{1}{L(\theta, \phi, z)}\frac{dL}{dz}(\theta, \phi, z)$$

approach a common value. Zaneveld and Pak (1972) have also given an alternate proof of the asymptotic state, and have shown that the volume scattering function may be calculated if the shape of the asymptotic radiance distribution is known.

V. The Inverse Problem

The main emphasis of radiative transfer studies in the ocean has always been on the description of the light field when the inherent optical properties are known. At present it appears possible to measure the submarine day light field with great accuracy (Smith et al., 1970, see also elsewhere in this volume). The inverse problem of finding the inherent optical properties when the submarine light field is known has taken on increased value in the light of these recent developments in instrumentation. Preisendorfer (1965) calls these the problems of the second class. Zaneveld and Pak (1972) have indicated a solution to the inverse problem when the light field is axially symmetric. In this section we will give an analytic solution to the inversion of the equation of radiative transfer for any light field.

First the radiance distribution is integrated with respect to the azimuth so that the distribution is made artificially axially symmetric. Let the axially symmetric distribution and the volume scattering function be represented by a series of Legendre functions:

$$L(\theta, z) = \sum_{n=0}^{\infty} A_n(z)P_n(\cos\theta)$$

$$\beta(\gamma) = \sum_{n=0}^{\infty} B_n P_n(\cos\gamma)$$

Expanding the volume scattering coefficient using the addition theorem for Legendre polynomials and substituting in the equation of radiative transfer yields the set of differential equations:

$$\frac{n}{2n-1}\frac{d}{dz}A_{n-1}(z) + \frac{n+1}{2n+3}\frac{d}{dz}A_{n+1}(z) + \left(c - \frac{4\pi}{2n+1}B_n\right)A_n(z) = 0$$

so that:

$$c - \frac{4\pi}{2n+1}B_n = -\frac{1}{A_n(z)}\left[\frac{n}{2n-1}\frac{d}{dz}A_{n-1}(z) + \frac{n+1}{2n+3}\frac{d}{dz}A_{n+1}(z)\right]$$

7

or:

$$c - \frac{4\pi}{2n+1} B_n = - \frac{\int_0^\pi \frac{\mathrm{d}L\,(\theta,\,z)}{\mathrm{d}z} P_n(\cos\theta)\cos\theta\sin\theta\,\mathrm{d}\theta}{\int_0^\pi L(\theta,\,z) P_n(\cos\theta)\sin\theta\,\mathrm{d}\theta}$$

It is clear that for large n:

$$c = \lim_{n\to\infty}\left\{c - \frac{4\pi}{2n+1} B_n\right\}$$

Once c has been obtained the following expression for the volume scattering function may be given in terms of the light field:

$$\beta(\gamma) = \sum_{n=0}^{\infty} \frac{2n+1}{4\pi} P_n(\cos\gamma)\left[c + \frac{\int_0^\pi \frac{\mathrm{d}L(\theta,\,z)}{\mathrm{d}z} P_n(\cos\theta)\cos\theta\sin\theta\,\mathrm{d}\theta}{\int_0^\pi L(\theta,\,z) P_n(\cos\theta)\sin\theta\,\mathrm{d}\theta}\right]$$

It has thus been shown that the light field and its derivative with depth at a point completely specify the beam attenuation coefficient and the volume scattering function.

VI. The Particle Index of Refraction

The determination of an average index of refraction for the suspended particles in a sample of sea water is of great enough interest to be considered here, even though not strictly within the theory of radiative transfer. An "average" index of refraction for a sample may be calculated if the particle size distribution is known by assuming an index of refraction and then calculating the volume scattering function using Mie (1908) theory. The index of refraction that reproduces the observed volume scattering is considered to be the index of refraction of the particles (Gordon and Brown, 1971). This method is cumbersome, however, and does not lend itself to the analysis of great numbers of samples.

A general method for the routine determination of the index of refraction of a collection of suspended particles has been given by Zaneveld and Pak (1973). The cumulative particle size distribution $g(D)$ is approximated by an exponential:

$$g(D) = N \exp(-AD)$$

where D is the diameter of a particle, and N and A are parameters to be determined from an observed particle size distribution. The particle size distribution is multiplied by the extinction cross-section for particles with a complex index of refraction (Deirmendjian, 1969; Van De Hulst, 1957) and integrated. If the approximation of non-absorbing particles is made, one then obtains the following relationship for the ratio of particle scattering coefficients at two wavelengths as a function of the particle size distribution parameters:

$$\frac{b_p(\lambda_1)}{b_p(\lambda_2)} = \frac{\dfrac{1}{A^2} + \dfrac{k_1^2 - A^2}{(A^2 + k_1^2)^2}}{\dfrac{1}{A^2} + \dfrac{k_2^2 - A^2}{(A^2 + k_2^2)^2}}$$

where

$$k_1 = \frac{2\pi}{\lambda_1}|n_p - m_w| \quad \text{and} \quad k_2 = \frac{2\pi}{\lambda_2}|n_p - m_w|,$$

n_p is the particle index of refraction and m_w is the index of refraction of water. Here one may apply the observation that the particle scattering coefficient is proportional to the volume scattering function at 45°.

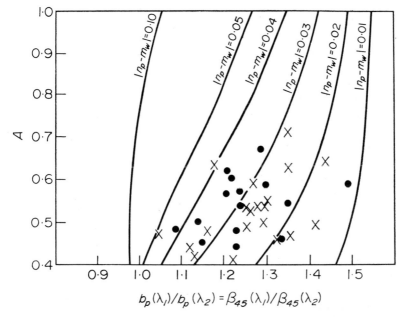

Fig. 3. The difference between the particulate and water indices of refraction $|n_p - m_w|$ as a function of the exponential particle size distribution parameter A and the ratio of light scattered at 45° for two wavelengths $\beta_{45}(\lambda_1)/\beta_{45}(\lambda_2)$. Dots and crosses mark samples taken from two stations off the coast of Ecuador. Dots mark station 4–2; crosses mark station 4–8. $\lambda_1 = 436\ m\mu$; $\lambda_2 = 546\ m\mu$.

Even if the constant of proportionality is not known one may postulate that:

$$\frac{b_p(\lambda_1)}{b_p(\lambda_2)} = \frac{\beta_{45}(\lambda_1)}{\beta_{45}(\lambda_2)}$$

provided that the constant of proportionality between b_p and β_{45} is independent of wavelength. This method has the further advantage of avoiding the necessity of obtaining absolute values for the volume scattering function at two wavelengths, if the ratio can accurately be determined. The resulting equation is summarized in Fig. 3. The method has been applied to a great number of stations in the Equatorial Pacific. The results are reported elsewhere in this volume by Pak and Plank (Chapter 11).

VII. Conclusions

The main effort of theoretical work has been on the behavior of a laser beam under water. This interesting problem has not yet been solved completely due to the highly complicated mathematics necessary. Best results have been obtained by means of Monte Carlo methods.

The classical problem has received less attention than it deserves. A new surge of work is expected when improved techniques for the measurement of the entire submarine daylight field become available. When the light field is measured in combination with light scattering and attenuation the various theoretical results could be tested. Considerations of the classical problem should be expanded to include vertically non-homogeneous oceans as well as turbulence effects.

Recently the asymptotic state has been examined more closely. A relation between the asymptotic attenuation coefficient and the inherent optical properties has been found by numerical and approximate analytical methods. An exact analytical expression has not yet been obtained.

New techniques for the calculation of the volume scattering function from the submarine daylight field have been developed. These techniques should provide an independent check on the various methods of measurement of the volume scattering function currently in use.

An average particle index of refraction can routinely be calculated from the ratio of light scattered at 45° at two wavelengths and the particle size distribution.

Acknowledgement

This research was supported by the Office of Naval Research through contract N000 14–67–A–0369–0007 under project NR 083–102.

REFERENCES

AGARD. (1970). Conference Proceedings No. 77 "Electromagnetics of the Sea". AGARD, Paris.
Ambarzumian, V. A. (1958). "Theoretical Astrophysics." Pergamon Press, New York.
Beardsley, G. F. and Zaneveld, J. R. V. (1969). *J. Opt. Soc. Am.*, **59**, 373–377.
Berreman, D. W. (1972). *J. Opt. Soc. Am.*, **62**, 502–510.
Bravo-Zhivotovskiy, D. M., Dolin, L. S., Luchinin, A. G. and Savel'yev, V. A. (1969). *Izv. Atmosph. Ocean. Phys.*, **5**, 83–87.
Brown, W. P. (1972). *J. Opt. Soc. Am.*, **62**, 45–54.
Chandrasekhar, S. (1950). "Radiative Transfer." Oxford University Press, London.
Chilton, F., Jones, D. D. and Talley, W. K. (1969). *J. Opt. Soc. Am.*, **59**, 891–898.
Consortini, A. and Ronchi, L. (1970). *Appl. Opt.*, **9**, 125–128.
Dave, J. V. and Gazdag, J. (1970). *Appl. Opt.*, **9**, 1457–1466.
Deirmendjian, D. (1969). "Electromagnetic Scattering on Spherical Polydispersions." Elsevier, New York, London, Amsterdam.
Devaux, C. and Herman, M. (1971). *Compt. Rend.*, **273B**, 849–852.
Dolin, L. S. and Savel'yev, V. A. (1971). *Izv. Atmosph. Ocean. Phys.*, **5**, 328–331.
Feinstein, D. L., Piech, K. P. and Leonard, A. (1970). *In* "Electromagnetics of the Sea". Paper No. 38. AGARD Conference Proceedings No. 77. AGARD, Paris.
Furutsu, K. (1971). *J. Opt. Soc. Am.*, **62**, 240–254.
Golubitskiy, B. M. and Tantashev, M. V. (1969). *Izv. Atmosph. Ocean. Phys.*, **5**, 428–430.
Gordon, H. R. and Brown, O. B. (1971). *Trans. Am. Geophys. Union*, **52**, 245.
Granatstein, V. L., Rhinewhine, M., Levin, A. M., Feinstein, D. L., Mazuriwski, M. J. and Piech, K. R. (1972). *Appl. Opt.*, **11**, 1217–1224.
Heggestad, H. M. (1971). *J. Opt. Soc. Am.*, **61**, 1293–1300.
Herman, M. and Lenoble, J. (1968). *J. Quant. Spectrosc. Radiat. Transfer*, **8**, 355–367.
Hessel, K. R. and LaGrone, A. H. (1970). *In* "Electromagnetics of the Sea". Paper No. 28. AGARD Conference Proceedings No. 77. AGARD, Paris.
Hodgson, R. T. (1972). "Fourier Imaging in a Hydrological Medium; a Test of the Linear Invariant Hypothesis." Ph.D. Thesis, Corvallis, Oregon State University.
Honey, R. C. and Sorenson, G. P. (1970). *In* "Electromagnetics of the Sea." Paper No. 39. AGARD Conference Proceedings No. 77. AGARD, Paris.
Ivanov, A. P. and Ganich, P. Ya. (1968). *Oceanology*, **8**, 762–768.
Ivanov, A. P., Kalinin, I. I., Kozlov, V. D., Skrelin, A. L. and Sherbaf, D. (1969). *Izv. Atmosph. Ocean. Phys.*, **5**, 116–118.
Jerlov, N. G. (1968). "Optical Oceanography." Elsevier, New York, London, Amsterdam.
Kattawar, G. W. and Plass, G. N. (1972). *J. Phys. Oceanog.*, **2**, 146–156.
Kerr, J. R., Titterton, P. J. and Brown, C. M. (1969). *Appl. Opt.*, **8**, 2233–2239.
Kochetkov, V. M. (1970). *Izv. Atmosph. Ocean. Phys.*, **6**, 342–348.
Lundgren, B. and Højerslev, N. (1971). Københavns Universitet, *Inst. Fysisk Oceanografi*, Report No. 14.
Lutomirski, R. F. and Yura, H. T. (1972). *Appl. Opt.*, **10**, 1652–1658.

Makarevich, S. A., Ivanov, A. P. and Il'ich, G. K. (1969). *Izv. Atmosph. Ocean. Phys.*, **5**, 40–43.

Mie, G. (1908). *Ann. Phys. (Leipzig)*, **25**, 377.

Minin, E. A. and Goncharov, E. G. (1970). *Izv. Atmosph. Ocean. Phys.*, **11**, 725–727.

Plass, G. N. and Kattawar, G. W. (1972). *J. Phys. Oceanog.*, **2**, 139–145.

Preisendorfer, R. W. (1965). "Radiative Transfer on Discrete Spaces." Pergamon Press, Oxford.

Preisendorfer, R. W. (1968). *J. Quant. Spectrosc. Radiat. Transfer*, **8**, 325–338.

Prieur, L. and Morel, A. (1971). *Cahiers Oceanogr.*, **23**, 35–47.

Romanova, L. M. (1968a). *Izv. Atmosph. Ocean. Phys.*, **4**, 679–684.

Romanova, L. M. (1968b). *Izv. Atmosph. Ocean. Phys.*, **4**, 175–179.

Romanova, L. M. (1969). *Izv. Atmosph. Ocean. Phys.*, **5**, 261–265.

Romanova, L. M. (1970). *Izv. Atmosph. Ocean. Phys.*, **6**, 281–285.

Romanova, L. M. (1971). *Izv. Atmosph. Ocean. Phys.*, **7**, 270–277.

Smith, R. C., Austin, R. W. and Tyler, J. E. (1970). *Appl. Opt.*, **9**, 2015–2022.

Sobolev, V. V. (1963). "A Treatise on Radiative Transfer." D. Van Nostrand, Princeton, N.J.

Sodha, M. S., Ghatak, A. J. and Malik, D. P. S. (1971). *J. Opt. Soc. Am.*, **61**, 1492–1494.

Sugimori, Y. and Hasemi, T. (1971). *J. Oceanogr. Soc. Jap.*, **27**, 73–80.

Timofeeva, V. A. and Neuymin, G. G. (1968). *Izv. Atmosph. Ocean. Phys.*, **4**, 747–757.

Timofeeva, V. A. (1971). *Izv. Atmosph. Ocean. Phys.*, **7**, 467–469.

Timofeeva, V. A. and Solomonov, V. K. (1970). *Izv. Atmosph. Ocean. Phys.*, **6**, 364–357.

Torrieri, D. J. and Taylor, L. S. (1972). *J. Opt. Soc. Am.*, **62**, 145–147.

Van de Hulst, H. C. (1957). "Light Scattering by Small Particles." Wiley, New York.

Wells, W. H. (1969). *J. Opt. Soc. Am.*, **59**, 686–691.

Yura, H. T. (1971). *Appl. Opt.*, **10**, 114–118.

Zaneveld, J. R. V. and Beardsley, G. F. (1969). *J. Opt. Soc. Am.*, **59**, 378–380.

Zaneveld, J. R. V., Hodgson, R. and Beardsley, G. F. (1970). *In* "Electromagnetics of the Sea". Paper No. 41. AGARD Conference Proceedings No. 77. AGARD, Paris.

Zaneveld, J. R. V. and Pak, H. (1972). *J. Geophys. Res.*, **77**, 1689–1694.

Zaneveld, J. R. V. and Pak, H. (1973). *J. Opt. Soc. Am.*, **63**, 321–324.

Zege, E. P. (1971). *Izv. Atmosph. Ocean. Phys.*, **7**, 86–92.

Chapter 7

Underwater Visibility and Photography

SEIBERT Q. DUNTLEY

Visibility Laboratory, Scripps Institution of Oceanography,
University of California, San Diego, La Jolla, California, U.S.A.

I. RANGE LIMITATIONS BY NATURAL WATERS

A self-luminous object cannot be seen or photographed at a limitless range through otherwise unlighted water, even if an ample amount of light from it reaches the observer or camera. Beyond a certain distance no image can be perceived, although a bright, diffuse glow may be easily discernible; the object is hidden by some of its own light that has been scattered toward the observer. This occurs even in the clearest water.

The reason for the obscuration is simple: an image is formed by a small amount of light that departs from the object in the exact direction of the eye or the camera lens and arrives there without having been scattered or absorbed. A much larger amount of light also departs from the object in adjacent directions. Some of it is scattered to the lens. This light does not contribute to the image, but it passes through the lens and floods the image plane. The ratio of scattered light to image-forming light increases rapidly with the distance of the self-luminous object, and at some range it exceeds the contrast threshold of the eye or the camera. The image cannot then be seen; only a glow of scattered light is discernible.

In order to speak about range limitations in a general but quantitative way, let us specify object distance in multiples of the length of water path required to reduce image-forming light by a factor of $1/e$, where $e = 2\cdot71828$ and $1/e = 0\cdot367879$. Let that path be called the attenuation length because it is equal to the reciprocal of the well known attenuation coefficient of water. Attenuation length may exceed 20 m in very clear oceans, or it may be 5 m in coastal zones and shrink to a few cm in turbid harbours.

The limiting range at which the image of a self-luminous object is visible in unlighted water varies from 15 to 20 attenuation lengths, depending upon the water's ratio of scattering to absorption. In the daylighted sea ordinary objects can be sighted at 3 to 6 attenuation lengths depending upon the nature of the object and whether the observer looks downward, horizontal or upward. Even shorter limiting ranges are found when these same objects are viewed or photographed by means of light from a submerged lamp located near the camera. In fact, vision or photography beyond 3 attenuation lengths requires the lamp to be separated from the camera, or prevented by a shield from shedding direct light on the path of sight, and/or that polarizers be used at both lamp and camera.

The reduction in limiting range from 15 to 20 attenuation lengths for self-luminous objects to 3 to 6 attenuation lengths in the case of ordinary objects lighted by submerged lamps near the camera is caused by back-scattering throughout the illuminated path of sight. Most of the lost range can be restored by placing the lamp close to the object, rather than in the vicinity of the camera. Long time exposures are often used when this is done, because a severe loss in total light is imposed by absorption in the long water path. Readers interested in a more complete and formal mathematical treatment of the foregoing principles of underwater visibility and photography are referred to the author's paper entitled "Light in the Sea" (Duntley, 1963).

II. RESOLUTION LIMITATIONS BY NATURAL WATERS

A. INTRODUCTION

It has been said that horizontal visibility through daylighted water is closely similar to horizontal visibility through the atmosphere, except for a factor of about 1000 in range. Yet many who have looked and photographed carefully through both media have the impression that resolution suffers differently in water than in air, even after allowing for the factor of 1000 in object distance. It is common practice to attribute resolution losses in either media to density gradients and to imply that water is a poorer optical medium than air. Is that really true? If so, are the deleterious density variations in the oceans attributable chiefly to temperature gradients or to salinity gradients, or to both? Are such gradients always, inescapably present everywhere in the oceans, or are the optically harmful ones found only in certain places or under particular circumstances? Is the resolution of optical images determined more by temperature and salinity gradients than by the presence of microscopic transparent biological material? What is the highest spatial frequency that is transmitted by natural waters?

My curiosity about the foregoing questions has been aroused many times, but most actively when I obtained good photographic quality in an underwater photograph taken through 9·6 scattering lengths (16 attenuation lengths) of water path.

B. IN-WATER PHOTOGRAPHY AT MODEL SCALE

The above mentioned picture was made in one of the indoor tanks at the Visibility Laboratory. The objects photographed were small dummies which looked like swimmers. They were moored in the tank so that time exposures could be made when necessary.

The tank used is only 5-m long, but dyes and scattering agents can be added to produce any desired number of attenuation lengths in that 5 m. In doing this we always match the shape of the volume scattering function to that of typical ocean water and set the ratio of b to c at any value we select.

It is well known that good quality underwater pictures can be made with lights located near the camera provided the object is only an attenuation length or two away. Figure 1 shows how the dummies looked through 2 attenuation lengths of lighted water, how much fainter they appear at 3 attenuation lengths, and that at 4 lengths they are barely discernible.

Various forms of light control and/or the use of polarizers enable the useful range to be lengthened somewhat, but beyond 5 or 6 attenuation lengths all efforts usually fail if the source of illumination and the camera are about equally distant from the object. Whenever the lamp can be placed near the object, however, pictures at longer range are possible, provided that sufficient exposure can be achieved.

Figure 2 shows the dummy swimmers at 12 attenuation lengths, or 7·2 scattering lengths. They were illuminated by a nearby carefully shielded 600-W tungsten halogen lamp. Linear polarizers were used at the lamp and at the camera to suppress back-scattered light from the

7*

illuminated water behind the swimmers, i.e. to darken the background. The scene was nearly as if the dummies were self-luminous in otherwise unlighted water. The picture was taken on Kodak Plus X film with a Nikon 35-mm camera having a lens 50 mm in focal length. The aperture was $f/2$. An exposure of several minutes was required.

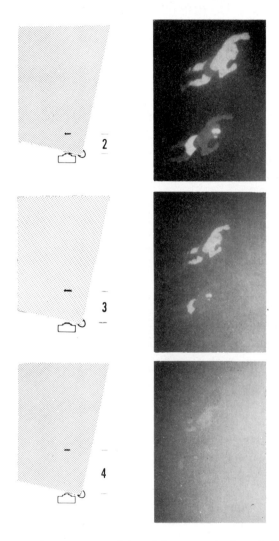

Fig. 1. Dummy swimmers moored in a laboratory tank and photographed at distances of 2, 3, and 4 attenuation lengths, respectively, by means of a submerged lamp mounted close to the camera, as shown in the diagrams on the left. Camera: Nikon, 50 mm, $f/2$ Lamp: 600-W tungsten halogen.

Fig. 2. Enlargement of dummy swimmers moored in a laboratory tank 12 attenuation lengths from the camera. Submerged lamp was mounted close to the swimmers. Linear polarizers were used at both lamp and camera to darken the water background. Camera and lamp: same as in Fig. 1.

Figure 2 is enlarged 4·7 times more than Fig. 1. There is no evidence of loss of image quality, yet published modulation transfer functions for similar water reproduced by Mertens in his recent book (1970) depict severe attenuation for spatial frequencies greater than one cycle per mrad! Such an MTF would have severely degraded Fig. 2. Clearly, the water in our tank had no such characteristic.

A Scripps graduate student (Morgan Morley) carefully duplicated the published modulation transfer function measurements in our laboratory tank and with the same type of water that we used in making Figs. 1 and 2. His results agreed with the published MTF curves! Thus, the mystery deepened.

Negatives of distant high contrast bar patterns were photographed in lighted water and subjected to microphotometer studies. They showed square-shouldered traces, even when the images were nearly buried in light scattered by the water path. Those square shoulders could not be reconciled with an MTF which rolls off at one cycle per mrad.

A private communication from Dr. Mertens brought the information that his newest MTF measurements in ocean water showed a roll-off at about 17 cycles per mrad, but even this higher value did not explain our photographic observations. There was still a mystery.

It was evident that we needed to measure the MTF of our water by some unambiguous method designed to show its high spatial frequency response with great precision. There must be no question concerning the angular extent of isoplanatism or the unsuspected presence of important imaginary terms in the optical transfer function. These requirements caused us to study the point spread function produced by our water and its two-dimensional Fourier transform, which is the optical transfer function.

C. POINT SPREAD PATTERNS

Studies of point spread patterns were begun in the same 5-m tank in which the dummies had been photographed. Later the experiment was transferred to a 15-m tank in order to test longer paths through ocean water containing no additives. Figures 3 to 6 show our apparatus as we used it in the 15-m tank.

Fig. 3. Beam of 514 nm light from an argon-ion laser is injected into ocean water in a 15-m laboratory tank. The laser beam passes through a submerged reversed microscope objective to form a tiny point source of light in the water.

Our experimental tank is equipped with an optical system by means of which we can inject a horizontal beam of 514 nm light from an Argon-ion gas laser into the water anywhere. Figure 3 shows how the submerged laser beam was passed through a reversed microscope objective to form a tiny point image in the water. The divergence of the light from that image could be changed by using different microscope objectives.

Outside a high quality optical glass window at the far end of the tank was the microscope shown in Fig. 4. Its sub-stage condenser was replaced with a reversed microscope objective arranged to image the distant point source in the plane of the microscope stage. The microscope then provided a greatly magnified view of this image through its eyepieces or on its viewing screen. The microscope also has a 35-mm camera, a large viewing screen and a multiplier phototube for measuring the total flux coming through the microscope. An opaque metal foil containing a small hole was placed on the microscope stage to limit the size of the measured field of view. Each of these modes of observation was used at appropriate times throughout the research.

Fig. 4. Microscope used outside 15-m laboratory tank to receive light from the submerged point source. Substage condenser was replaced by a reversed microscope objective to image the point source in the plane of the microscope stage. Microscope has eyepieces, camera, monitor screen and multiplier phototube photometer.

We are pleased to find that classical, circular diffraction patterns were formed by this system when the tank contained no water. After it was filled with clear, filtered tap water, for which $1/c = 14$ m, an equally good diffraction pattern appeared when the optically flat window in the tank was precisely normal to the optical axis. This was true at all source distances.

Temperature structures which developed when the tank was allowed to stand quietly overnight made no discernible effect on the diffraction pattern. Clearly, there was no significant roll-off of the MTF within the pass band of the optical system, which is determined by the diameter of its entrance pupil. This diameter was set by the sub-stage reversed microscope objective at 12 mm. With the laser operating in a low-coherence mode this corresponded with a spatial cut-off frequency in excess of 20 cycles per mrad.

Soluble dyes and magnesium aluminum hydroxide scattering agents were added to the tap water until the path length corresponded to 25 attenuation lengths. This addition greatly reduced the amount of light reaching the observation window, but it had no effect on the quality of the diffraction pattern. There was no evidence of an MTF roll-off.

We refilled the tank with water collected from the ocean well beyond the surf line at a depth of about 6 m. The water was biologically rich. It contained a considerable particle burden of both organic and minerogenic types. Its attenuation length was less than 2 m. The diffraction pattern appeared exactly as it had before. Other sea water samples were obtained. Some were from deeper water further from shore. All gave the same result; the diffraction pattern was not disturbed. We were convinced that if an MTF roll-off does, in fact, exist it must occur at some higher spatial frequency than 20 cycles per mrad.

D. HIGHER SPATIAL FREQUENCIES

Figure 5 shows how the microscope was positioned at the end of the 15-m tank to explore higher spatial frequencies than those passed by the 12-mm diameter sub-stage reversed microscope objective. Massive concrete pads set deep in the ground protruded through the floor of the building to support the tank. Two heavy 30-cm steel I-beams bridged between such pads at the end of the tank to support two optical benches. The microscope can be seen supported by one bench and a diffraction-limited objective lens about 120 mm in diameter is mounted on the other optical bench close to the optically flat window at the nearer end of the tank. An adjustable iris diaphragm was

Fig. 5. Long focal length, diffraction-limited lens in use at the 100-mm diameter window of the 15-m laboratory tank. It imaged the submerged point source in the plane of the microscope stage. No substage lens was used.

provided at the objective. Several other high-quality lenses were also used during the experiments. The sub-stage microscope objective was removed so that the image of the submerged point source formed by the large objective lens would be formed in the plane of the microscope stage. A beautiful circular diffraction pattern was displayed on the viewing screen shown in Fig. 6. It was possible to count more than 30 maxima.

Great difficulty was experienced in aligning the optically flat window at the end of the tank. Auto-collimation techniques were not good

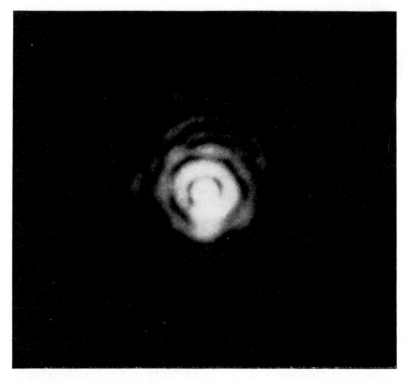

Fig. 6. Circular diffraction pattern photographed on the monitor screen of the microscope. Ring imperfections were subsequently corrected by aligning the tank window perfectly with incoming wave fronts. Pattern shows that diffraction limited image was found by the total system, including 15 m of ocean water.

enough. With water in the tank the circular fringes were broken or misshapen unless the window was very perfectly aligned with the incoming wavefronts. This was accomplished best by adjusting the window while watching the diffraction pattern. With aperture diameters of 50 mm or less, excellent patterns could be achieved by this technique. At greater diameters, particularly at the largest size (100 mm) used, minor imperfections in the ring structure were inescapable. Presumably it is necessary to use a diffraction limited lens designed to form its own window with the water in order to obtain a perfect diffraction pattern at large aperture diameters.

All of the observations described earlier in this paper were repeated using the larger lens diameters. Most of the tests were performed, however, in a single night using a fresh sample of coastal ocean water for which the 15-m water path contained 24 attenuation lengths. No degradation of the diffraction pattern was observed due to this water or to any water, fresh or salt, natural or filtered. It did not matter whether the water was stirred, circulated by pumps, or allowed to stand for many hours. Thus, even with an 8-fold increase in spatial frequency, no MTF roll-off was found. All spatial frequencies of image-forming light out to the limit tested were found to be attenuated alike.

Our experience with the critical alignment of the window in the tank suggests an explanation for the MTF roll-offs at a few spatial cycles per mrad that have been reported by various investigators who have inferred their results from in-water photographs of bar patterns, or from photoelectric devices using rotating spoke patterns etc. It appears that the high frequency portions of the MTF's they obtained do not relate to any property of natural waters, but characterize the optical systems they employed.

E. ABNORMAL REFRACTIVE EFFECTS

The preceding sections of this paper relate to normal conditions in ocean waters and other waters. Abnormal circumstances, such as a submerged heat source in the vicinity of the objective lens, can produce refractive effects that destroy the diffraction pattern and produce a vastly enlarged time-varying point spread pattern for which the optical transfer function exhibits severe attenuation, even at small spatial frequencies.

Large transparent biological scatterers in abnormal concentrations can produce similar effects. Figure 7 is a photograph of such a case. Half a liter of very highly concentrated biological material was injected into the optical path about one meter from the window in the end of the tank. Instantly the circular diffraction pattern was replaced by a very much larger, fluctuating point spread pattern, as shown in Fig. 7.

This interesting observation made us decide to explore other abnormal circumstances, but in a more controlled way. We chose to confine the abnormality within a submerged cylindrical plastic tube, shown in Fig. 8. It is 1-m long and about 20 cm in diameter. It has high quality optically flat windows at each end. Our 5-m tank was carefully cleaned and filled with fresh filtered tap water. The 12-mm diameter objective lens was used. The 1-m closed cell was mounted directly in front of the measurement window and provided with a system of hoses so that the water in it could be circulated or changed at will. Initially, the new cell was filled with water from the tank so that the temperature within and without the cell would be duplicated precisely. The diffraction patterns were then observed and photographed, both with the cell in

place and with it moved aside. The presence of the cell made no difference in the diffraction pattern.

A plastic 700-ml bottle was mounted above the water in the tank
and connected to the submerged cell through a hose. Water from the

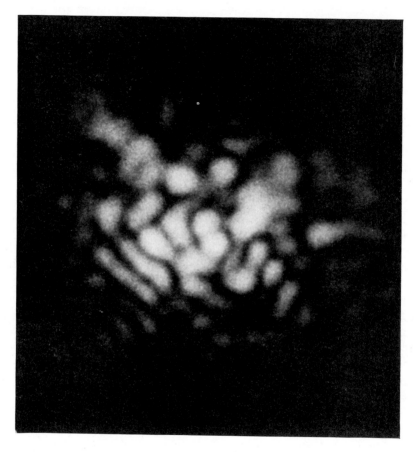

Fig. 7. Rapidly moving break-up of the circular diffraction pattern shown in
Fig. 6 when 1 liter of highly concentrated biologically rich water was injected
into the water path close to the window of the laboratory tank.

bottle entered the cell through a port in its bottom, flowed upward on a
diagonal path, and left the cell through a hose connection at its top.
It was found that passing a 700-ml charge of tank water through the
cell produced no discernible effect in the pattern produced by the
microscope.

The first experiment in the tank concerned temperature gradients. A small amount of tank water was warmed slightly and placed in the 700-ml elevated reservoir. By mixing water direct from the tank with the slightly warmed water, it was possible to produce a 700-ml charge of water differing in temperature by any desired amount with respect to the temperature of the water in the tank. In initial trials, water having a temperature 0·1°F higher than that of the water in the tank

Fig. 8. Submergible 1-m cell having optically flat windows. Samples of biologically rich waters, heated waters, high-salinity waters etc. can be passed through the path of sight without contaminating the water in the large tank.

and the cell was caused to flow from the elevated bottle through the cell. This temperature difference produced gradients in the cell at least an order of magnitude higher than those expected in the oceans, other than at the sea surface. These temperature gradients produced no observable effect on the diffraction pattern. Higher and higher temperature gradients were tried until a slight defect in the diffraction pattern was observed with a temperature difference of ½°F. This slight effect did not destroy the rings of the Airy pattern; the disturbance was very subtle. Next a 700-ml volume of water having the temperature difference of 1°F was passed through the cell and this produced only a barely more noticeable effect. Even water 2°F warmer than that in the tank did not seriously affect the appearance of the Airy pattern. Thus, it was concluded that within the range of spatial frequencies passed by a 12-mm aperture lens, temperature gradients enormously greater than those found in the ocean produce no significant effect.

The effect of biological scatterers have long been suspected to be the source of most optical resolution loss in natural water. The first such sample was a very concentrated biological sample from a stagnant fresh water pool near our campus. A fresh sample of this material in a closed container was immersed in the laboratory tank until its temperature matched that of the water in the tank. Then a 700-ml sample was placed in the elevated container and allowed to circulate through the 1-m cell. The optical effect of this very concentrated sample of fresh water biology was dramatic. The Airy disc was destroyed and replaced by a complex pattern that varied rapidly as the biologically rich water passed through the cell. Whereas, even enormous temperature gradients had disturbed the diffraction pattern almost none at all, the same volume (700 ml) of biologically rich water destroyed the circular diffraction pattern completely and produced a large time-varying point spread pattern which exceeded the field of view of the microscope.

The effect of the first biological experiment was so extreme that we decided to test a much lower concentration of biological material. For this purpose, a sample of ocean water was collected from the Scripps pier and stored in the experimental tank until its temperature matched that of the water in the tank. Then a 700-ml portion was caused to pass through the submerged cell. The optical effect of the ocean water was very much less than that of the fresh water from the stagnant pond. The circular diffraction pattern was disturbed only slightly.

Since the ocean water, obviously, contained salts, it was queried whether salinity gradients, rather than biological scatterers, had been responsible for the initial disturbance of the diffraction pattern that was observed when 700 ml of sea water was introduced into the cell. To explore this, the cell was flushed and filled again with filtered tap water. The high quality circular diffraction pattern appeared, just as before. A solution of table salt in a concentration of $\frac{1}{2}$ g per liter was made and brought to temperature equilibrium with water in the laboratory tank. Then 700 ml of this solution was placed in the elevated container and permitted to flow through the cell. This produced salinity gradients in the test cell hundreds of times greater than any which occur naturally in the oceans, apart from possible effects at the sea surface. These huge salinity gradients produced virtually no detectable effect on the diffraction pattern. It appears that neither temperature gradients nor salinity gradients, even of large magnitude, are of consequence in the spatial frequency range we tested, whereas biological scatterers in abnormally high concentrations do produce a point spread function.

Our experiments with abnormal refractive effects in water have led us to believe that neither temperature nor salinity gradients ordinarily found in the oceans attenuate high spatial image frequencies appreciably, even on very long optical paths including tens of scattering lengths. The same is true of normal concentrations of marine biology. From this standpoint it appears that, scattering length for scattering length, ocean water is usually a better optical medium than air, provided that loss of power by absorption can be neglected. This conclusion may not apply, however, where fresh water mixes with ocean water or where major thermal or biological abnormalities exist.

References

Duntley, S. Q. (1963). *J. Opt. Soc. Am.*, **53**, 214–233.

Mertens, L. E. (1970). "In-Water Photography." John Wiley, New York.

Chapter 8

Polarization Measurements in the Sea

ALEXANDRE IVANOFF

*Laboratoire d'Océanographie Physique de l'Université de Paris VI,
Equipe de recherche associée au C.N.R.S., France*

I. INTRODUCTION

Whereas polarization of the light from a clear sky was discovered as early as in 1811 by the French physicist Arago, the underwater daylight polarization, anticipated by scientists such as Le Grand (1939), was observed for the first time in 1954 by the American biologist Waterman (1954). This last discovery illustrates how much more difficult it is to make observations or measurements underwater.

Le Grand anticipated also that the polarization of underwater light is rather similar to that of skylight. Today we may add that both these two fields of research have developed similarly. Both started by observations and improvements of instruments, and by a theoretical approach at first limited to Rayleigh scattering. Then scientists attempted to include scattering by larger particles, and to take into account the multiple scattering. In both cases, but especially for underwater polarization, the theory is far from being completed, although it is progressing more rapidly since the use of computers. But the theoretical attempts are out of the scope of this paper. We shall limit ourselves to polarization measurements, and to a qualitative explanation of the observed phenomena.

II. Polarization of Light Scattered from a Parallel Beam of Artificial Light

Before reviewing our present knowledge concerning the polarization of underwater daylight we think it would be useful to consider first the polarization of the radiation scattered from a parallel beam of artificial light, since in these conditions the results depend only upon the optical properties of the medium, but not upon the radiance distribution and the multiple scattering, as it does when considering the underwater daylight polarization. In other words, the polarization of the flux scattered from a parallel beam of artificial light enables the isolation of the effect of the optical properties of the medium. Whereas the polarization by Rayleigh scattering is well known (**e**-vector normal to the scattering plane; degree of polarization equal to

$$P_{max} \frac{\sin^2 \alpha}{1+\cos^2 \alpha}$$

where α is the scattering angle and P_{max} the maximum value of the degree of polarization, obtained at right angles to the incident beam, equal to 1 for spherical particles but diminishing with the anisotropy of the particles), the polarization by larger particles is still poorly analyzed. When applying the Mie theory to monodisperse suspensions, one gets rather complicated results, the value of the degree of polarization fluctuating with the scattering angle, the **e**-vector being sometimes normal and sometimes parallel to the scattering plane (see for instance the work by Pavlov and Grechuchnikov (1965). Great progress has been accomplished since computer technics have enabled the application of the Mie theory to polydisperse suspensions. Quite recently Morel (1972) analyzed in such a manner the scattering by spherical particles of different diameters, assuming a Junge's particle size distribution, $N = ad^{-m}$, where N is the number of particles whose diameter is larger than d, m a constant characterizing the distribution, and a the number of particles whose diameter is larger than 1 μm. Actually, according to many measurements, the particle size distribution in seawater follows more or less approximately such a law, with m values varying from 2 to 4·5 (see also Chapter 2). Figure 1 represents the results obtained by Morel for $m = 3·5$ and a refractive index (relative to that of sea water) equal either to 1·05 or to 1·10. Continuous lines represent the volume scattering functions; points correspond to the fraction of scattered light whose **e**-vector is normal to the scattering plane; triangles or squares correspond to the fraction of scattered light whose **e**-vector is parallel to the scattering plane. In one case (that of triangles) the

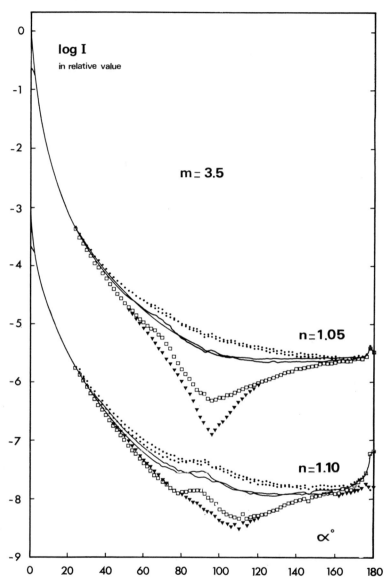

Fig. 1. Calculated volume scattering functions (from Morel, 1972) for spherical particles distributed according to Junge's law, with $m = 3.5$ and $n = 1·05$ or $1·10$.

limits of the particle size distribution are $0 \cdot 2 \leqslant \dfrac{\pi d}{\lambda} \leqslant 50$ (which, for $\lambda = 0 \cdot 5 \, \mu\text{m}$, corresponds approximately to $0 \cdot 022 \, \mu\text{m} \leqslant d \leqslant 6 \, \mu\text{m}$), whilst in the other case (that of squares) these limits are $0 \cdot 2 \leqslant \dfrac{\pi d}{\lambda} \leqslant 200$ (or $0 \cdot 022 \, \mu\text{m} \leqslant d \leqslant 24 \, \mu\text{m}$). It appears that whereas both the value of the refractive index and the limits of the particle size distribution have only a small effect on the volume scattering function (except at very small angles), their effect on the polarization (actually on the intensity of the fraction of light whose **e**-vector is parallel to the scattering plane) is considerable. Adding larger particles ($6 \, \mu\text{m} \leqslant d \leqslant 24 \, \mu\text{m}$) decreases the degree of polarization, whereas adding smaller particles increases it. Table 1 gives several of the results (for $m = 3 \cdot 5$) obtained by Morel (personal communication):

TABLE 1. Computation of maximum value of the degree of polarization. Junge distribution with $m = 3 \cdot 5$ (after Morel, personal communication).

Refractive index (relative to sea water) n	Limits of $\dfrac{\pi d}{\lambda}$	Maximum value of the degree of polarization (**e**-vector normal to the scattering plane)	Corresponding scattering angle
1·05	0·2–50	0·962	
	1–50	0·958	
	10–50	0·920	$\approx 96°$
	0·2–200	0·849	
	1–200	0·827	
	10–200	0·640	
1.10	0·2–50	0·789	$\approx 104°$
	0·2–200	0·676	

It should be noted also that the maximum value of the degree of polarization occurs at scattering angles greater than 90°, and varying considerably with the refractive index of the particles. The influence of the shape of the particles remains unknown, as in these calculations the particles are supposed spherical. There are practically no experimental studies supporting these theoretical considerations. As far as we know, only Hatch and Choate (1930) have demonstrated some forty years ago that the degree of polarization of light scattered at right angles by different suspensions of non-uniform particulate minerogenic substances decreases when their arithmetic mean diameter increases (Fig. 2). Morel (1965, 1970), during his extensive study of the volume scattering function of sea water, did polarization measurements at different scattering angles, and did not observe any significant deviation of the angle of maximal polarization from 90°, but the measurement of the intensity of the fraction of light whose **e**-vector is parallel to the scattering plane was probably not precise enough, as this intensity is very low. Between 1959 and 1964 Ivanoff made many measurements of the degree of polarization $P_{90°}$ of light scattered at

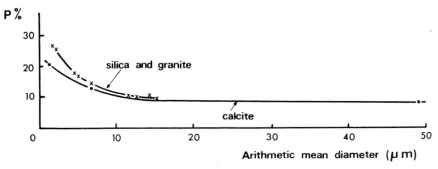

Fig. 2. Variation of the degree of polarization of light scattered at right angles with the arithmetic mean diameter of the particles, according to Hatch and Choate (1930).

right angles by sea water, either studying sea water samples in laboratory (Ivanoff, 1959, 1961; Ivanoff et al., 1961), or using an underwater remotely controlled polarimeter (Ivanoff et al., 1961). Although the scattering coefficient $\beta_{90°}$ (which depends mainly upon the total amount of suspended material) and the degree of polarization $P_{90°}$ of light scattered at right angles (which varies with the particle size distribution, with their refractive index and with their shape) are theoretically two independent parameters, it has been observed that for sea water the degree of polarization generally decreases when the scattering coefficient increases, the maximum value of $P_{90°}$ approaching the one corresponding to optically pure water (0·835), the minimum value (in turbid coastal waters) being around 0·4. This is illustrated in Fig. 3, representing all the results obtained by Ivanoff (1961) during a cruise in the Western Mediterranean (the scattering coefficient $\beta_{90°}$ has been assumed as being equal to 1 for approximately optically pure water). Now, the flux scattered by sea water may be considered as being the sum of the flux scattered by the molecules and of the flux scattered by the suspended particles. Consequently the degree of polarization of the total scattered flux is a function of the degrees of polarization corresponding to the molecular scattering and to the particle scattering, the influence of the latter increasing with the amount of particles. So the above mentioned experimental result can be explained by assuming that the degree of polarization of sea water is determined at least partly by the relative importance of the particle scattering as compared to the molecular scattering (in the same way as the volume scattering function

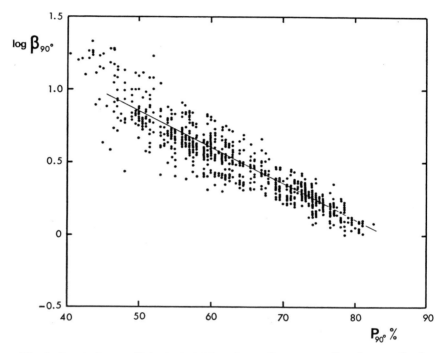

Fig. 3. Scattering coefficient at right angles and corresponding degree of polarization for all the sea water samples studied by Ivanoff (1961) during a cruise in the Western Mediterranean.

of sea water is dependent upon the ratio of the particle scattering coefficient to the molecular scattering coefficient, Morel, 1965). However, the large variations of $P_{90°}$ for a fixed value of $\beta_{90°}$ (Fig. 3) and in particular for a high value of $\beta_{90°}$ (in which case molecular scattering is negligible) indicate that the degree of polarization by particles may vary greatly, whereas their volume scattering function varies relatively little.

It would be very important to determine the two polarized components of the particle volume scattering function, and to see how much the polarization properties vary with the considered suspended particles.

Very probably they may vary greatly. For instance Ivanoff observed exceptional cases, he called "polarizing layers", when $P_{90°}$ and $\beta_{90°}$ increase simultaneously, indicating the presence of strongly polarizing particles (Fig. 4 illustrates such a "polarizing layer", observed in the Western Mediterranean at depths of 40 to 80 m). So, although we have at times been too enthusiastic, for instance when proposing a $P-\beta$ diagram (Ivanoff, 1959a) in order to characterize water masses, we are

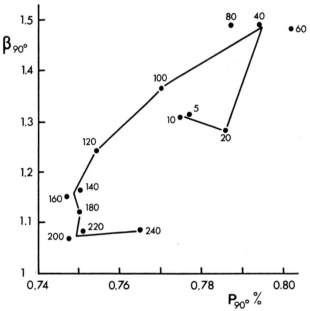

Fig. 4. A "polarizing layer" observed by Ivanoff (unpublished) at depths of 40 to 80 m, in the Western Mediterranean.

still convinced that it would be quite fruitful to make an extensive study of the polarizing properties of sea waters. Independently of their importance in underwater daylight polarization, they would very probably give important informations concerning the suspended material, and perhaps help to distinguish water masses. For instance we remember that some ten years ago, when working in the straits of Gibraltar, we had been quite unable to distinguish the Atlantic water going in and the Mediterranean water flowing out by their scattering properties, whereas we observed a significant difference between their polarizing properties.

II. Polarization of Underwater Light

A. INTRODUCTION

If a scattering medium is not illuminated by a parallel beam of light but rather by a conical beam, the degree of polarization is still maximum approximately at right angles to the axis of the beam, but its value decreases as the aperture of the conical beam increases. The effect has been calculated by Ivanoff and Lenoble (1957) in the case of Rayleigh anisotropic scattering, and Ivanoff (1957a) found a rough agreement between the observed values of the underwater daylight polarization, and those calculated from the observed radiance angular distributions, when taking into consideration the above-mentioned effect. So, even if polarization and angular distribution of underwater daylight were not just two different consequences of the same mechanism, which is multiple scattering of daylight by sea water, the radiance distribution would explain the decrease of the degree of polarization with the diffuseness of the underwater light, that is to say with: increasing turbidity of the water, increasing depth, increasing penetration of the considered wavelength, increasing albedo of the bottom etc. However in reality the value of the underwater polarization is not merely a consequence of the radiance distribution. After the daylight has penetrated into the ocean, multiple scattering redistributes simultaneously radiances and polarization in a manner which depends upon the optical properties of the sea water and upon those of the suspended particles. As a result, spatial distribution of underwater radiances and that of underwater polarization are closely connected, but they are not direct consequences of each other. Such a close relation between the underwater polarization and the angular radiance distribution leads to two general remarks:

(1) The sea, the sky and the sun have, all together, only one element of symmetry, which is the vertical plane going through the sun. The direction of the sun is not an axis of symmetry. However when the sky is sufficiently cloudy, the direction of the zenith is approximately such an axis. It is strictly so at depths where the asymptotic radiance distribution is reached.

(2) Both underwater polarization and radiance distribution will behave in different manners according to the depth:

> —At shallow depths, their aspect may be complicated by the effect of the skylight and that of the surface of the sea; but on the other hand, the intensity of the direct sunlight is then predominant, which is a simplification.

—At great depths, when the asymptotic radiance distribution is reached, everything is, to the contrary, very simple. In particular the state of polarization should be independent of the depth. Figure 5 from Timofeeva (1961) shows clearly how, in a milky solution, the degree of polarization becomes constant as soon as the asymptotic radiance distribution is reached. On this figure are represented one above the other the variations with the depth of the radiance L (in relative value) and those of the corresponding degrees of polarization P, for different values of the zenith angle θ (the milky solution being illuminated by a parallel beam of solar light normal to its surface). At depths greater than 30 cm the asymptotic radiance distribution is reached, whereas the degree of polarization becomes constant.

—At intermediate depths, the pattern is progressively changed by multiple scattering from just below the surface of the sea to the depth of the asymptotic distribution. This modification is more or less fast according to the optical properties of the medium. The only way to study quantitatively the underwater polarization is firstly to determine exactly the scattering and the polarizing properties of the medium, and then to solve the problem of multiple scattering in presence of polarized light. The volume scattering function is nowadays generally well-known, and the methods developed since Chandrasekhar enable the treatment of the multiple scattering, at least in unpolarized light and with simplifying assumptions concerning the radiance distribution. In particular, good results have been obtained concerning the asymptotic radiance distribution. However, these calculations disregard the polarization of the underwater light which in reality would be inexistent only if the asymptotic radiance distribution was approximately spherical. The problem is that it is not so simple to introduce polarization into the theory of multiple scattering. However this should be done, and consequently polarizing characteristics of particles should be known as well as their volume scattering function.

Until physicists, theorists and computers solve these difficult problems, we must be satisfied with a qualitative interpretation of the experimental data. We shall briefly repeat these data, as we do not believe it useful to go into details, which can be found in the references at the end of the chapter. We shall limit ourselves to a general outlook, and endeavour to point out the main features. The story of the polarization measurements in the sea is short, and the scientists involved in it are very few.

From 1954 to 1957 a pioneer work was done by Waterman (1954, 1955), by Ivanoff (1955, 1956) and by Waterman and Westell (1956), using visual or photographic instruments operated by divers. In 1957

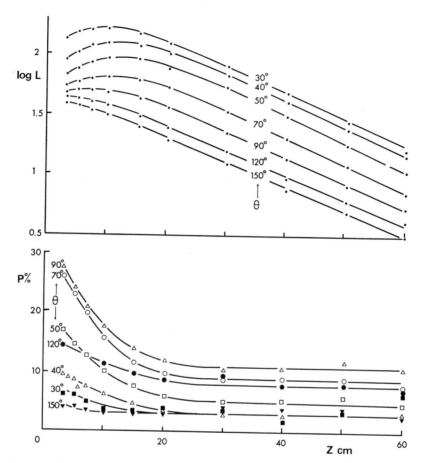

Fig. 5. Some results obtained by Timofeeva (1961) in a milky solution illuminated by a parallel beam of sun light normal to its surface. L is the radiance (in relative value), P the degree of polarization, Z the depth, and θ the zenith angle.

a large step forward was accomplished, as Ivanoff pointed out the connection between underwater daylight polarization and the angular distribution of submarine illumination (1957a) and designed the first photoelectric polarimeter (1957b). Effects of transparency, depth and wavelength were satisfactorily studied with this instrument by Ivanoff and Waterman (1958b), but only in horizontal lines of sight. The existence of elliptically polarized light, already observed in 1954 by Waterman, was explained by Ivanoff and Waterman (1958a). In 1959, a second photoelectric polarimeter working also only in horizontal lines of sight was used by Sasaki et al. (1959), whose results corroborate those obtained previously.

Since 1961, most of the polarization measurements have been done by Timofeeva and her co-workers both in the sea and in milky solutions studied in the laboratory. Measurements at sea have been accomplished with a photoelectric polarimeter designed by Timofeeva and Kaïgorodov (1963), which enabled an extensive study of the angular distribution of the degree of polarization at different depths and for different wavelengths (Timofeeva, 1962). Timofeeva et al. (1966), in studying milky solutions in the laboratory, discovered the existence of neutral points analogous to those well-known in optics of the atmosphere. This has been corroborated some years later by measurements at sea and during an extensive study of the spatial distribution of the e-vector in milky solutions, which enabled Timofeeva (1969) to illustrate the effect of the optical properties of the solution on the orientation of the e-vector and on the positions of the neutral points. Recently, Timofeeva (1970) made also an attempt to relate the value of the degree of polarization to the optical properties of the medium. Meanwhile, improved electrical polarimeters have been designed by Kaïgorodov and Neuymin (1968), and by Lundgren (1971). We shall not describe these instruments, nor the others mentioned above, as they are explained in detail in the references we give. Although multiple scattering by large particles could theoretically produce elliptical polarization, it seems that in the sea daylight is polarized linearly, except the very particular case analyzed by Ivanoff and Waterman (1958a), which shall be briefly discussed later on.

The experimental study of partially linearly polarized light consists in measuring its degree of polarization and in determining the predominant orientation of the e-vector. We shall begin with the value of the degree of underwater light polarization.

B. DEGREE OF POLARIZATION

As already stated above, the most complete description of the angular distribution of the degree of underwater light polarization was published in 1962 by Timofeeva. It was completed in 1966 by the observation in the laboratory of neutral points (Timofeeva et al., 1966) and, in 1970 by a more precise study, also done in the laboratory, using milky solutions (Timofeeva, 1970).

8

Figure 6, left side, represents the angular distribution of the degree of polarization in relative value, as obtained in the Black Sea, with a clear sky, and a solar zenith angle approximately equal to 60°, at an optical depth equal to 8, in different vertical planes at an angle φ from

i : 60°

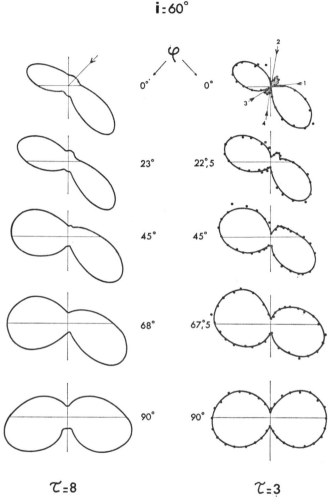

$\tau = 8$ $\tau = 3$

Fig. 6. Examples of angular distribution of the degree of polarization (in relative value) in different vertical planes at an angle φ from the plane through the sun or through the incident beam. The left side represents the results obtained by Timofeeva (1962) in the Black Sea, with a clear sky, a solar zenith angle of approximately 60°, at an optical depth of 8. The right side represents the results obtained by Timofeeva (1970) in a coloured milky medium, at an optical depth of 3, for an angle of incidence of the light of 60°.

the plane through the sun. Figure 6, right side, represents in the same way the angular distribution of the degree of polarization, as obtained with a better precision in the laboratory with coloured milky media, at an optical depth equal to 3, for an angle of incidence of the light upon the surface of the medium equal to 60°. In the vertical plane going through the incident beam ($\varphi = 0$) there are shown four neutral points analogous to those already well known in optics of the atmosphere.

The lack of symmetry around the direction of the sun (or that of the incident beam) appears very clearly, and is quite normal. The only element of symmetry is the vertical plane through the sun, and this appears in the two lowest diagrams of the figure ($\varphi = 90°$). In the sun's

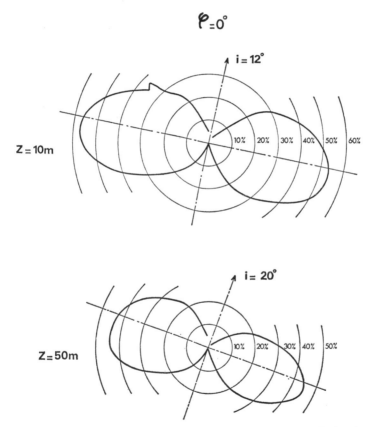

Fig. 7. Examples of angular distribution of the degree of polarization in the sun's bearing, determined by Lundgren (1971) (personal communication) near Sardinia. Z is the depth, i the solar zenith angle.

bearing, the distribution of the degree of polarization is not symmetrical with respect to the apparent direction of the sun. This is less apparent in Fig. 7, which shows some results obtained recently by Lundgren (personal communication) near Sardinia, in a vertical plane through the sun which was, however, rather close to the zenith. On the contrary, the difference between the two lobes of the diagram is striking in Fig. 8, representing the results obtained by Ivanoff (1957a) around Corsica, at a depth of 15 m, still in the vertical plane

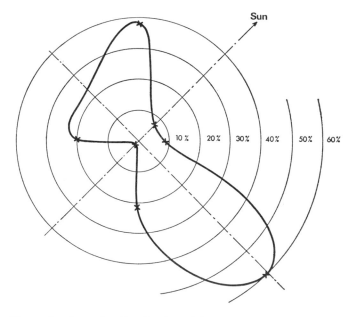

Fig. 8. Example of angular distribution of the degree of polarization in the sun's bearing, determined by Ivanoff (1957a) near Corsica, at a depth of 15 m.

through the sun. This may be an error. There may also be an effect of the skylight polarization, as the sea was so flat calm on this day that when diving it was possible to recognize the faces of the people on board the ship. But very probably the interference of the skylight polarization disappears as soon as the surface of the sea is rough.

Apparently, at small depths the degree of polarization is maximum in directions which are at least approximately normal to the direction of the sun (after its refraction at the sea surface). In Fig. 9, on which Timofeeva (1962) has plotted for different optical depths the degree of polarization against the "scattering angle", appears a shift towards angles greater than 90°. However, the "scattering angle" used in this figure is the angle A between the line of sight and the apparent direction of the sun. This angle is approximately the scattering angle only at shallow depths and when the sun is shining, in other words when the radiance distribution is very sharp in the direction of the sun. At greater depths, as it is well known, the direction of maximal radiance shifts towards the zenith, and the "scattering angle" could be under these conditions the angle between the line of sight and the direction of maximal radiance. However as the radiance distribution has no axis of symmetry until its asymptotic value is reached and as moreover the light does not come from a sole direction, it appears somewhat senseless to speak about a "scattering plane" or a "scattering angle". This is probably the reason for the dispersion of the points in Fig. 9.

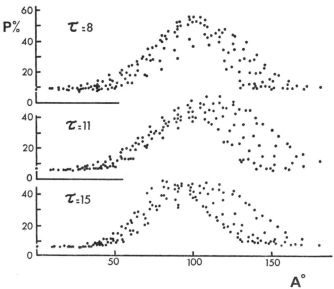

Fig. 9. Degree of polarization and corresponding angle A between the line of sight and the apparent direction of the sun, for different optical depths τ. Reproduced from Timofeeva (1962).

It is important to know which should be the orientation of the line of sight in order to get the maximum value of the degree of polarization. The answer is not at all evident. Ivanoff and Waterman (1958b) have admitted that the degree of polarization is maximum in a horizontal direction normal to the sun's bearing. The lowest diagram of the Fig. 6, right side, is in agreement with such an assumption, but the lowest diagram of the same figure, left side, is not. The distributions on the

right in Fig. 6 are probably more exact than those on the left. Furthermore, the above assumption is supported by the trend of the lowest
curve of Fig. 27, Chapter 10 in the reference quoted. Assuming a
shallow depth, let us consider the effect of different factors: firstly, the
turbidity of the water. Waterman and Westell (1956) already have
observed that the degree of polarization decreases at the same time as
the visibility, but "visibility" is as vague as "turbidity". In order to
estimate the latter, Ivanoff and Waterman (1958b) used a Secchi disk,
which is a very simple but inaccurate device. Figure 10 shows the results

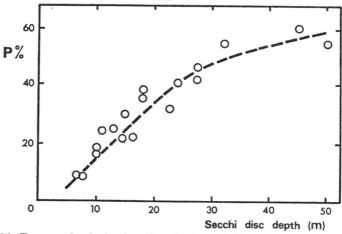

Fig. 10. Degree of polarization (in a horizontal direction normal to the sun's
bearing) measured by Ivanoff and Waterman (1958b) around the Bermuda
Islands at 9 m depth, and corresponding Secchi disc depth.

they obtained around the Bermuda Islands, at a depth of 9 m, in a
horizontal direction normal to the sun's bearing. Reaching 0·6 in very
clear waters, the approximately maximum value of the degree of polarization may be less than 0·1 in rather turbid coastal waters. Referring
to Fig. 3, we may recall that when illuminating with a parallel beam of
light, the degree of polarization of the light scattered at right angles
by sea water varies between approximately 0·8 and 0·4. This enables
us to appreciate the effect of the diffuseness of the daylight, even at
shallow depths and when the sun is shining. Note that the turbidity
acts doubly: it decreases the polarizing power of the medium, and it
increases the diffuseness of the submarine illumination.

Timofeeva (1961), working in the laboratory with milky solutions,
has studied the variations of the degree of polarization with the
absorption and the scattering coefficients. Figure 11 represents the

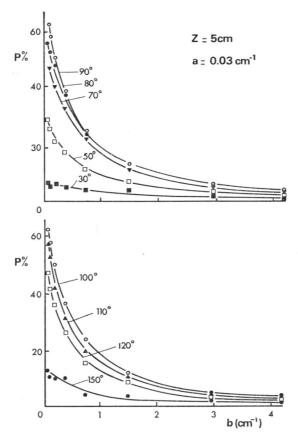

Fig. 11. Degree of polarization as a function of the scattering coefficient b, for different zenith angles. Measured by Timofeeva (1961) in a milky solution illuminated normally to its surface, at 5 cm depth. The absorption coefficient a of the solution was equal to 0.03 cm⁻¹.

results she obtained at a depth of 5 cm with an incident beam normal to the surface of the medium, for different zenith angles, the absorption coefficient being equal to 0.03 cm⁻¹, and the scattering coefficient varying from nearly zero to 4 cm⁻¹. It appears that when the scattering coefficient increases, the degree of polarization at first diminishes very rapidly, then more and more slowly. Recently, Timofeeva (1970) established a relation between the degree of polarization and the ratio k/c of the radiance attenuation function over the attenuation coefficient. However, although the volume scattering function of milky media is similar to that of sea water, it should be verified that their polarizing properties are also alike. Moreover, as we have seen above, there are reasons for considering that polarizing properties of suspended particles are more diversified than are their volume scattering functions. The reality is generally more complex than laboratory models. But this does not diminish their usefulness, if one keeps in mind that they are merely a simplifying approach. According to the general rule that the degree of polarization decreases as the diffuseness of the illumination increases, the underwater light polarization decreases when the sky is cloudy. At the same time the angular distribution of the degree of polarization tends to be symmetrical around the vertical. This has been observed by several investigators. For example, Fig. 12 represents

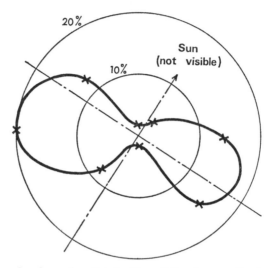

Fig. 12. Example of angular distribution of the degree of polarization with an overcast sky. Measured by Ivanoff (1957a) in the Channel at 15 m depth.

the results obtained by Ivanoff (1957a) in the sun's bearing (but the sun was not visible) in the Channel, at a depth of 15 m. The maximum value of the degree of polarization is 0·2, whereas, when the sun was shining, it was about 0·3, at the same depth. In the same manner, as the diffuseness of the underwater illumination increases for the most penetrating wavelengths, the degree of underwater polarization is minimum for these wavelengths. This has been first demonstrated for horizontal lines of sight by Ivanoff and Waterman (1958b), then established for any direction by Timofeeva (1962). As the results of the latter are presented as three-dimensional models, we have reproduced in Fig. 13 the measurements done by Ivanoff and Waterman in the Sargasso Sea, at a depth of 16 m, in the sun's bearing ($\varphi = 0$) and at right angle to it ($\varphi = 90°$). Still in the same manner, when the depth

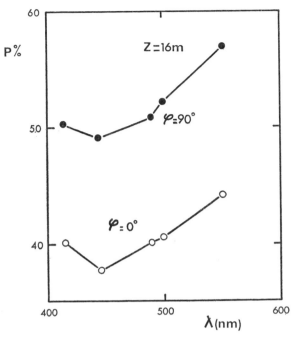

Fig. 13. Effect of wave-length on the degree of polarization, measured by Ivanoff and Waterman (1958b) in the Sargasso Sea, at 16 m depth, in horizontal directions parallel and normal to the sun's bearing.

is increasing, the diffuseness of the submarine illumination increases also (especially if the sun is shining), and at the same time the degree of polarization decreases. Moreover both the radiance distribution and the polarization pattern tend to be symmetrical around the vertical. When this is achieved, the radiance distribution has reached its asymptotic state, and the polarization pattern varies no more with the depth, as already stated earlier on (Fig. 5). Qualitatively it is very simple. But quantitatively, as we have already pointed out, we are still very far from solving such a problem.

The variations of the polarization distribution in the sea with depth have been studied by Timofeeva (1962), but here again, unfortunately her results are presented as three-dimensional models. Figure 14 represents some of the results obtained for horizontal lines of sight by Ivanoff and Waterman (1958b) in the Sargasso Sea, for different azimuths. Broken lines correspond to measurements done with a 500 nm narrow band filter, whereas solid lines correspond to measurements done without any coloured filter. Note that in the sun's bearing the degree of polarization increases again at depths greater than 80 m. This is

8*

probably because the direction of maximal radiance approaches the zenith with increasing depth. Curves A and B must join each other along a horizontal line when the asymptotic radiance distribution is reached, as the degree of polarization is then independent of the azimuth.

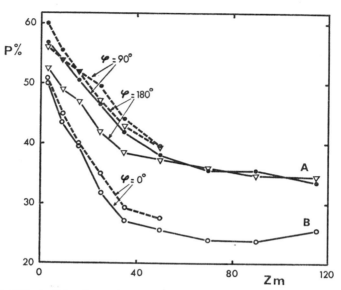

Fig. 14. Effect of depth on the degree of polarization for horizontal directions and different azimuths φ from the sun's bearing. Measurements by Ivanoff and Waterman (1958b) near Bermuda Islands. Broken lines correspond to measurements done with a 500 nm narrow band filter, solid lines correspond to measurements done without any coloured filter.

No polarization measurements have been done in the sea at depths where the asymptotic radiance distribution is reached, and consequently we do not know the asymptotic polarization distribution, nor its dependency upon the optical properties of the medium. According to measurements done by Timofeeva (1961) in milky solutions, the asymptotic polarization distribution is approximately symmetrical with respect to the horizontal plane, and the maximum value of the degree of polarization (corresponding to approximately horizontal lines of sight) depends upon the ratio b/k of the scattering coefficient over the radiance attenuation function (P_{max} increases when the ratio b/k decreases, and tends to a limit when b/k tends to zero). It would be interesting to find again these results by a theory of multiple scattering in presence of polarized light, and to take into account the influence of the polarizing properties of the medium.

C. ORIENTATION OF THE PREDOMINANT e-VECTOR

Let us consider now the orientation of the predominant e-vector. According to the classic theory of light scattering, the e-vector of scattered light is the projection of the e-vector of incident light on a plane normal to the direction of scattering. Consequently, if the incident light is unpolarized, the orientation of the predominant e-vector of the scattered light is the line of intersection of a plane normal to the direction of scattered light with a plane normal to the direction of incident light; in other words the predominant e-vector is normal to the scattering plane. This very simple situation holds up in underwater polarization so long as the incident light has one very predominant direction (that is to say as long as one can define a scattering plane), and moreover only if the incident light is unpolarized. This is approximately the case at shallow depths and only when the sun is shining. Then the direction of the sun (after its refraction on the sea surface) may be called "direction of incident light", and this incident light is unpolarized. In these conditions the orientation of the predominant e-vector is given by a trigonometrical formula established by Ivanoff and Waterman (1958a) in their study of the elliptical polarization, which precisely is limited to very shallow depths. This formula explains rather well the distribution of the e-vector at shallow depths when the sun is shining. For instance in a horizontal line of sight normal to the vertical plane going through the sun, the angle between the e-vector and a horizontal plane is merely equal to the angle of refraction of the direct sunlight. But this is no longer true at increasing depths, or when the weather is cloudy.

Actually, the problem is much more complicated as soon as the "incident light" does not come from a sole direction but rather is characterized by a radiance distribution (in which case, as we have pointed out above, the "scattering plane" has no signification), and moreover as soon as this incident light comprises already scattered light, and consequently is itself partially polarized. The problem of multiple scattering in the presence of polarized light has been rather well investigated in optics of the atmosphere (Sekera, 1957), but very little as far as we know in optics of the sea. So, once again, we must be satisfied with the experimental data. The only extensive study of the e-vector distribution has been done by Timofeeva (1969), both in the laboratory using milky solutions, and in the sea. She demonstrated that the trigonometrical formula mentioned above is not sufficient, and that optical properties of the medium must be considered (which is quite obvious, as multiple scattering is involved). For instance, in a horizontal line of sight normal to the sun's bearing the angle between the

e-vector and a horizontal plane would be equal to the angle of refraction only so long as the optical depth is less than 5. At greater depths this angle decreases, as it has been observed previously (Waterman, 1954), and as it should be, because when the asymptotic distribution is reached it is equal to zero.

Figure 15, from Timofeeva (1969), represents the **e**-vector distribution in a milky solution (ψ is the angle between the **e**-vector and the horizontal line perpendicular to the line of sight) as a function of the zenith angle θ, for different values of the angle φ between the vertical planes passing through the incident beam (angle of incidence equal to 60°) and through the line of sight, and for an optical depth equal to 3.

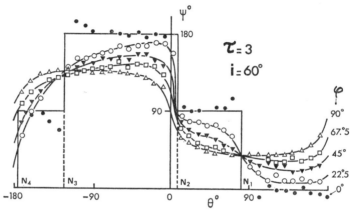

Fig. 15. **e**-vector distribution in a milky solution (from Timofeeva, 1969). ψ is the angle between the **e**-vector and the horizontal line perpendicular to the line of sight; θ is the zenith angle; φ is the angle between the vertical planes passing through the incident beam (angle of incidence is equal to 60°) and through the line of sight; optical depth $\tau = 3$.

Figure 16, also from Timofeeva (1969), represents the variations of the degree of polarization with the zenith angle θ, in the vertical plane passing through the incident beam ($\varphi = 0$). In this plane appear the four neutral points N_1, N_2, N_3 and N_4, whose positions, as demonstrated by Timofeeva, depend upon the angle of incidence on the surface of the milky solution, and upon the ratio k/c of the radiance attenuation function over the attenuation coefficient. Figure 15 shows that for directions lying between N_1 and N_2 or between N_3 and N_4 the **e**-vector is in a vertical plane ($\psi = 90°$), whereas for the other lines of sight located in the vertical plane passing through the incident beam it is perpendicular to this plane ($\psi = 0$ or $\psi = 180°$). In order to maintain

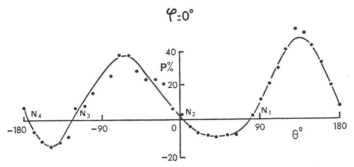

Fig. 16. Degree of polarization as a function of the zenith angle θ, in the vertical plane passing through the incident beam (angle of incidence equal to 60°). Measured by Timofeeva (1969) in a milky solution. $N_1 N_2 N_3$ and N_4 are the four neutral points. The degree of polarization is considered as being negative when the **e**-vector is in a vertical plane instead of being perpendicular to it (see Fig. 15).

the continuity of the curve in Fig. 16, the degree of polarization is considered as being negative when the **e**-vector is in a vertical plane, instead of being perpendicular to it. But actually, as emphasized by Timofeeva *et al.* (1966), negative values of the degree of polarization have no sense, as the **e**-vector may have any orientation. Figure 17, still from Timofeeva (1969), represents the **e**-vector distribution in the sea, for an optical depth equal to 6. There is an evident similarity with Fig. 15. It appears that the angle between the predominant **e**-vector and the plane defined by the line of sight and the direction of the incident beam ("scattering plane") actually varies between 90° and 0. This is an effect of the multiple scattering, which makes possible any orientation of the **e**-vector. Let us note that for lines of sight located in the

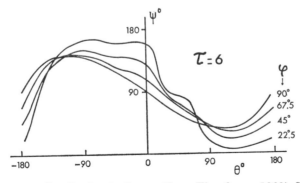

Fig. 17. **e**-vector distribution in the sea (from Timofeeva, 1969). Symbols have the same meaning as in Fig. 15.

vertical plane passing through the incident beam ($\varphi = 0$), the **e**-vector is either horizontal, which is the usual case, or located in the vertical plane, which is the case between the pairs of neutral points. In fact, as the vertical plane passing through the incident beam is a plane of symmetry, these two different orientations of the **e**-vector are the only two possible.

In spite of the valuable results obtained with laboratory models, only a complete theory of multiple scattering by large particles and in presence of polarized light could explain correctly the underwater light field and its state of polarization, and explain the influence of all the parameters such as: the sky radiance distribution (and in particular the sun's elevation), the absorbing, scattering and polarizing properties of the medium, the depth etc. If marine animals sensitive to polarized light can really use it for their orientation, they should have an extraordinary computer, but many facts are extraordinary in nature.

D. ELLIPTICAL POLARIZATION

A few words should be said about the existence in the sea of elliptically polarized light. Such a light has been observed by Waterman (1954), and its origin has been explained by Ivanoff and Waterman (1958a) by a total reflection at the sea surface of the linearly polarized light scattered from below. Although their theoretical analysis is limited by several simplifying assumptions, the predictions they have calculated are in good agreement with the observations. Thus, elliptical light is detectable: at shallow depths and when the sea is calm, at large zenith angles of the sun, in directions just outside the critical angle, in azimuths excluding the bearings of the sun and of the anti-sun. The elliptical light rotates clockwise to the right of the sun, and anti-clockwise to its left. Under the best conditions maximum ellipticity ratios of about 0·3 could be present, and the elliptically polarized flux could reach 10% of the total flux, whereas its ratio to linearly polarized flux could be more than 0·5.

Since Ivanoff and Waterman no one has been interested in the elliptically polarized underwater light, whose existence is strictly limited. However, the existence of clockwise and anti-clockwise natural underwater lights is exciting for the imagination, as complex organic molecules probably appeared for the first time in the sea. Some twenty years ago, Miller succeeded in obtaining such molecules by illuminating water with short wavelengths, in the presence of ammoniac and methane. One could wonder whether an elliptical polarization of the light could influence the structure of these molecules in two different manners according to the direction of rotation of the oscillation.

REFERENCES

Hatch, T. and Choate, S. P. (1930). *J. Franklin Inst.*, **210**, 793–804.

Ivanoff, A. (1955). *C. R. Acad. Sci.*, **241**, 1809.

Ivanoff, A. (1956). *J. Opt. Soc. Amer.*, **46**, 362.

Ivanoff, A. (1957a). *Ann. Geophys.*, **13**, 22–53.

Ivanoff, A. (1957b). *Bull. Inf. COEC*, **IX**, 491.

Ivanoff, A. (1959a). *J. Opt. Soc. Amer.*, **49**, 103.

Ivanoff, A. (1959b). Preprints Int. Oceanogr. Congr., pp. 553–555.

Ivanoff, A. (1961). I.U.G.G. Symposium on Radiant Energy in the Sea, **10**, 45–51.

Ivanoff, A., Jerlov, N. and Waterman, T. H. (1961). *Limnol. Oceanogr.*, **6**, 129–148.

Ivanoff, A. and Lenoble, J. (1957). *C. R. Acad. Sci.*, **245**, 329.

Ivanoff, A. and Waterman, T. H. (1958a). *J. Mar. Res.*, **16**, 255–282.

Ivanoff, A. and Waterman, T. H. (1958b). *J. Mar. Res.*, **16**, 283–307.

Kaïgorodov, M. N. and Neuymin, G. G. (1968). Trans. (Trudy) *Inst. Mar. Hydrophys. Ukr. Acad. Sci.*, **36**.

Le Grand, Y. (1939). *Ann. Inst. Oceanogr.*, **XIX**, 393–436.

Lundgren, B. (1971). Københavns Universitet, *Rep. Inst. Fys. Oceanogr.*, **17**, 1–34.

Morel, A. (1965). *Ann. Geophys.*, **21**, 281–284.

Morel, A. (1970). *AGARD Con. Proc.*, **77**, 30–1–30–9 and 47–6–47–7.

Morel, A. (1972). *C. R. Acad. Sci.*, **274**, 1387.

Pavlov, V. M. (1965). Trans. (Trudy) *Inst. Oceanology*, **77**, 41–52.

Pavlov, V. M. and Gretchuchnikov, B. N. (1965). Trans. (Trudy) *Inst. Oceanology*, **77**, 53–66.

Sasaki, T., Okami, N., Watanabe, S. and Oshiba, G. (1959). *Rec. Oceanogr. Works Jap.*, **5**, 91–97.

Sekera, Z. (1957). *In* "Polarization of Skylight", Encyclopedia of Physics, Geophysics II, Vol. XLVIII, pp. 288–328. Springer, Berlin.

Timofeeva, V. A. (1961). *Bull. (Izv.) Acad. Sci. USSR, Geophys. Ser.*, **5**, 766–774.

Timofeeva, V. A. (1962). *Bull. (Izv.) Acad. Sci. USSR, Geophys. Ser.*, **12**, 1843–1851.

Timofeeva, V. A. (1969). *Bull. (Izv.) Acad. Sci. USSR, Atmosph. Oceanic Phys. Ser.*, **5**, 1049–1057.

Timofeeva, V. A. (1970). *Bull. (Izv.) Acad. Sci. USSR, Atmosph. Oceanic Phys. Ser.*, **6**, 513–522.

Timofeeva, V. A. and Kaïgorodov, M. N. (1963). *Okeanologia*, **3**, 506–516.

Timofeeva, V. A. and Neuymin, G. G. (1968). *Bull. (Izv.) Acad. Sci. USSR, Atmosph. Oceanic Phys. Ser.*, **4**, 1305–1323.

Timofeeva, V. A., Vostroknutov, A. A. and Koveshnikova, L. A. (1966). *Bull. (Izv.) Acad. Sci. USSR, Atmosph. Oceanic Phys. Ser.*, **2**, 1259–1266.

Waterman, T. H. (1954). *Science N.Y.*, **120**, 927–932.

Waterman, T. H. (1955). Papers in Marine Biology and Oceanography, pp. 426–434, Pergamon Press, Oxford.

Waterman, T. H. and Westell, W. E. (1956). *J. Mar. Res.*, **15**, 149–169.

Some additional references can be found in the publication by Lundgren (1971), and, concerning Russian articles on optics of the sea, in the review by Timofeeva and Neuymin (1968).

Chapter 9

Optics of Turbid Waters
(Results of Laboratory Studies)

V. A. TIMOFEEVA

Black Sea Branch of Marine Hydrophysical Institute,
Ukrainian Academy of Sciences, Katzively, Simeiz, Crimea, USSR

I. INTRODUCTION

Experimental studies of the light field in turbid media are discussed and applied to the optics of the sea under natural irradiation. Particular emphasis is placed upon the work carried out at the Marine Hydrophysical Institute of the Ukrainian Academy of Sciences where the first studies of this kind were made.

Studies of the light field in water with a view to develop a method to calculate the field when the optical properties of the water and the irradiation conditions are given, is one of the basic tasks for the optical research in turbid waters and sea water in particular.

There are three ways of solving this problem:

(*a*) *Theoretical calculations.* In a turbid medium light is not only absorbed, but also scattered. However, the problem is complicated to a considerable extent by the fact that, for example in the sea, any volume of water scatters not only the direct light from the sun and the sky but also light scattered into it from the surroundings. Taking account of the multiple scattering and the strong forward scattering in the theoretical calculations of the light field in the sea is extremely difficult.

(b) *Field work.* This is a very labour-consuming and expensive method of investigation. As a rule, during field measurements of light fields it is not possible to measure all the necessary optical characteristics of the water simultaneously. Therefore results obtained in one area are not representative for another. In addition, an observer can neither give nor change the conditions of the experiments in a natural environment. Finally, it is often difficult to compare results obtained by different workers.

(c) *Laboratory studies.* The light field in the sea as in any turbid medium is determined by three optical characteristics: the absorption coefficient, the beam attenuation coefficient and the volume scattering function. It does not matter if the scattering medium is sea water, clouds, etc. provided that the parameters mentioned above have the same values, the light fields are similar. Thus to make meaningful studies of any aspects of the light field in a medium, one should know the optical characteristics of it. Measurements of such characteristics in seas and oceans are one of the main tasks of hydrooptical expeditions, but it is more practical to carry out studies of the relations between the characteristics of light field and medium under laboratory conditions on model media.

Let us dwell briefly upon laboratory investigations of the light field in turbid media as applied to optics of the sea.

The development of this research occurred in particular in the U.S.S.R. thanks to fundamental achievements on the theory of radiative transfer. The first correct explanation of the phenomena producing the colour of the sea was given by Shuleikin (1922) who deduced a formula taking into account light scattering by water molecules and large particles (air bubbles, suspended matter) and absorption by molecules as well as dissolved substances. He also introduced the parameter $\chi = b/c$ which corresponds to probability of photon survival by scattering, and which is an important concept in modern theory. Shuleikin tested his theoretical results under laboratory conditions by preparing a medium which simulated the optical properties of the sea. Furthermore Shuleikin (1924) calculated volume scattering functions for large suspended particles, which permitted him (Shuleikin, 1933) to perform a quantitative analysis of the light field in a turbid medium by a special grapho-analytical method. He discovered that when a scattering medium of great thickness is irradiated by a parallel vertical beam the light flux observed in the direction towards the light source first decreases exponentially up to a certain depth and then more slowly. The energy of beams scattered in various directions first increases with depth reaching some maximum and then diminishes.

Below a certain depth the distribution of the relative energy of these beams becomes stationary.

Gershun's works (Gershun, 1936a,b,c) on theoretical photometry played a very important role in the development of optics of turbid media. Based on his experience during many years he presented a general theory of the light field in which his photometric ideas were considerably developed. New photometric concepts were introduced suitable for describing scattering media. This led to the development of new methods and specialized instruments, e.g. the scalar irradiance meter.

At the Black Sea Branch of the Marine Hydrophysical Institute of the Uk.S.S.R. an experimental setup of instrumentation for making laboratory investigations has been created (Timofeeva, 1953, 1956, 1957a,b, 1960; Timofeeva and Kaïgorodov, 1963; Timofeeva and Koveshnikova, 1966). Usually experiments were conducted on colourless or coloured milky media with volume scattering functions similar to that of normal sea water but sometimes other artificial media were used (Timofeeva, 1971a). The medium was irradiated either by a wide beam of sunlight (Timofeeva, 1956) or artificial light (Timofeeva and Solomonov, 1970; Timofeeva and Opaets, 1971). The results obtained under laboratory conditions were tested by the theory and by measurements in the sea.

Laboratory investigations on turbid media have also been made by the following workers: Ivanoff and Waterman (1958a), Isacchi and Lenoble (1959), Blouin and Lenoble (1962) and Ivanov (1969).

Before presenting the results of the research on turbid media we shall give a list of symbols.

τ optical depth
a absorption coefficient
b scattering coefficient
c total attenuation coefficient
k asymptotic radiance attenuation coefficient

All values of the coefficients are given for the base e.

$\chi = \dfrac{b}{c}$ probability of photon-survival

$\epsilon = \dfrac{k}{c}$

ρ parameter characterizing the volume scattering function of the particles of a medium (see eq. 3).
L_0 radiance just below the surface of a medium in the direction of the light source.

L radiance inside a medium
E_d downwelling irradiance
E_u upwelling irradiance
E_{od} downwelling scalar irradiance
E_{ou} upwelling scalar irradiance
E_o scalar irradiance

$R = \dfrac{E_u}{E_d}$ irradiance reflectance

$\bar{\mu}$ average cosine
p degree of polarization

To explain other symbols we give Fig. 1 in which the vertical plane through the sun coincides with the plane of the paper.

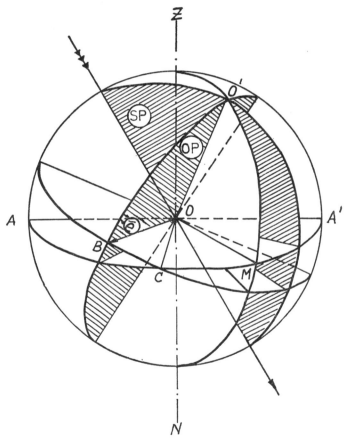

Fig. 1. Diagram describing the notation. O is a point under the surface of medium.

OO′   optical axis of the polarimeter

$\bar{e}$ electrical vector (OB)

OP   oscillations plane or polarization plane (a plane through the optical axis of the instrument and through $\bar{e}$)

SP   scattering plane (a plane through the refracted and scattered beams)

CO   reference line for the angle ψ (the horizontal line in a plane perpendicular to the optical axis of the polarimeter)

ψ the angle between the reference line CO and the electrical vector $\bar{e}$ ($180° - \angle BOC$) $0° \leqslant \psi \leqslant 180°$

φ azimuth angle (the angle between the vertical plane through the light source and the vertical plane through the optical axis of the polarimeter, i.e. $\angle AOM$)

θ zenith angle, i.e. $\angle ZOO'$ $0° < \theta < 180°$

II. Radiance

A. IRRADIATION OF A MEDIUM BY PERPENDICULAR BEAMS

(1) *Change in the spatial distribution of radiance with depth*

We understand the spatial distribution of radiance as radiance in various directions around a given point. In 1948, for the first time one managed to obtain (Timofeeva, 1950, 1951a) the spatial distribution of radiance in turbid media at various depths.

Polar diagrams are given in Fig. 2, each of them being a section of a body of radiance distribution of some point of milky medium illuminated by a vertical flux of solar beams, by the plane passing through a symmetry axis. Optical depths τ used in the diagrams are denoted by numbers above and by numbers below lengths of radii-vectors for $\theta = 0°$ in nominal units (the length of a radius-vector in the direction of $\theta = 180°$ for $\tau = 0$ is taken as a unit). Only those radii-vectors are given in diagrams for which real measurements were performed.

The body formed by the radiance vectors at a point at the medium surface irradiated by a vertical sun beam is a sharply delineated cone with an apex angle of 0·5°. The corresponding body just below the surface has a typical mushroom-like shape.

As the optical depth increases the shape of the radiance distribution alters passing in consecutive order all stages of Fig. 2. The attenuation of direct light with depth is accompanied by a growth of light scattered in other directions. Light scattered at small angles grows first of all. As depth increases the lower part of the diagram head increases continuously, i.e. the quantity of light scattered forward at large angles

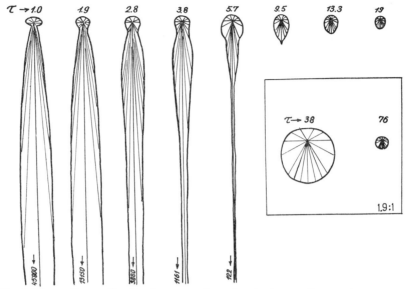

Fig. 2. The radiance distribution at various depths in a milky medium for $\epsilon = 0.04$ (Timofeeva, 1951a).

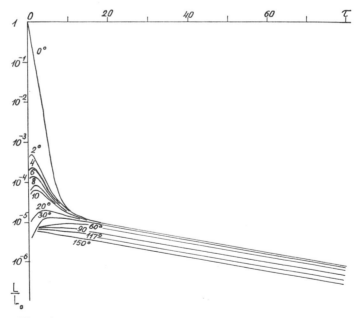

Fig. 3. The change with depth of radiance in various directions in a milky medium (Timofeeva, 1951a).

increases. The upper part of the "head" practically does not change at first and then starts decreasing. For great optical depths the distribution comes close to a circle in a given medium.

For further analysis the same data are given in a different form in Fig. 3. Relative radiance L/L_o in a logarithmic scale is plotted against the optical depth τ. The vertical angles θ are denoted by numbers near the curves.

As shown by the group of curves the light distribution at small depths differs qualitatively from that at great depths. In the first case the role of incident radiation is decisive of the formation of the light field. As depth increases the radiance distribution continuously changes its shape. The light regime at these depths is called the transient light regime. At great depths the light field is dictated not by the conditions of illumination but by the optical properties of the medium and the shape of radiance distribution does not change with increasing depth. The light regime at these depths is usually called the deep light regime or asymptotic radiance distribution and the limiting form of the radiance distribution—the stationary one.

(a) *Transient light regime.* It follows from Fig. 3 that the decrease of radiance for $\theta = 0°$ to a certain depth is represented by a straight line, i.e. it follows the exponential law $e^{-\tau}$. Below this depth the decrease of radiance rapidly approaches the deep regime value.

As depth increases the radiance in other directions first increases and reaches a maximum value at different depths τ_{max} for different directions. Below this depth the radiance decreases. Such a trend of the curves supports Shuleikin's theoretical conclusions (1933). Initial values of all curves are very small but they are not equal to 0 (Timofeeva, 1953).

(b) *Deep light regime.* It follows from Fig. 3 that beginning with a certain depth the radiance decrease in any direction follows the exponential law but with an attenuation coefficient which is relatively small. The law of radiance attenuation can be written as follows

$$L = L' \exp\left(-\epsilon(\tau - \tau')\right) \tag{1}$$

where L'-radiance occurs at a depth τ' at which light is maximally scattered.

Shuleikin (1933) was the first to point to the presence of an asymptotic distribution of radiance and also to the fact that radiance of maximally scattered light must decrease with increasing depth more slowly than that of directed beams. His conclusions are fully supported by these experiments.

The effect of various parameters of the medium on the character of the deep light regime is theoretically investigated by Rozenberg

(1959). His conclusions were tested theoretically under laboratory conditions (Ivanov and Il'ich, 1965, 1967) and experimentally by Timofeeva (1961a). Her method of determination of a real attenuation coefficient for turbid media is consistent with the basic idea by Gershun with due regard to the theory of transfer and applicable, generally speaking, to the sea.

(2) *Change in the downwelling irradiance with depth*

Naturally, in all cases when the radiance distribution is known one can also calculate such integral characteristics as E_d, E_u and E_o. We give the analysis of one characteristic E_d only. Graphs of the function $E_d = f(\tau)$ are shown in Fig. 4 for various values of ϵ with incident

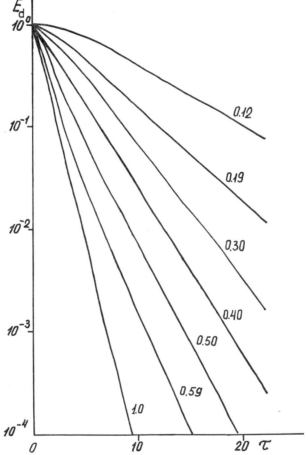

Fig. 4. The change with depth of irradiance E_d in milky media with various ϵ (Makarevich and Robilko, 1969).

light normal to the surface of a semi-infinite medium. The data are obtained from measurements in milky media. As for the discussed curves $L(\tau)$ we have transient and deep regimes. At small τ the tangent to the curves has different inclinations to the abscissae at different points. At great τ the mentioned inclination does not depend upon depth and is dictated only by the type of light scattering function and by the value of ϵ, i.e.

$$E_d = E'_d \exp\left(-\epsilon(\tau - \tau')\right) \tag{2}$$

where E'_d is the irradiance at a depth τ'. With small τ all the curves must be convex. (Ivanov, 1969) attributes concavity of some curves in Fig. 4 to significant divergence of the light beam during the experiment. For $\epsilon \to 1$ the change in irradiance with depth is described by an exponential function $\exp(-\tau)$ at all depths.

(3) Parameter ϵ

Parameter $\epsilon = k/c$ is an important optical characteristic of a turbid medium. Its value depends on the type of light scattering function and on the value of χ.

A great number of experimental and theoretical works have been dedicated to the study of relations between ϵ and χ. Values of ϵ for a spherical light scattering function and also for strong forward light (without taking polarization into account) were calculated by Ambarzumian (1942). Later, a relation between ϵ and χ was obtained by Poole (1945) also for spherical light scattering (disregarding polarization). For Rayleigh scattering the calculation of ϵ was performed by Sobolev (1956), Rozenberg (1959), Herman and Lenoble (1964), Prieur and Morel (1971). Lenoble (1956) made calculations for strong forward scattering and Blouin and Lenoble (1962) tried to obtain relations (ϵ, χ) experimentally. However, their measurements were reliable only in the range $\chi > 0.9$. A later attempt to check experimentally the mentioned relation was carried out only for $\chi > 0.75$ by Herman and Lenoble (1964).

The strong forward light scattering and the necessity for the taking polarization of light into account make the theoretical calculations highly complicated. The need for high sensitivity of the measuring apparatus hampers experiments with small χ.

A relation between ϵ and χ for the whole range $0 < \chi < 1$ and for strong forward scattering was experimentally obtained by Timofeeva and Gorobetz (1967).

The dependence of ϵ upon χ obtained in coloured milk media is given on the left in Fig. 5 (curve 1). For each χ the values ϵ are found highly reliable. Some scatter of the experimental points can be explained by the difference in milk in the various experiments. Curve 2 is obtained in a milky medium (skimmed milk) (Herman and Lenoble, 1964) and curve 3 in a water suspension of afcolac with particle size of about 0·2 μm. Both curves lie above curve 1 which concerns a medium with larger particles.

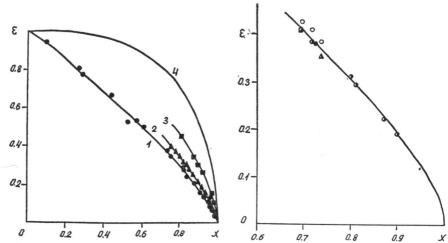

Fig. 5. The dependence of ε on χ in various media (on the left) and in the sea (on the right) (Timofeeva, 1971b). 1, milky medium; 2, milky medium (Herman and Lenoble, 1964); 3, medium with afcolac (Herman and Lenoble, 1964); 4, medium with Rayleigh scattering; ◑, Berlin lakes (Schellenberger, 1969); ○, Lake Pend Oreille (Tyler, 1960); ▲, Baltic Sea (Jerlov and Nygård, 1968); ◪, Lake Baikal (Sherstyankin, 1970).

Curve 4 corresponds to spherical scattering and is prepared from the data mentioned.

As appears from Fig. 5 parameter ε decreases with the increase of χ, i.e. the share of scattering in the total light attenuation. At small χ, the less stretched the light scattering function is: i.e. the finer particles are, the weaker is the dependence and at large χ the more marked is the dependence (ε, χ).

Curve 1 is rather simply expressed analytically:

$$\epsilon = (\rho^{\chi}(1-\chi)^{\chi})^{\frac{1}{2}} \tag{3}$$

When χ → 1 this formula turns into the formula presented by Timofeeva (1950, 1953, 1957c) for intensely scattering media and later found theoretically by Rozenberg (1958).

Value ρ for most milk samples proved to be equal to 0·25.

A section of curve (ϵ, χ) which is most interesting for natural water is given magnified on the right in Fig. 5. The curve is calculated using formula (3) for $\rho = 0.25$. The data from measurements performed under natural conditions are plotted by points. In spite of the fact that these points are obtained by investigators in most diverse waters (from the very turbid water of Berlin lakes to the extremely clear water of the Lake Baikal) all of them are close to the curve (the maximum deviation is about 4%). This points to the fact that: (i) the mean light scattering functions of natural waters are, in fact, similar: (ii) formula (3) is suitable for natural water.

Further investigations show that the empirical formula (3) is suitable both for artificial turbid media and for natural water within a wide range of particle dimensions from the largest particles to the colloid ones ($0.19 \leqslant \rho \leqslant 1.8$).

Formula (3) describes well the theoretical curves obtained by Prieur and Morel (1971) and Zege (1971) if the corresponding values of parameter ρ are used.

Thus, Fig. 5 permits us to calculate rather easily χ-values from ϵ-values and, on the contrary, to calculate parameter ϵ from known probability of photon-survival in a turbid medium. However, the determination of χ in turbid media (including the sea) is rather complicated. Efforts have been made (Timofeeva, 1957c, 1961a; Rozenberg, 1958; Levin and Ivanov, 1965; Pelevin, 1965) to find on one hand principal possibilities of separate determination of coefficients of absorption and scattering necessary for the determination of χ and on the other hand, to develop the simplest and handiest methods of such determinations. Much has already been done in this respect and yet we have no simple and handy methods for determination of χ. To

TABLE 1. Relation between probability of photon-survival $\chi(= b/c)$ and parameter $\epsilon(= k/c)$.

ϵ	0	0·01	0·02	0·03	0·04	0·05	0·06	0·07	0·08	0·09
0·0	1	0·999	0·998	0·996	0·994	0·990	0·986	0·982	0·977	0·972
0·1	0·966	0·960	0·953	0·946	0·939	0·932	0·924	0·916	0·908	0·900
0·2	0·892	0·883	0·875	0·866	0·857	0·848	0·839	0·830	0·820	0·811
0·3	0·802	0·792	0·783	0·773	0·763	0·753	0·744	0·734	0·724	0·714
0·4	0·704	0·694	0·684	0·674	0·663	0·653	0·643	0·632	0·622	0·612
0·5	0·601	0·591	0·580	0·570	0·559	0·548	0·538	0·527	0·516	0·505
0·6	0·494	0·483	0·472	0·461	0·450	0·439	0·427	0·416	0·405	0·393
0·7	0·382	0·370	0·359	0·347	0·335	0·324	0·312	0·300	0·288	0·276
0·8	0·264	0·251	0·239	0·227	0·214	0·202	0·189	0·176	0·163	0·150
0·9	0·137	0·124	0·111	0·098	0·084	0·070	0·056	0·043	0·029	0·014
1·0	0									

TABLE 2. Relation between parameter $\epsilon(= k/c)$ and probability of photon-survival $\chi(= b/c)$.

χ	0	0·01	0·02	0·03	0·04	0·05	0·06	0·07	0·08	0·09
0·0	1	0·993	0·986	0·979	0·972	0·965	0·957	0·950	0·943	0·936
0·1	0·928	0·921	0·913	0·906	0·898	0·890	0·883	0·875	0·867	0·859
0·2	0·852	0·843	0·835	0·827	0·819	0·811	0·803	0·795	0·786	0·778
0·3	0·770	0·761	0·753	0·745	0·736	0·728	0·719	0·710	0·702	0·693
0·4	0·684	0·675	0·667	0·658	0·649	0·640	0·631	0·622	0·613	0·604
0·5	0·595	0·585	0·576	0·567	0·558	0·548	0·539	0·530	0·520	0·511
0·6	0·501	0·492	0·482	0·472	0·463	0·454	0·443	0·433	0·424	0·414
0·7	0·404	0·394	0·384	0·374	0·364	0·353	0·343	0·333	0·323	0·312
0·8	0·302	0·291	0·280	0·270	0·259	0·248	0·237	0·225	0·214	0·202
0·9	0·190	0.178	0.165	0.152	0·139	0·125	0·110	0·093	0·075	0·052
1·0	0									

determine parameter ϵ one must measure two values: the coefficient c (by beam transmittance studies in the laboratory or in the sea) and the coefficient k (its measuring presents no principal and practical difficulties).

It follows from the above that it is expedient to consider parameter ϵ as a basic optical characteristic of natural water.

Functions $\epsilon(\chi)$ and $\chi(\epsilon)$ are given in Table 1 and 2.

B. IRRADIATION OF A MEDIUM BY OBLIQUE BEAMS

(1) *Transient light regime*

(a) *Variations of radiance with depth in the vertical plane of the sun.* Studies of the effect of asymmetry on the angular distribution of radiance were carried out by Timofeeva (1950, 1953) in turbid media illuminated by solar beams and scattered skylight. A diagram of the radiance distribution obtained at a depth of $\tau \approx 1$ in milky medium with $\epsilon \approx 0$ is given in Fig. 6 (on the left). Measured radius-vectors are plotted in the diagram. The direction of refracted beams is shown by arrows. Figure 6 (on the right) illustrates how a similar polar diagram alters with the increase of depth; the direction of maximum radiance approaches the normal to the surface of the medium. At great depths the distribution of radiance becomes symmetrical around the normal to the medium surface (near asymptotic state).

The change in the direction of the main light flux in the sea was discovered by Jerlov and Liljequist (1938). They found that the zenith angle of maximum radiance was 29° at a depth of 5 m, 27° at 15 m, 22° at 30 m and 19° at 40 m respectively. This approach to zenith has also been observed by Whitney (1941), Jerlov and Fukuda (1960), Tyler (1960), Jerlov and Nygård (1968).

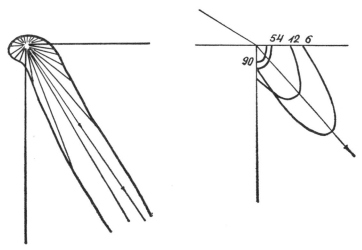

Fig. 6. Alteration of the diagram of radiance distribution with depth at oblique illumination of a milky medium. $\varphi = 0°$, $h_\odot = 66°$ (Timofeeva, 1953).

Measurements under natural conditions are in good agreement with the results of laboratory investigations. Curves of radiance in the vertical plane of the sun versus depth are given in Fig. 7. They are obtained in the sea by Timofeeva (on the left) and in a lake by Tyler (on the right).

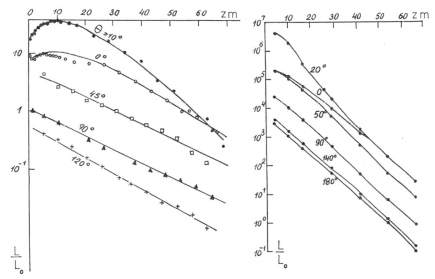

Fig. 7. The change of radiance with depth for different vertical angles in the vertical plane of the sun ($\varphi - 0°$) in the sea (on the left) (Timofeeva, 1950, 1951b, 1957a) and in Lake Pend Oreille (on the right) (Tyler, 1960).

Well marked maxima are observed in the left-hand diagrams. The exponential dependence of radiance upon depth for some directions is evident. Thus, the theoretical conclusions drawn by Shuleikin (1933) were in 1950 supported experimentally both under laboratory conditions and in the sea.

(b) *Spatial distribution of radiance.* Radiance was measured (Timofeeva *et al.*, 1966) in a hemisphere for nine azimuthal planes (for every 22·5°) and for every 10° of zenith angle. Curves showing the angular distribution of radiance in various azimuthal planes in a milky medium are presented in Fig. 8. Curves are given both for angles φ and for angles $(\varphi + 180°)$. Such a representation is permissible since the light field is symmetrical around the plane $\varphi = 0°$ ($\theta < 0°$ in the second hemisphere).

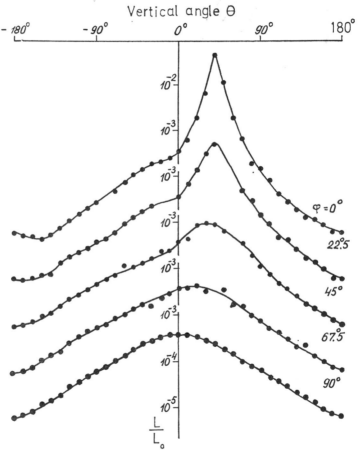

Fig. 8. Spatial distribution of radiance inside a milky medium by oblique irradiation. $\epsilon = 0.42$; $\tau = 3$; $h_\odot = 30°$ (Timofeeva and Koveshnikova, 1966).

Relative radiance (relative to the radiance just below the medium surface in the direction of the sun) is plotted against the $\dfrac{L}{L_o}$ —axis.

It follows from Fig. 8 that obliquity of the illumination substantially affects the angular distribution of radiance, especially in the vertical plane of the sun. Radiance in the vertical plane of the sun is maximal in the apparent direction of the sun and minimal in the opposite direction. As φ increases, the asymmetry decreases gradually, maximal radiance approaches zenith and at $\varphi = 90°$ the angular distribution of radiance is symmetrical around the direction $\theta = 0°$. Naturally,

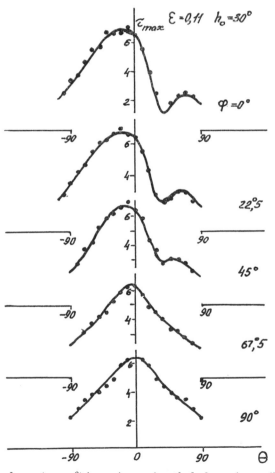

Fig. 9. Dependence $(\tau_{\max}, \theta)$ in various azimuthal planes in a milky medium by oblique irradiation. (Timofeeva and Opaets, 1971).

the radiance value decreases when moving off from the sun's vertical plane.

(c) *Maximal radiance and the depth of the maximum.* As mentioned above, with increasing depth in turbid media, radiance in a direction θ first increases then becomes maximum L_{max} at a depth of τ_{max} after which it starts decreasing. In Fig. 9 the dependence of τ_{max} upon θ is given for various azimuthal planes obtained in milky media illuminated by a projector (50 cm in diameter). It appears from Fig. 9 that τ_{max} also changes with the change of θ in different ways in various

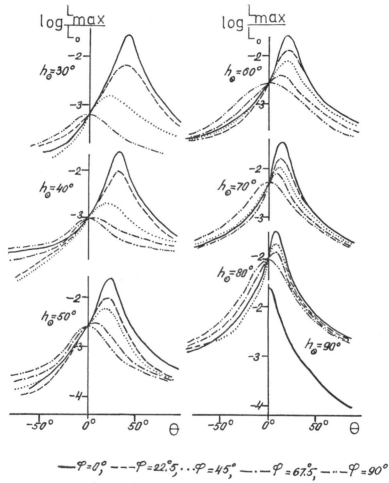

Fig. 10. Relation $\left(\dfrac{L_{max}}{L_o}, \theta \right)$ in various azimuthal planes at various $h_\odot$ in a milky medium. $\epsilon = 0\cdot11$ (Timofeeva and Opaets, 1971).

azimuthal planes. In the vertical plane of the sun two different maxima are found. Generally speaking, the curve is symmetrical neither around the refracted beam nor around the normal to the medium surface because of the effect of asymmetry of illumination.

The value of the curve maximum for $\theta > 0°$ and its position depend strongly upon φ. With the increase of φ the curve $(\tau_{\max}, \theta)$ alters so that at $\varphi = 90°$ it becomes symmetrical around the direction $\theta = 0°$.

The trend of the curves $\left(\dfrac{L_{\max}}{L_o}, \theta\right)$ in the same azimuthal planes when a turbid medium is illuminated by oblique beams $(h_\odot = 30°)$ is well indicated in Fig. 10 and needs no special elucidation.

(d) *The dependence of the maximal radiance and the depth of the maximum on* $h_\odot$. Families of curves showing angular distribution $\dfrac{L_{\max}}{L_o}$ and $\tau_{\max}$ respectively, in various azimuthal planes at different $h_\odot$ are given in Figs. 10 and 11. The effect of illumination asymmetry on these values as well as the decrease of this effect with the increase of $h_\odot$ are clearly seen in these figures.

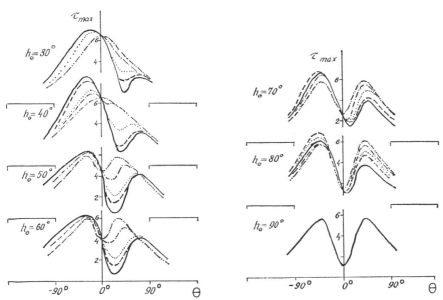

Fig. 11. Relation $(\tau_{\max}, \theta)$ in various azimuthal planes for various $h_\odot$ in a milky medium. $\epsilon = 0.11$ (Timofeeva and Opaets, 1971).

(e) *The effect of parameter* ϵ *on the maximal radiance and on the depth of the maximum.* The parameter ϵ of the average sea water is

3–4 times greater than in the case shown in Figs. 10 and 11. Therefore we give Tables 3 and 4 permitting us to determine $\dfrac{L_{max}}{L_o}$ and τ_{max} practically in all directions in turbid media of sea water type (and consequently in the sea too) with $\epsilon = 0.3$. The solar elevation, $h_\odot$, is between 30° and 90°. The effect of parameter ϵ on relation (τ_{max}, θ) is below discussed in more detail for a particular case of illumination of the medium when $h_\odot = 90°$.

The family of curves (τ_{max}, θ) obtained by Timofeeva and Opaets (1971) in various media is shown in Fig. 12. Since the light field under the given illumination is symmetrical around the normal to the surface

TABLE 3. Values of $\left(-\log\dfrac{L_{max}}{L_o}\right)$ in various directions at the depth τ_{max} of maximum radiance (Timofeeva and Opaets, 1971).

$h_\odot$	φ	−70°	−50°	−30°	−10°	0°	10°	20°	30°	40°	50°	70°	90°
30	0°	—	—	4·06	3·78	3·60	3·39	3·00	2·46	1·72	2·20	3·30	—
	22.5°	—	—	—	—	3·60	3·38	3·08	2·63	2·52	3·00	3·48	—
	45°	—	3·92	3·82	3·82	3·60	3·46	3·32	3·19	3·18	3·30	3·60	—
	67.5°	—	3·83	3·76	3·66	3·60	3·54	3·48	3·46	3·50	3·55	3·74	—
	90°	—	3·74	3·66	3·60	3·60	3·60	3·62	3·64	3·68	3·72	3·86	—
40	0°	—	—	—	3·80	3·50	3·06	2·34	1·70	2·20	2·95	3·50	3·80
	22.5°	—	—	—	3·80	3·50	3·06	2·50	2·24	2·50	3·10	3·60	3·90
	45°	—	—	—	—	3·50	3·25	3·00	2·90	3·00	3·65	3·93	—
	67.5°	—	—	—	—	3·50	3·32	3·26	3·32	3·44	3·57	3·82	—
	90°	4·00	3·85	3·67	3·54	3·50	3·54	3·60	3·67	3·76	3·85	4·00	—
50	0°	—	4·13	3·90	3·48	3·18	2·72	2·08	1·60	2·36	2·88	3·30	—
	22.5°	—	3·97	3·80	3·45	3·18	2·80	2·30	2·08	2·54	3·00	3·55	—
	45°	—	3·86	3·70	3·36	3·18	2·92	2·68	2·70	2·94	3·21	3·60	—
	67.5°	—	3·74	3·57	3·30	3·18	3·04	3·00	3·12	3·30	3·46	3·67	—
	90°	—	—	—	—	3·18	3·20	3·28	3·37	3·40	3·50	3·62	—
60	0°	—	3·65	3·55	3·22	2·90	2·28	1·68	2·22	2·76	3·12	3·56	—
	22.5°	—	3·65	3·54	3·18	2·90	2·40	2·08	2·40	2·84	3·21	3·61	—
	45°	—	3·65	3·50	3·17	2·90	2·62	2·44	2·72	3·09	3·32	3·65	—
	67.5°	—	3·60	3·44	3·10	2·90	2·77	2·83	3·07	3·26	3·43	3·70	—
	90°	—	—	—	—	2·90	2·97	3·13	3·30	3·43	3·56	3·77	—
70	0°	—	3·80	3·55	3·07	2·60	1·80	2·34	3·17	3·44	3·63	—	—
	22.5°	—	3·70	3·50	3·03	2·60	2·08	2·46	3·04	3·37	3·57	—	—
	45°	—	3·63	3·42	2·94	2·60	2·23	2·50	2·88	3·18	3·40	3·62	—
	67.5°	—	3·58	3·36	1·87	2·60	2·36	2·60	3·03	3·32	3·44	3·56	—
	90°	—	—	—	—	2·60	2·68	2·93	3·22	3·44	3·56	3·74	—
80	0°	—	3·80	3·50	2·77	2·10	1·72	2·50	3·08	3·45	3·67	—	—
	22.5°	—	3·67	3·38	2·76	2·10	1·82	2·56	3·07	3·35	3·52	3·73	—
	45°	3·84	3·60	3·30	2·64	2·10	2·08	2·67	3·08	3·30	3·48	3·70	—
	67.5°	3·70	3·55	3·22	2·52	2·10	2·18	2·67	3·08	3·30	3·48	3·70	—
	90°	—	—	—	—	2·10	2·28	2·80	3·17	3·43	3·57	3·74	—
90	0–180°	—	—	—	—	1·70	2·08	2·70	3·13	3·44	3·70	4·12	4·40

TABLE 4. Values of depths at which radiance in some direction is maximal at various altitudes of the sun (Timofeeva and Opaets, 1971).

$h_\odot$	φ \ θ	$-50°$	$-30°$	$-10°$	$0°$	$10°$	$20°$	$30°$	$40°$	$50°$
30	0°	0·3	1·1	2·3	2·9	2·5	1·0	1·6	1·2	0·4
	22·5°	0·3	1·3	2·4	2·9	2·5	1·3	1·8	1·4	0·5
	45°	0·4	1·4	2·6	2·9	2·4	1·9	1·8	1·2	0·4
	67·5°	0·3	1·4	2·6	2·9	2·7	2·4	1·8	1·1	0·4
	90°	0·3	1·6	2·7	2·9	2·7	2·2	1·6	1·0	0·3
40	0°	0·1	0·5	1·8	2·9	2·0	1·2	2·0	1·3	0·5
	22·5°	0·1	0·7	1·9	2·9	1·7	1·5	2·2	1·6	0·6
	45°	0·2	0·9	2·4	2·9	2·3	2·2	2·1	1·4	0·4
	67°·5	0·3	1·2	2·6	2·9	2·9	2·4	1·6	0·9	0·4
	90°	0·7	1·9	2·9	2·9	2·9	2·6	1·9	1·3	0·7
50	0°	0·3	1·1	2·4	2·9	1·4	1·3	1·6	0·7	0·2
	22·5°	0·2	1·0	2·9	2·9	1·4	1·5	1·7	0·9	0·4
	45°	0·2	1·1	3·0	2·9	2·1	2·1	1·7	0·8	0·3
	67·5°	0·4	1·6	2·8	2·9	2·7	2·3	1·7	1·0	0·5
	90°	0·4	1·6	2·7	2·9	2·7	2·2	1·6	1·0	0·4
60	0°	0·1	0·8	3·0	2·9	0·8	1·7	1·7	0·7	0·1
	22·5°	0·1	0·8	2·8	2·9	0·9	1·8	1·4	0·6	0·2
	45°	0·2	1·2	3·1	2·9	1·7	2·3	1·4	0·7	0·2
	67·5°	0·3	1·4	2·8	2·9	2·0]	2·3	1·4	0·6	0·2
	90°	0·5	1·6	2·6	2·9	2·6	2·2	1·6	1·0	0·5
70	0°	0·1	0·7	2·6	1·9	0·9	1·9	1·4	0·8	0·3
	22·5°	0·2	1·3	2·7	1·9	1·1	2·0	1·6	1·0	0·4
	45°	0·2	1·2	2·5	1·9	1·5	2·3	1·6	0·8	0·3
	67·5°	0·1	1·2	2·4	1·9	1·9	2·2	1·6	1·0	0·5
	90°	0·2	1·4	2·4	1·9	2·4	2·2	1·4	0·6	0·2
80	0°	0·1	1·1	3·1	1·3	1·7	2·0	1·3	0·7	0·3
	22°·5°	0·1	1·0	3·1	1·3	1·6	2·7	2·2	1·2	0·5
	45°	0·2	1·2	2·5	1·3	1·8	2·5	1·8	0·9	0·3
	67·5°	0·4	1·5	2·4	1·3	1·8	2·4	1·6	0·8	0·3
	90°	0·4	1·7	2·5	1·3	2·5	2·6	1·7	0·9	0·4
90	0–180°	0·2	1·5	2·0	1·0	2·0	2·5	1·5	0·7	0·2

of medium only half of the curves are given in the figure. The upper curve is obtained by extrapolation of curves (τ_{max}, ϵ) to $\epsilon = 0$. Values of parameter ϵ are inserted near the curves.

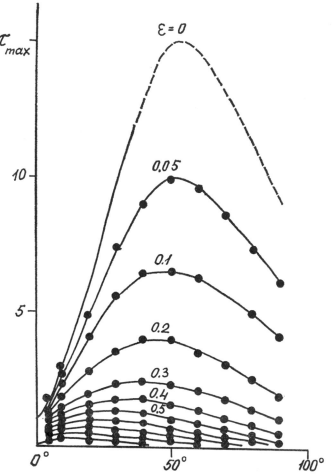

Fig. 12. The effect of ϵ on τ_{max} for various θ in milky media (Timofeeva and Opaets, 1971).

It follows from Fig. 12 that at $h_\odot = 90°$ τ_{max} first increases with increasing zenith angle to a maximum value and then tends towards 0. With the increase of ϵ, τ_{max} decreases in all directions (i.e. the depths of maximal radiances tend to the surface of medium) the more slowly the greater is ϵ, and the maximum (τ_{max}, θ) shifts to small θ, tending to 0° at $\epsilon = 1$.

(2) *Deep light regime*

    (a) *Dependence of the shape of the diagram of asymptotic radiance distribution on the value of parameter* ϵ. In a hypothetical purely scattering medium ($\epsilon = 0$) the diagram of the asymptotic radiance distribution shown in polar coordinates is a circle with the pole in its centre. Timofeeva (1957b) showed that with very small ϵ (up to $\epsilon \leqslant 0.2$) the diagram is also close to a circle but with the pole displaced (away from the centre of the circle). Families of curves are given in Fig. 13. They

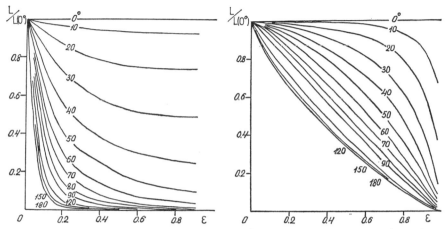

Fig. 13. Dependence of relative radiance $L(\theta)/L(0)$ on the parameter ϵ in various directions in a deep regime for milky media (on the left) and for media with Rayleigh scattering (on the right). Zenith angles are denoted by numbers on the curves (Timofeeva and Solomonov, 1970).

show the change of relative radiance $L(\theta)/L(0)$ in different directions with the increase of ϵ in a milky medium as well as in a medium with Rayleigh scattering (according to published data). One can see that light scattered at great angles decreases very rapidly with the increase of ϵ in the milky medium (on the left) especially the upward scattered light. With further increase of ϵ the decrease of radiance becomes somewhat slower. The trend of corresponding Rayleigh curves is quite different. Consequently the shape of the volume scattering function strongly affects the angular distribution of radiance in the deep regime. There is good reason to believe that Fig. 13 is valid for average sea water. Therefore it is useful to give Table 5 allowing us to determine the shape of the diagram of asymptotic radiance distribution in turbid media of sea water type at $0 < \epsilon < 1$.

(b) *Dependence of irradiance reflectance upon parameter* ϵ. The ratio E_u/E_d is defined as irradiance reflectance R. It is known that in the sea for $\tau > 1.5$, R in fact does not depend on depth (Schellenberger, 1969). Therefore a relation between R and ϵ found for the deep-water

TABLE 5. $L(\theta)/L(0°)$ for various zenith angles θ and $\epsilon(= k/c)$ in the deep regime (Timofeeva, 1971a).

θ \ ϵ	0	0·1	0·2	0·3	0·4	0·5	0·6	0·7	0·8	0·9	1
0	1	1	1	1	1	1	1	1	1	1	1
10	1	0·975	0·965	0·955	0·940	0·935	0·935	0·925	0·925	0·925	0
20	1	0·930	0·865	0·813	0·780	0·760	0·750	0·740	0·740	0·740	0
30	1	0·847	0·715	0·630	0·580	0·540	0·510	0·495	0·490	0·490	0
40	1	0·735	0·545	0·445	0·375	0·325	0·290	0·273	0·255	0·245	0
50	1	0·640	0·393	0·300	0·240	0·185	0·145	0·125	0·110	0·095	0
60	1	0·535	0·290	0·190	0·135	0·095	0·065	0·055	0·045	0·035	0
70	1	0·460	0·212	0·125	0·070	0·047	0·030	0·022	0·017	0·010	0
80	1	0·380	0·143	0·080	0·045	0·030	0·020	0·013	0·011	0·005	0
90	1	0·295	0·100	0·050	0·025	0·015	0·010	0·006	0·003	—	0
100	1	0·240	0·070	0·035	0·013	0·009	0·006	0·003	—	—	0
110	1	0·200	0·056	0·025	0·009	0·007	0·004	0·002	—	—	0
120	1	0·180	0·045	0·020	0·007	0·004	0·003	—	—	—	0
130	1	0·160	0·036	0·015	0·005	0·004	—	—	—	—	0
140	1	0·145	0·028	0·012	0·005	0·003	—	—	—	—	0
150	1	0·135	0·025	0·010	0·004	0·002	—	—	—	—	0
160	1	0·130	0·020	0·009	0·004	0·002	—	—	—	—	0
170	1	0·126	0·016	0·009	0·003	0·002	—	—	—	—	0
180	1	0·120	0·015	0·008	0·003	0·002	—	—	—	—	0

regime is probably also valid for smaller depths. Irradiances E_u and E_d necessary for the calculation of R were determined by the method of graphical integration from the known angular distribution of radiance in the deep-water regime, in particular from the data of Table 5 obtained in a milky medium. The radiance in directions $\theta > 90°$ in strongly absorbing media (at $\epsilon > 0·6$) is small and measurements of it are not quite reliable. Therefore E_u in this case was determined not by integration but by interpolation through the use of graphs (E_u, ϵ).

In Fig. 14 curves (R, ϵ) derived for three media are given: milk, colloid with latex, and medium with Rayleigh scattering. It is clear that even a slight increase of ϵ in milky medium leads to a sharp fall of R in the range $0 < \epsilon < 0·2$. With further increase of ϵ, coefficient R decreases more slowly tending to 0 at $\epsilon = 1$. In media in which latex is present, curve (R, ϵ) is more gently sloping and becomes quasi-linear for Rayleigh scattering in the range $0 < \epsilon < 0·9$. As expected for media with equal ϵ the more forward scattering the less is irradiance reflectance.

The most interesting (from viewpoint of applicability to natural water) section of curve (R, ϵ) is shown in Fig. 14 on an enlarged scale. Results of measurements obtained in natural waters different in transparency (the same waters which were mentioned in connection with

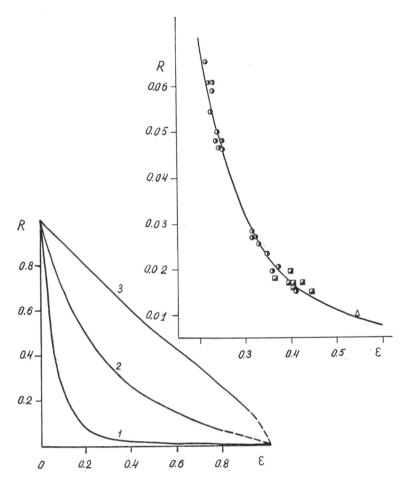

Fig. 14. Dependence of irradiance reflectance on ϵ in various media (Timofeeva, 1971b). 1, milky medium; 2, medium with latex; 3, medium with Rayleigh scattering; ◑, waters of Berlin ($\lambda = 480$–670 nm) (Schellenberger, 1969); ◪, Lake Baikal, $\lambda = 380$–480 nm (Sherstyankin, 1970); △, Lake Baikal, $\lambda = 380$ nm (Sherstyankin, 1970).

Fig. 5) are also presented here. As we see, plots referring to natural waters (including the point for the near ultraviolet) are near the "milk" curve. Thus we conclude that this dependence is valid for natural water as well.

(c) *Dependence of average cosine on parameter* ϵ. The average cosine μ can be defined by

$$\bar{\mu} = \frac{E_d - E_u}{E_o} \qquad (4)$$

and can also be calculated from the formula

$$\mu = \frac{1-\chi}{\epsilon} = \frac{1-b/c}{k/c} = \left(\frac{a}{k}\right) \tag{5}$$

Both these formulas were used for processing the experimental data. The obtained practically equal results are averaged. The corresponding curves are given in Fig. 15 (for milky media, media with latex and those with Rayleigh scattering). In milky media $\bar{\mu}$ first increases very rapidly with the increase of ϵ, and then more slowly tending to the limit. On the contrary, in media with Rayleigh scattering the growth of $\bar{\mu}$ is slow to begin with and then rather rapid. The curve for latex occupies an intermediate position. At all ϵ, the stretched shape of asymptotic polar diagrams, and thus the magnitude of $\bar{\mu}$ is substantially greater for milky media than for those with latex and even more so for media with Rayleigh scattering.

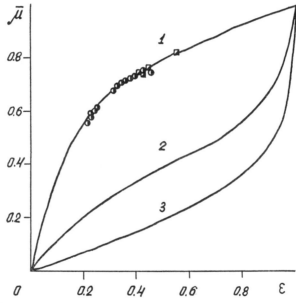

Fig. 15. Dependence of the average cosine on ϵ in various media. 1, milky medium; 2, medium with latex; 3, medium with Rayleigh scattering; ◑, Berlin Lakes (Schellenberger, 1969); ◪, Lake Baikal (Sherstyankin, 1970).

Taking Fig. 5 and eq. (5) into account one can conclude that the dependence (μ, ϵ) derived for milky media is also valid for natural water (Table VI).

TABLE 6. Dependence of average cosine $\bar{\mu}$ on $\epsilon(= k/c)$.

ϵ	0	0·01	0·02	0·03	0·04	0·05	0·06	0·07	0·08	0·09
0·0	0	0·060	0·080	0·117	0·155	0·190	0·223	0·254	0·285	0·313
0·1	0·340	0·366	0·390	0·413	0·434	0·455	0·474	0·492	0·509	0·526
0·2	0·541	0·556	0·570	0·583	0·596	0·608	0·620	0·631	0·641	0·652
0·3	0·661	0·671	0·680	0·688	0·696	0·704	0·712	0·719	0·727	0·734
0·4	0·740	0·747	0·753	0·759	0·765	0·771	0·776	0·782	0·787	0·792
0·5	0·798	0·803	0·807	0·812	0·817	0·821	0·826	0·830	0·835	0·839
0·6	0·843	0·847	0·851	0·856	0·860	0·864	0·867	0·871	0·875	0·879
0·7	0·883	0·887	0·890	0·894	0·898	0·902	0·905	0·909	0·913	0·917
0·8	0·920	0·924	0·928	0·932	0·935	0·939	0·943	0·947	0·951	0·955
0·9	0·958	0·962	0·966	0·970	0·974	9·979	0·983	0·987	0·991	0·996
1·0	1									

III. POLARIZATION

In turbid media including the sea, the polarization is usually linear. A historical review of the research on the polarization of natural light in the sea is given by Pavlov (1965) and by Lundgren (1971) therefore we shall not dwell on it. Laboratory investigations of polarization of natural daylight in turbid media in the Black Sea Branch of the Marine Hydrophysical Institute started in 1958. The instruments, the experimental installations and the methods of measurements have been described previously (Timofeeva, 1961b; Timofeeva and Kaïgorodov, 1963; Timofeeva et al., 1966).

A. IRRADIATION OF A MEDIUM BY PERPENDICULAR BEAMS

(1) *Variations of the spatial distribution of the degree of polarization with depth.*

A complete distribution of the degree of polarization in a turbid medium (a milky medium) was obtained by Timofeeva (1961b,c). The result is shown in the upper graph of Fig. 16. The abscissa is the optical depth and the ordinate is the degree of polarization and the logarithm of the relative radiance in various directions. The lower family of curves is the polarization, the upper family is the radiance and the numbers on the curves are the zenith angles. As is seen, the degree of polarization first decreases with increasing depth rather rapidly and then more slowly and finally reaches a value which is independent of depth. The decrease of p with depth is also shown in the lower graph of Fig. 16 giving results obtained in the sea (see also Ivanoff and Waterman, 1958b). An inspection of the radiance and polarization curves in Fig. 16 reveals that the depth at which the polarization reaches its final value coincides with the depth at which asymptotic radiance distribution occurs. It follows from this: (1) the degree of polarization

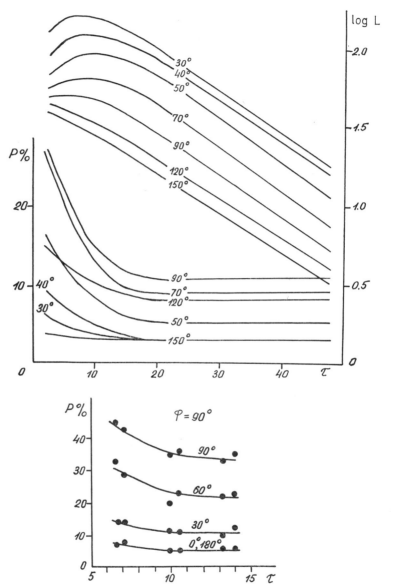

Fig. 16. Dependence of the radiance and the degree of polarization on depth in various directions in a milky medium (above) and relation (p, τ) in the sea.

decreases with increasing diffuseness of the light which is in agreement with Ivanoff's (1957) conclusions and (2) the asymptotic state is characterized not only by a stationary shape of the radiance distribution but also by a stationary angular distribution of the degree of polarization. This supports the theoretical conclusions by Sobolev (1956) and Rozenberg (1959).

With vertical irradiation the polarization plane proved to be perpendicular to the scattering plane at all depths.

The angular distribution of the degree of polarization at various optical depths is shown in Fig. 17. It is evident that the angle at which the polarization is maximal increases with depth tending towards a limiting value characteristic for the asymptotic distribution of polarization degree. The smaller ϵ is, the greater is the increase of the angle. With decreasing depth the angle of maximum polarization tends towards the limiting value at $\tau = 0$.

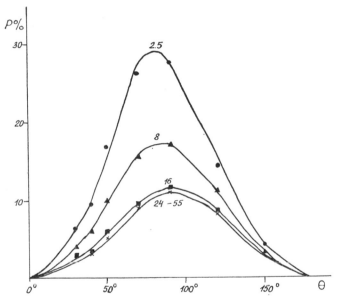

Fig. 17. Dependence of the angular distribution of the degree of polarization on depth in a milky medium. Optical depths are shown by numbers near curves.

Thus the angular distribution of the degree of polarization in a turbid medium can generally not be described by the Rayleigh equation according to which $\theta_{p_{\max}} = 90°$.

B. IRRADIATION OF A MEDIUM BY OBLIQUE BEAMS

Nowadays the similarity of the phenomena of scattering in various turbid media (like planet atmospheres, natural waters, opal glass, photographic emulsions etc.) is beyond doubt. Therefore it should be expected that just as in the atmosphere and in colloids (vapours, smokes) there exist also in liquid media "neutral points" (at which $p = 0$) and fields of "negative" polarization (polarization plane parallel to the scattering plane). This expectation is supported by experimental material (Timofeeva et al., 1966; Timofeeva, 1969, 1970).

(1) *The transient light regime*

(a) *Spatial distribution of ē and p.* In Fig. 18 two families of curves are given. They represent the spatial distributions of the angle ψ of the electrical vector ē (above) and the polarization degree p (below) measured simultaneously with the radiance distribution that was presented in Fig. 8. The number by each curve in Fig. 18 is the angle φ of the azimuth plane in which the measurements were made for a full meridional circle. The variation of ψ with φ in zenith and nadir is due to the fact that the reference line CO (Fig. 1) is fixed relative to the instrument. The comparatively great scatter of the points for the vertical plane through the sun is probably due to insufficient symmetry around this plane of the irradiation in the experimental installation. Such an

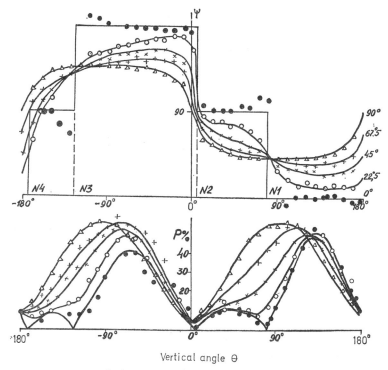

Fig. 18. Relations (ψ, θ) and (p, θ) in various azimuthal planes in a milky medium by oblique irradiation.

asymmetry, judging by Fig. 18, affects most of all the angle ψ near the neutral points where ψ changes rapidly with small changes in φ.

The neutral points 1–4 by analogy with optics of the atmosphere (see Fig. 26 lower right-hand diagram) are called Brewster (No. 1), Babinet (No. 2) and Arago (No. 3) points respectively; the fourth point is not named. The angle ψ in the almucantars (θ constant) through (for $0 < \varphi < 90°$) Brewster's point and through (for $90° < \varphi < 180°$) Arago's point is practically independent of the azimuth angle and in the given medium at the given depth it is numerically equal to the angle of refraction j. For directions in the vertical plane through the sun ψ is either equal to 0° (or 180°) or 90°. The first case with the polarization plane perpendicular to the scattering plane is called "positive polarization" and the second with the polarization plane parallel to the scattering plane "negative polarization". In the horizontal direction the angle ψ has its greatest value at $\varphi = 90°$. Knowing these peculiarities of the distribution of the vector $\bar{e}$ one can by measuring ψ obtain (regardless of the radiance distribution) not only the angle of the vertical plane through the sun but also the sun elevation. The few works dedicated to the study of the angle of the polarization plane of natural light in the sea rest upon the assumption that the polarization plane is perpendicular to the scattering plane (Waterman and Westell, 1956; Ivanoff and Waterman 1958a; Pavlov and Gretchushnikov, 1965). In the formulas used in these studies the angle ψ depends only on geometrical factors, as the sun elevation $h_\odot$ (more precisely j) and the angle of the line of sight. Factors like the optical properties of the water, the optical depth etc. were not taken into account.

For the atmosphere many investigators also assume that $OP \perp SP$, but both in the atmosphere and in turbid media this is not always true (see Rozenberg, 1963b; Timofeeva et $al.$, 1966). We shall now consider the correctness of calculating ψ under the above assumption.

On the basis of simple (geometric) considerations Timofeeva (1969) derived the expression for the angle ψ_1, between OP and SP:

$$tg\psi_1 = \frac{\sin \psi tgj \, \text{cosec} \, \theta + \cos \psi \, \text{cosec} \, \varphi(1 - tgj \, ctg\theta \cos \varphi)}{\sin \psi \, \text{cosec} \, \varphi(1 - tgj \, ctg \, \theta \cos \varphi) - \cos \psi tgj \, \text{cosec} \, \theta} \quad (6)$$

Besides the angles φ, θ and j characterizing the geometry of the experiment eq. (6) involves the angle ψ which is associated with the physics of the phenomenon. If one assumes that $\psi_1 = 90°$, i.e. $OP \perp SP$ we obtain

$$tg\psi = \frac{tgj \sin \varphi}{\sin \theta - tgj \cos \varphi \cos \theta} \quad (7)$$

In particular at $\theta = 90°$

$$tg\psi = tgj \sin \varphi \qquad (8)$$

This formula is given by Waterman and Westell (1956) and by Pavlov and Gretchushnikov (1965) for horizontal line of sight. The general formula for ψ presented by Waterman and Westell (1956) differs from (7) in that the subtrahend in the denominator is missing—an error that Ivanoff and Waterman (1958a) pointed out. Using Fig. 18 and eq. (6) Timofeeva (1969) calculated the angle ψ_1 for various θ and φ. It turns out that with increasing φ a gradual turn of OP relative to SP occurs for almost all angles θ. Actually the angle ψ can take any value between 0° and 90°. The greatest deviations from 90° are observed for directions in the vertical plane through the sun the smallest for directions perpendicular to it. It should be be noted that during the experiment carried out by Timofeeva disturbing effects (horizontal non-uniformity, silvery clouds etc.) were not present to which observed deviations in the atmosphere have been ascribed. Thus the mentioned turn of OP is caused by something else.

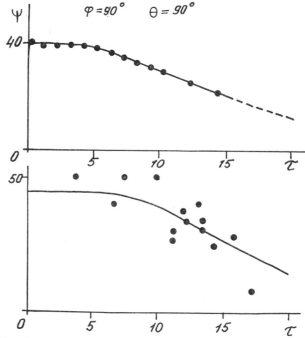

Fig. 19. The change of angle ψ with depth in a milky medium (above) and in the sea (below).

Therefore, in the general case, ψ_1 is different from 90° and formulas (7) and (8) does not correspond to the real spatial distribution of $\bar{e}$ neither in artificial media nor in the sea. Besides, formula (8) cannot be recommended for practical use due to the fact that, as shown below, the angle ψ at $\theta = 90°$ depends on the depth at which measurements are performed. This dependence is studied by Timofeeva (1969) in media with $0.16 \leqslant \epsilon < 0.99$. It was noticed that ψ in fact does not depend on ϵ (to within 3%) where τ is constant. Therefore values obtained for different ϵ are averaged.

Relations (ψ, τ) obtained under laboratory conditions (above) and in the sea (below) are given in Fig. 19. As one can see, only up to $\tau = 4$–6 the angle ψ for the given direction is really equal to j (in this case 40°). With increasing depth ψ decreases rather rapidly tending towards 0° in the deep regime. This dependence of ψ on τ should not be ignored when measuring polarization in the sea.

Figure 20 shows a family of curves similar to those of Fig. 18 but obtained in the sea. Measurements were made by a photoelectric

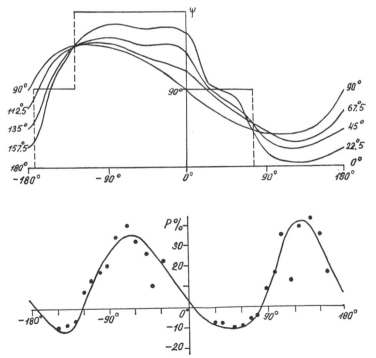

Fig. 20. Relation (ψ, θ) in various azimuthal planes (above) and relation (p, θ) in the vertical plane of the sun (below) in the sea.

polarimeter (Timofeeva and Kaïgorodov, 1963) in clear and calm weather. The qualitative similarity between the curves in Figs. 18 and 20 is evident, so the results of the laboratory investigations are supported by the measurement under natural conditions.

In order to analyse the curves of the lower family of Fig. 18 it is convenient to present them as polar diagrams showing the variation of the degree of polarization with θ in various azimuth planes (Fig. 21, to the left). For comparison similar curves obtained in the sea are also given. The neutral points are, as before, marked with numbers 1 to 4 and the fields of "negative" polarization are shaded.

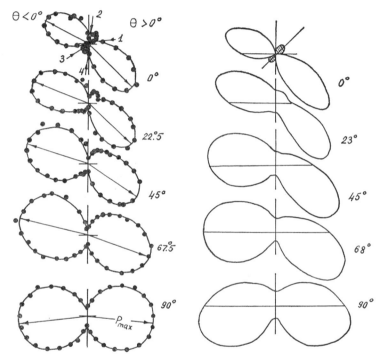

Fig. 21. Relation (p, θ) in various azimuthal planes in a milky medium (on the left) and in the sea (on the right). Numbers denote angles φ.

Figure 21 shows that the variation of the degree of polarization with the zenith angle undergoes a systematic change with the azimuth angle. The polar diagrams are neither symmetrical around the refracted beam (for $\varphi = 0°$) nor around the normal to the medium surface for any φ except for $\varphi = 90°$. Such an asymmetry in earlier observations in the sea was explained by some authors as an effect of the polarized skylight.

But under laboratory conditions the media were irradiated from only one direction. A comparison shows a very good qualitative agreement between the diagrams on the right and left hand sides of Fig. 21. Thus the main cause for the asymmetry of the polarization is not the sky-light but the asymmetry of the irradiation about the normal to the surface. One should note that because of the absence of sufficiently reliable data on the angle ψ, earlier results from measurements of the degree of polarization of natural light (Timofeeva, 1962) were illegally averaged at $\varphi = 0°$ and the areas of "negative" polarization remained unnoticed. Later (Timofeeva et $al.$, 1966) after obtaining the necessary accuracy it became evident that "neutral points" and fields of "negative" polarization also exist in the sea. Therefore corresponding changes were introduced into the upper right-hand diagram. Thus the results of laboratory measurements made it possible to predict the trend of the curves from in $situ$ measurements. An especially vivid impression of the spatial distribution of $\bar{e}$ and p in a turbid medium is produced by the maps in Fig. 22 constructed from experimental data like those of Fig. 18. The coordinate grids are given in transverse azimuthal projections. The symmetry axis to the right is the zenith-nadir line Z–N, the meridional circles correspond to the azimuthal planes separated by 22·5° and the parallel circles are the almucantars of the optical axis of the polarimeter. The direction of the refracted beams is shown by an arrow and the fields of "negative" polarization by a dotted line.

To the left the symmetry line of the grid is the direction of the refracted beams. Here the "meridional" planes are the scattering planes.

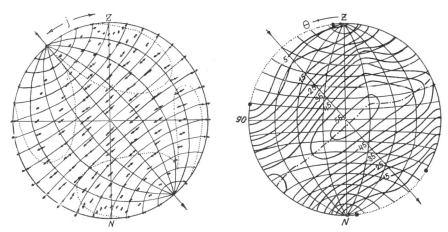

Fig. 22. On the left: distribution of vector $\bar{e}$ in a milky medium around a given point (if we look at it from outside); on the right: isolines of the degree of polarization in the same medium.

The angle of the polarization plane ($\overline{e}$) is shown by arrows the length of which are proportional to the degree of polarization. Deviations from perpendicularity of $\overline{e}$ to the scattering plane are obvious in the regions around the sun and antisun points but are observed in other directions too where the degree of polarization is not as small as in the regions just mentioned. The regions where the deviation is measurable is surrounded by a dotted line.

Numbers on the isolines in the right-hand map show the degree of polarization. The isolines are usually perpendicular to the refracted beams but not everywhere. For instance, in the region $20° < \theta < 60°$ and $112 \cdot 5° < \varphi < 180°$ a sharp deviation of the isolines from the general trend is observed. This occurs probably because of the asymmetry of the irradiation on the surface of the medium. Fields of small and great polarization are clearly defined in the maps. The points of maximal polarization are connected with a dot-dash line. Unfortunately because the phenomenon is so complex it has been impossible so far to explain precisely all the deviations of the isolines from the basic trend.

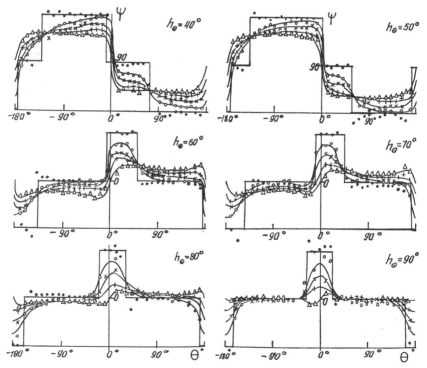

Fig. 23. Relation (ψ, θ) in various azimuthal planes at various $h_\odot$ in milky media.

(b) *The influence of the sun elevation on the angle of the polarization plane and on the degree of polarization.* The influence of $h_\odot$ on the spatial distribution of $\bar{e}$ and p has been studied by Timofeeva (1969, 1970). The experimental data were obtained in milky media as far as possible with equal ϵ and at equal optical depths τ but with different $h_\odot$. The corresponding families of curves are given in Figs. 23 and 24. With the increase of $h_\odot$ the asymmetry of the light field decreases markedly, the angle ψ tends to 0° for all θ except those corresponding to the fields of "negative" polarization and the degree of polarization tends towards a distribution symmetrical around $\theta = 0°$ in all azimuthal planes. For $h_\odot = 90°$ the value of the polarization is small in zenith and nadir and is not very reliable since the measurements were performed at the sensitivity limit of the equipment.

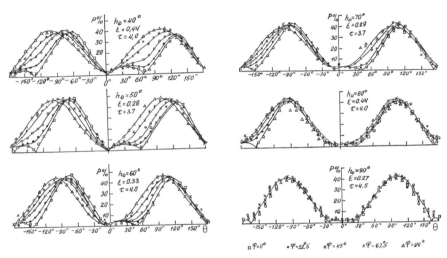

Fig. 24. Relation (p, θ) in various azimuthal planes at various $h_\odot$ in milky media.

The effect of the sun elevation on the polarization of daylight in the sea was studied in horizontal directions by Waterman and Westell (1956). However due to the great spread of the observed data (Fig. 25 to the right) the authors could only draw the following conclusion: there is a clear dependence of p on $h_\odot$ in the vertical plane through the sun but not in the plane perpendicular to it. Using Figs. 23 and 24 one can demonstrate the influence of $h_\odot$ on ψ and p in any direction. In Fig. 25 (middle and to the left) relations for horizontal directions are given as examples. As we see, the angle ψ is zero, regardless of $h_\odot$ (at least for $h_\odot \geqslant 30°$) in directions in the vertical plane through the sun.

The degree of polarization first increases slowly with increasing $h_\odot$, then more quickly and finally approaches its limiting value for $h_\odot = 90°$. In the perpendicular direction ψ is close to $50°$ for $h_\odot = 0$ and becomes equal to $0°$ for $h_\odot = 90°$. Here the degree of polarization remains constant and naturally equal to the limiting value of p at $\varphi = 0°$.

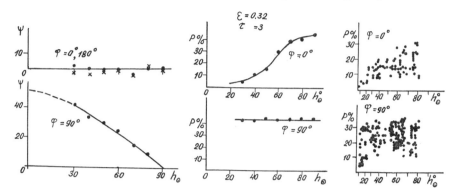

Fig. 25. Dependence of ψ and p on $h_\odot$ in a milky medium (Timofeeva, 1970) and in the sea (Waterman and Westell, 1956) (on the right).

Figure 26 presents schematically the positions of the neutral points (determined from Figs. 23 and 24) relative to the direction of the refracted beams for various $h_\odot$. One can see that with increasing $h_\odot$ the angular distance between the neutral points around the sun direction decreases and similarly in the antisun direction. The directions of maximal polarization are denoted by θ_{max}.

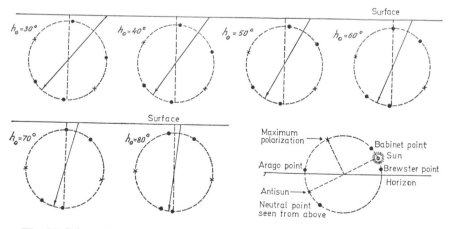

Fig. 26. Schematic representation of the "neutral points" in milky media (Timofeeva et al., 1966) and in the atmosphere (below on the right) (Rozenberg, 1963a).

It follows from Fig. 24 that the maximal polarization in a medium depends very little both on φ (this dependence being less the greater is $h_\odot$) and on $h_\odot$. On the other hand the angle θ_{max} for which the degree of polarization is greatest depends strongly on both φ and $h_\odot$ as shown in Fig. 27. The angles φ are denoted by numbers near the curves. Note also that the θ_{max}-axis is broken. Obviously, the effect of the asymmetry of the irradiation on θ_{max} decreases with increasing $h_\odot$ and the distance between the curves becomes equal to zero at $h_\odot = 90°$, $\theta_{max} = 98°$. This is the angle at which Shifrin (1951) observed the greatest polarization of light scattered from large particles. At $\tau = 3$ the role of multiple scattering is still small and one can expect that the polarization is determined mainly by the light scattered once from individual

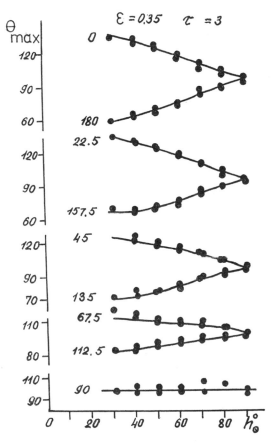

Fig. 27. Dependence of the direction of maximal polarization on $h_\odot$ in various azimuthal planes.

particles. Knowing the angle of scattering at which the polarization is maximal it is easy to derive (Timofeeva, 1969) the following formula for calculation of θ_{max} in any azimuthal plane and for any $h_\odot$:

$$\cos \theta_{max} = -\sin 98° \sin j \cos \varphi + \cos 98° \cos j \tag{9}$$

The curves presented in Fig. 27 are calculated using this formula. The curves agree very well with the experimental points, and so the assumption in (9) seems to be good in this case. With decreasing ϵ and with increasing τ some approach of the curves in each pair is also observed. The reason is quite clear: both variations act to make the light field more symmetric similarily to an increase of $h_\odot$ (the role of multiple scattering increases).

(c) *Dependence of ψ and p on τ and ϵ.* During the study of these relations Timofeeva obtained many experimental results all of which

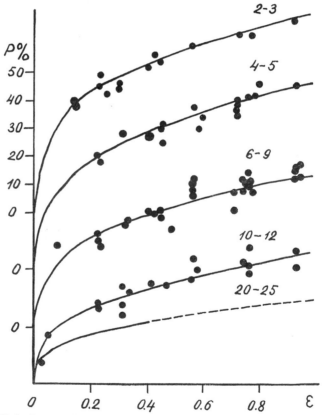

Fig. 28. Relation (p, ϵ) at various optical depths in a milky medium by oblique irradiation.

cannot be mentioned here. Let us give as an example some experimental data for the horizontal direction at $\varphi = 90°$. The relation (p, ϵ) is given in Fig. 28 where the origins of the three upper curves have been shifted along the ordinate axis, the scale remaining unchanged. The interval of optical depths corresponding to the experimental points are denoted by numbers by each curve. The comparatively great spread of the points is attributed to the presence of significant amounts of haze in the atmosphere during the measurements (the sun was the light source). Sometimes the haze was so thick that the decrease of the radiance relative to a clear day could be seen by the naked eye. Under such conditions the angle of divergence of the light field naturally increases and the polarization is diminished. Fig. 28 gives an idea of the accuracy of the obtained data.

An analysis of the whole material showed that a good first-order approximation of the dependence of the degree of polarization on ϵ is the formula

$$p = p_o \epsilon^{\delta} \qquad (10)$$

where $\delta = f_1(\varphi)$, $0.33 \leqslant \delta \leqslant 0.50$, and $p_o = f_2(\varphi, \theta, \tau)$ (Timofeeva, 1970). Let us calculate the spatial distribution of the degree of polarization by means of formula (10) for a concrete case (functions f_1 and f_2 have been defined by Timofeeva, 1970). Figure 29 presents experimental points obtained in a medium with $\epsilon = 0.58$ and at an optical depth $\tau = 5$ and curves calculated from formula (10). The curves agree with the points with sufficient accuracy. A somewhat greater inaccuracy is observed for small values of ϵ (of the order of 0.2).

The dependence of the positions of the neutral points on τ as well as on ϵ is discussed by Timofeeva (1969). It is shown that the positions of the Brewster and Babinet points do, in fact, not depend on τ (at least up to $\tau = 8$–10). Probably in the given medium the direct light predominates over the scattered light down to these depths. Deeper the neutral points must shift and at great depths be transformed into two circles symmetrical around the normal to the surface. With an increase of ϵ the Brewster and Babinet points converge, the Brewster point moving more than the Babinet point. At the same time the positions of the points becomes more sensitive to variations of ϵ. For large ϵ, i.e. in a weakly scattering medium, for instance tap water, the field of negative polarization between the points is quite narrow, and the neutral points are arranged symmetrically around the sun and the antisun directions; the Babinet point is displaced relative to the zenith so that the whole picture is quite similar to that for the sky. Because the positions of at least the Brewster and the Babinet points change

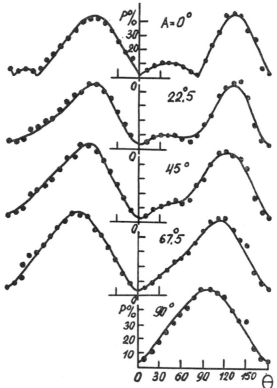

Fig. 29. Comparison of experimental points and the calculated curves (p, θ) for various φ.

little with depth (at least in media with $\epsilon = 0\cdot3$ and $0\cdot6$) but depend on ϵ, especially at small sun elevations it is possible if only approximately to judge the value of ϵ in a turbid medium by the positions of the neutral points.

(2) The deep light regime

The relation between the degree of polarization and the value of the parameter ϵ in the deep regime is given in Fig. 30.

From Fig. 30 the following conclusions are drawn:

(a) With decreasing ϵ the degree of polarization diminishes rather quickly for zenith angles close to $90°$. For small angles, for instance $\theta = 30°$, the degree of polarization is in fact, independent of ϵ down to about $\epsilon = 0\cdot075$ and then diminishes quickly to zero.

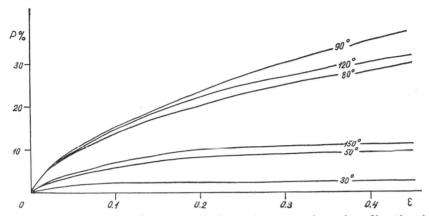

Fig. 30. Dependence of the degree of polarization upon ε in various directions in deep regime in a milky medium. θ-values are denoted by numbers on the curves.

(b) When $\epsilon \to 0$ the degree of polarization tends towards zero in all directions. In other words light would be unpolarized in a hypothetical purely scattering medium which is in good agreement with the theory.

(c) When $\epsilon \to 1$ the degree of polarization for all directions tends towards a limit. The curves in Fig. 30 are probably valid for various wavelengths. Thus one can expect that the polarization is higher at wavelengths where ϵ is larger. Let us consider sea water for example. Since the absorption coefficient varies more with wavelength than the scattering coefficient, the relation between ϵ and wavelength will mainly be determined by the value of the absorption coefficient. The smaller the absorption coefficient is the smaller is ϵ and p, i.e. for the most penetrating light the degree of polarization is smallest and greater in the short-wave part and, especially, in the long-wave part of the spectrum. This conclusion is in good agreement with the data from measurements in the sea (Ivanoff and Waterman, 1958b, Timofeeva, 1965).

The plane of oscillations is in the deep regime always perpendicular to the plane normal to the surface of the medium and passing through the line of sight.

In conclusion of the review we shall note that the good agreement between the results of laboratory measurements and those obtained under natural conditions permits one to recommend wider use of the method of laboratory investigations in optics of turbid waters and in other fields of research associated with the study of light in scattering media.

218   V. A. TIMOFEEVA

REFERENCES

Ambarzumian, V. A. (1942). *Izv. Akad. Nauk SSSR, Ser. Geogr. Geofiz.*, **6**, 97–103.
Blouin, F. and Lenoble, J. (1962). *Rev. Opt.*, **43**, 615–620.
Gershun, A. A. (1936a). Svetovoe pole, ONTI, NKPT, Leningrad, Moscow.
Gershun, A. A. (1936b). *Tr. Gos. Opt. Inst.*, **11**, 3–17.
Gershun, A. A. (1936c). *Tr. Gos. Opt. Inst.*, **11**, 43–56.
Herman, M. and Lenoble, J. (1964). *Rev. Opt.*, **11**, 552–572.
Isacchi, S. and Lenoble, J. (1959). *Rev. Opt.*, **38**, 217–237.
Ivanoff, A. (1957). *Ann. Geophys.*, **13**, 22–53.
Ivanoff, A. and Waterman, T. (1958a). *J. Mar. Res.*, **16**, 255–282.
Ivanoff, A. and Waterman, T. (1958b). *J. Mar. Res.*, **16**, 283–307.
Ivanov, A. P. (1969). Optics of Scattering Media, p.h. Science and Techn., Minsk.
Ivanov, A. P. and Il'ich, G. K. (1965). *J. Appl. Spectrosc.*, **2**, 356–361.
Ivanov, A. P. and Il'ich, G. K. (1967). *Izv. Akad. Nauk SSSR, FAaO*, **3**, 662–666.
Jerlov, N. G. and Liljequist, G. (1938). *Svenska Hydrograf. Biolog. Komm. Skrifter, Ny Ser. Hydrogr.*, **14**, 1–15.
Jerlov, N. G. and Fukuda, M. (1960). *Tellus*, **12**, 348–55.
Jerlov, N. G. and Nygård, K. (1968). Københavns Universitet, *Rep. Inst. Fys. Oceanog.*, **1**, 1–6.
Lenoble, J. (1956). *Nature (London)*, **178**, 756–757.
Levin, I. M. and Ivanov, A. P. (1965). *Opt. Spectrosk.*, **18**, 920–923.
Lundgren, B. (1971). Københavns Universitet, *Rep. Inst. Fys. Oceanog.*, **17**, 1–34.
Makarevich, S. A. and Robilko (1969). *In* "Optike rasseivayushchikh sred" (A. P. Ivanov, ed.), pp. 175–176. Izd. Nauka i tekhnika, Minsk.
Pavlov, V. M. (1965). *Tr. Okeanologii*, **77**, 41–53.
Pavlov, V. M. and Gretchushnikov, B. N. (1965). *Tr. Okeanologii*, **77**, 53–66.
Pelevin, V. N. (1965). *Izv. Akad. Nauk SSSR, FAaO*, **I**, 539–545.
Poole, H. H. (1945). *Sci. Proc. Roy. Dublin Soc.*, **24**, 29–42.
Prieur, L. and Morel, A. (1971). *Cah. Oceanogr.*, **23**, 35–47.
Rozenberg, G. V. (1958). *Opt. Spektrosk.*, **5**, 440–449.
Rozenberg, G. V. (1959). *Opt. Spektrosk.*, **7**, 407–416.
Rozenberg, G. V. (1963a). *In* "Spectroskopii svetorasseivayushchikh sred", 5–36, Izd-vo Akad. Nauk BSSR, Minsk.
Rozenberg, G. V. (1963b). Sumerki Gos. Izdatel' stvo fiz. mat. liter., Moscow.
Schellenberger, G. (1969). *Acta Gydrophisica*, **14**, 207–235.
Sherstyankin, (1970). Unpublished manuscript.
Shifrin, K. S. (1951). "Rasseyanei sveta v mutnoi srede", Izd. tekhn. teoretich lit., Moscow-Leningrad.
Shuleikin, V. V. (1922). *Izv. in-ta Fiziki Biofiz.*, **2**, 119–136.
Shuleikin, V. V. (1924). *Philos. Mag.*, **48**, 307–320.
Shuleikin, V. V. (1933). *Geofizika*, **3**, 145–180.
Sobolev, V. V. (1956). Perenos luchistoi energii v atmosfere zvezd i planet. Gos. tekhizdat, Moscow.
Timofeeva, V. A. (1950). Dissertatsiia " Slozhnoe rasseyanie sveta", Akad. Nauk SSSR, Moscow.
Timofeeva, V. A. (1951a). *Dokl. Akad. Nauk SSSR*, **76**, 677–680.
Timofeeva, V. A. (1951b). *Dokl. Akad. Nauk SSSR*, **76**, 831–833.

Timofeeva, V. A. (1953). *Tr. Morskogo Gidrofiz. Inst. Akad. Nauk SSSR*, **3**, 35–81.
Timofeeva, V. A. (1956). *Tr. Morskogo Gidrofiz. Inst. Akad. Nauk SSSR*, **7**, 153–160.
Timofeeva, V. A. (1957a). *Tr. Morskogo Gidrofiz. Inst. Akad. Nauk SSSR*, **2**, 97–104.
Timofeeva, V. A. (1957b). *Dokl. Akad. Nauk SSSR*, **113**, 556–559.
Timofeeva, V. A. (1957c). *Izv. Akad. Nauk SSSR, Ser. Geofiz.*, **2**, 265–272.
Timofeeva, V. A. (1960). *Tr. Morskogo Gidrofiz. Inst. Akad. Nauk SSSR*, **20**, 100–105.
Timofeeva, V. A. (1961a). *Opt. Spectrosk.*, **10**, 533–534.
Timofeeva, V. A. (1961b). *Izv. Akad. Nauk SSSR, Ser. Geofiz.*, **5**, 766–774.
Timofeeva, V. A. (1961c). *Dokl. Akad. Nauk SSSR*, **140**, 361–363.
Timofeeva, V. A. (1962). *Izv. Akad. Nauk SSSR, Ser. Geofiz.*, **12**, 1843–1851.
Timofeeva, V. A. (1965). *In* "Gidrofizich. i gidrokhimich. issled.", pp. 56–59. Naukova dumka, Kiev.
Timofeeva, V. A. (1969). *Izv. Akad. Nauk SSSR, FAaO*, **10**, 1049–1057.
Timofeeva, V. A. (1970). *Izv. Akad. Nauk SSSR, FAaO*, **6**, 611–616.
Timofeeva, V. A. (1971a). *Izv. Akad. Nauk SSSR, FAaO*, **6**, 688–691.
Timofeeva, V. A. (1971b). *Izv. Akad. Nauk SSSR, FAaO*, **7**, 1326–1329.
Timofeeva, V. A. and Gorobetz, F. I. (1967). *Izv. Akad. Nauk SSSR, FAaO*, **3**, 291–296.
Timofeeva, V. A. and Kaïgorodov, M. N. (1963). *Okeanologiya*, **3**, 506–516.
Timofeeva, V. A. and Koveshnikova, L. A. (1966). *Izv. Akad. Nauk SSSR, FAaO*, **3**, 320–323.
Timofeeva, V. A. and Opaets, L. S. (1971). *In* "Morskikh gidrofizich. issl.", Vol. 54, pp. 85–95, Morskogo Gidrofiz. Inst. Akad. Nauk USSR, Sevastopol.
Timofeeva, V. A. and Solomonov, V. K. (1970). *Izv. Akad. Nauk SSSR, FAaO*, **5**, 513–522.
Timofeeva, V. A., Vostroknutov, A. A. and Koveshnikova, L. A. (1966). *Izv. Akad. Nauk. SSSR, FAaO*, **12**, 1259–1266.
Tyler, J. (1960). *Bull. Scripps. Inst. Oceanogr.*, **7**, 363–412.
Waterman, T. H. and Westell, W. E. (1956). *J. Mar. Res.*, **15**, 149–169.
Whitney, L. V. (1941). *J. Mar. Res.*, **4**, 122–131.
Zege, E. P. (1971). *Izv. Akad. Nauk SSSR, FAaO.* **7**, 121–132.

Chapter 10

Some Applications of the Optical Tracer Method

HASONG PAK and WILLIAM S. PLANK

School of Oceanography, Oregon State University, Corvallis, Oregon, U.S.A.

I. INTRODUCTION

The optical method of tracing ocean circulation basically requires the assumption that the optical properties under consideration are a conservative property of the water. There are processes in the sea water such as biological production, sinking and dissolving of particles, breakdown and flocculation of particles, chemical precipitation and others which tend to make the particulate matter in the sea a non-conservative property, and the application to the tracing of water flow becomes inaccurate. The non-conservative tendency of the particulate matter should be carefully accounted for whenever the application of the optical tracer method is attempted. For these reasons the particulate matter in the sea water is sometimes considered a semi-conservative property of water masses and its application to tracing the water masses can be made with certain limitations.

There are many cases in the literature which report the results of successful applications of light scattering measurements to tracing water masses (Jerlov, 1953, 1958, 1959, 1964; Ketchum and Shonting, 1958; Joseph, 1959; and many others), and their successes are, in general, dependent on the degree of conservativeness of the optical property which they use. The present paper introduces some cases of water mass tracing by optical parameters such as light scatterance and beam transmittance, and the direct measurements of suspended particle distributions.

II. Interaction of the Cromwell Current and the Galapagos Islands

The upper layer circulation pattern in the vicinity of the Galapagos Islands shows a quite complicated picture contrary to the common impression of the simple and relatively steady south equatorial current responding to the steady southeasterly trade wind. The complications result from (1) the convergence of the surface waters from both hemispheres which have distinctive water mass characteristics, (2) interruption of the flows by the Galapagos Islands, and (3) the internal interaction between the surface and the Equatorial undercurrent.

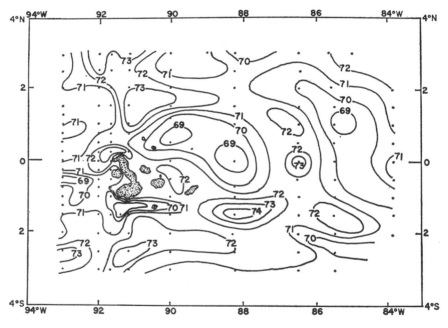

Fig. 1. Dynamic topography in dynamic meter × 10 at 100 m relative to 600 m.

The surface water of the South Equatorial current is characterized by low temperature and high salinity (less than 20°C and larger than 34‰), while the surface water from the Northern hemisphere (probably from the Equatorial Counter Current) is characterized by higher temperature and lower salinity. Their boundary is known as the

Equatorial front, and it slopes so that warm, less saline water overrides the cold saline water. The large scale frontal position may deviate considerably from a zonal configuration, and may contain perturbations. These perturbations may be described as intrusions of one water into the other as a tongue of one characteristic water mass and vice versa. This perturbation effect is the first factor that makes the circulation pattern complicated. The effects of the interruption of the flows by the Galapagos Islands is intuitively obvious. We expect some form of eddies around the islands causing the flow pattern to become more complex. We also expect particles of terrigenous origin to be carried downstream. The interaction between the islands and the currents may be indicated by the distribution of the eroded particles.

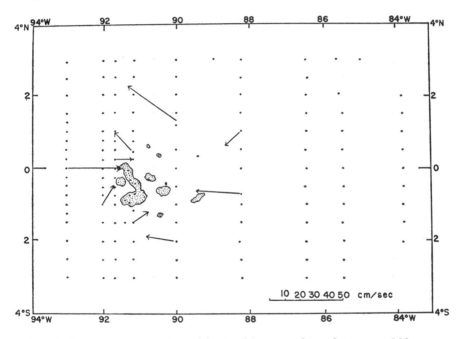

Fig. 2. Current vectors obtained by tracking parachute drogues at 100 m.

The internal interaction between the south equatorial current and the equatorial undercurrent will also contribute to a more complicated flow pattern. Between the equatorial undercurrent and the oppositely directed south equatorial current, a large vertical velocity shear is expected, and the position of the maximum shear will be complicated due to the effects of the Islands.

The complicated nature of the near surface circulation is shown in

the dynamic topography at 100 m depth relative to 600 m (Fig. 1).
The figure shows a number of closed cells forming flows in both east
and west directions. The same picture may be seen from Fig. 2, where
parachute drogue data at 100 m are represented by velocity vectors.

We can find fair agreement in the direction of the flow between the
dynamic topography in Fig. 1 and the velocity vectors in Fig. 2 determined by the drogue observations. This agreement lends credence to
the complicated flow pattern shown in the dynamic topography. It is
quite probable that our lack of knowledge about the Equatorial undercurrent on the east side of the Galapagos Islands results mainly from
its complicated nature relative to that on the west side of the Galapagos
Islands.

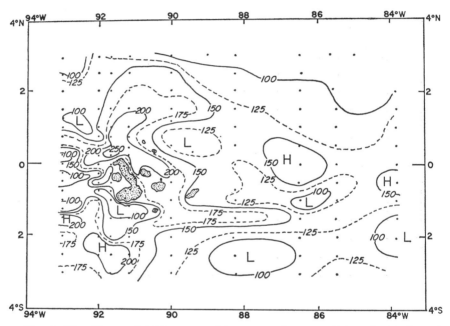

Fig. 3. The Equatorial Thermostad in meters between 170 and 190 cl ton^{-1}
isanosteric surfaces.

During the Yaloc-71 Cruise, from October 16, 1971 to December 7,
1971, the optical oceanography group of Oregon State University
aboard R/V *Yaquina* made optical and hydrographic observations at
152 stations for the purpose of studying the interaction of the Equatorial
undercurrent with the Galapagos Islands and the extension of the
Equatorial undercurrent on the east side of the islands. Temperature,
salinity, light scattering and particle size distributions were measured

at each station covering about 13 points in the upper 600 m of water.
Parachute drogues at 100 m were tracked at nine stations by radar
ranging from a ship's position fixed by the satellite navigator.

Results of the Yaloc-71 cruise data will be presented to show the
extension of the Equatorial undercurrent, also called the Cromwell
Current, to the east side of the Galapagos Islands.

At a depth at which the Equatorial undercurrent is commonly
found, say 100 m, the circulation pattern seems to be subject to the
various complicating effects described earlier and considerable difficulties are confronted in defining the current. The picture becomes
much simpler on the lower side of the Cromwell Current (depths of
200 or 300 m), where the water characteristics are much more homogeneous and the effects of the complicating processes quite limited.
Thus the flow pattern is very well delineated on the 250 m surface in
terms of the hydrographic data as well as light scattering and particle
concentration. The Cromwell Current, with its high velocity core, is
characterized by large vertical eddy diffusivity and accordingly it is
characterized by horizontal maxima or minima on level surfaces. For
an example, in a level above the Cromwell Current, it is characterized
by temperature minimum and in a level below the Cromwell Current
it is characterized by the maximum temperature, and the Cromwell

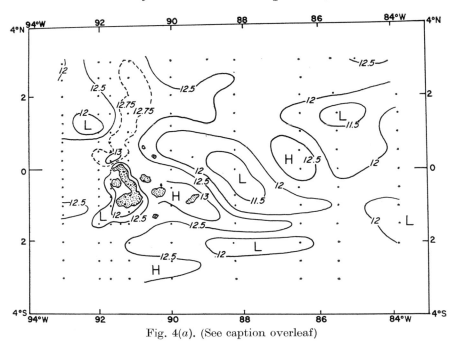

Fig. 4(a). (See caption overleaf)

10

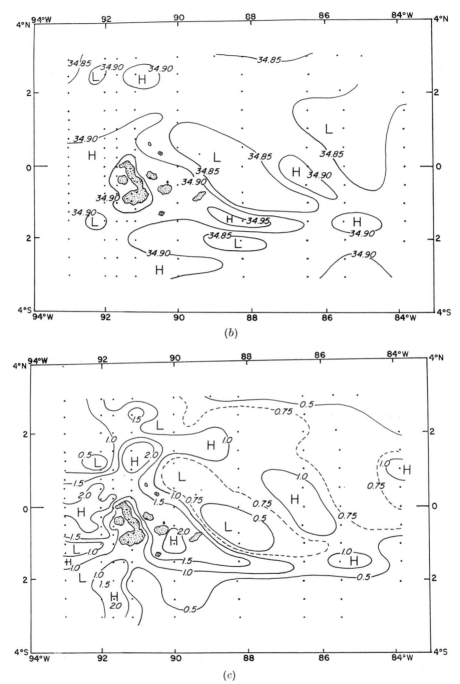

Fig. 4. Distribution of (a) temperature (°C), (b) salinity (0/00), and (c) oxygen (ml l⁻¹) on the 250 m surface.

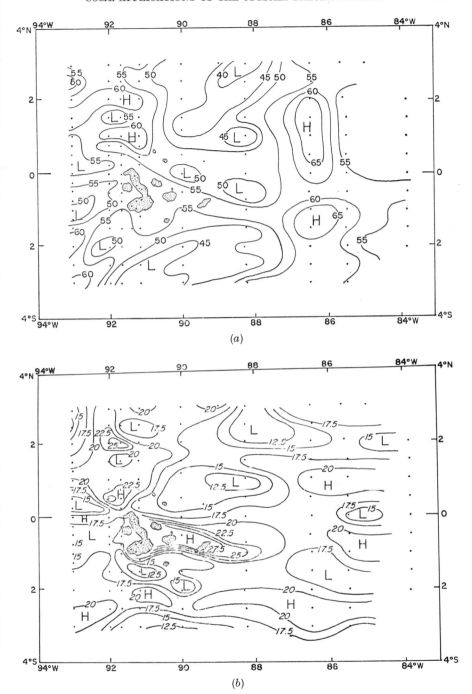

Fig. 5. Distribution of (a) light scattering on the 250 m surface (m-ster)⁻¹ and (b) suspended particulate matter integrated from 50 m to 250 m (2 × 10⁵ mg m⁻²).

Current core is located at the thermocline. Owing to the large eddy diffusivity within the Cromwell Current, the current water is also identified by a relatively thicker layer of water with temperature or density corresponding to the core of the current (Cromwell, 1953; Wooster and Cromwell, 1958; Knauss, 1960; Montgomery and Stroup, 1962). White (1969) defined this layer as the distance between the isanosteric surfaces, 170 to 190 cl ton^{-1}, and called this the Equatorial thermostad. The distribution of the Equatorial thermostad determined by the Yaloc-71 data is shown in Fig. 3.

Since most isolines dip down below the level of its core, the Cromwell Current is expected to be characterized in the 250 m surface by a maximum in temperature, salinity, and oxygen. Figure 4 shows that this is indeed the case. The core of the Cromwell Current also seems to be indicated by a maximum in light scattering and particle concentration on the east side of the Galapagos Islands (Fig. 5). The distribution of light scattering and particle concentration corresponding to the Cromwell Current on the east side of the islands appear to be an indication of the interaction between the islands and the Cromwell Current.

III. Characterizing Water Masses

The inherent optical properties of sea water have been studied with the intent of characterizing water masses, especially in connection with water masses which are spreading (Jerlov, 1953, 1958, 1959). As was pointed out earlier, the inherent optical properties become non-conservative mainly due to biological activities in the euphotic zone and settling through the water column. There are only a few cases in which the application of the inherent optical properties to the tracing of water masses has been successful; Antarctic intermediate water (Jerlov, 1959), Red sea water spreading in the Indian Ocean (Jerlov, 1953), Mediterranean water spreading in the Atlantic Ocean (Jerlov, 1953, 1961), upwelling of deep water (Joseph, 1959; Pak et al., 1970), river plumes (not really a water mass) (Jerlov, 1958; Pak, 1970) and some others which demonstrate the optical properties of a particular body of water as characterizing the water in contrast to others and correlating with other known oceanographic parameters (Kullenberg, 1968; Joseph, 1955; Spilhaus, 1965; Morrison, 1970; Ivanoff, 1959; Kozlyaninov and Ovchinnikov, 1961; Paramonov, 1964; Neuymin and Paramonov, 1965; Pickard and Giovando, 1960) and Timofeeva and Neuymin (1968) who reviewed many other works by Russian investigators. There are difficulties involved with characterizing a water mass by inherent optical properties with the same accuracy that is achieved by using temperature or salinity.

Application of optical properties in water mass studies can still be made for a good many important problems in oceanography, provided the limitations imposed by non-conservativeness are taken account of. For instance, Jerlov (1953) compared light scattering profiles of the different oceans and indicated, that Antarctic Bottom water in the Atlantic Ocean is associated with a particle maximum which is quite different from that in the other oceans. In the coastal upwelling zone

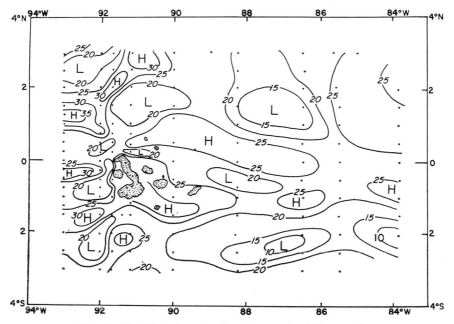

Fig. 6. "Bulk" index of refraction on the 250 m surface.

off the Oregon Coast a temperature inversion is often found near the bottom of the permanent pycnocline. Light scattering measurements showed that the inversion water has a maximum in light scattering. The optical data were important in interpreting the temperature inversion as an extension of near shore surface water (Pak *et al.*, 1970). This water had become rich in nutrients and higher in salinity because of upwelled subsurface water and was then warmed by solar radiation and supplied with large light scattering particles from biological production. The water moved offshore by Ekman transport and sank to near the permanent pycnocline due to its high salinity. The application of the optical properties in characterizing water masses will be improved greatly if the light scattering particles can be identified in terms of size distributions and mineralogical composition.

Measurements of the particle size distribution are possible on a routine basis by means of the Coulter-Counter. The lower limit of particle sizes measurable with the Coulter-Counter is determined by problems of electrical noise and clogging. Particles may be classified as organic or inorganic. It is known that the index of refraction varies appreciably between these two types. Using the Mie theory, the "bulk" index of refraction for a given sample can be computed if the particle size distribution and the volume scattering functions at two different wavelengths of light are known (Zaneveld and Pak, Chapter 6). With this information, particle size distribution and bulk index of refraction, suspended particles may be classified by shape of the particle size distribution and its bulk index of refraction. This method has been applied to the data collected near the Galapagos Islands (Fig. 6). It clearly shows an island wake resulting from the interaction of the Cromwell current with the islands. This wake appears as a zone of maximum index of refraction.

IV. BOTTOM NEPHELOID LAYER

The study of deep nepheloid layers has become an important area of research in which optical tracers are extremely useful. These layers containing high concentrations of particulate matter were first reported by Jerlov (1953) who made light scattering measurements on samples obtained with hydrographic bottles. This method has also been used extensively at Oregon State University (Plank et al., 1972). Data have also been collected using photographic light scattering measurements (Ewing and Thorndike, 1965; Hunkins et al., 1969; Eittreim et al., 1969; Ewing and Connary, 1970) and measurements of light scattering at a fixed angle using an in situ nephelometer (Plank et al., 1972).

Information concerning total concentration and type of particulate matter has been obtained by centrifuging large samples (Jacobs and Ewing, 1969) and size distributions of particulate matter within nepheloid layers have been determined by means of the Coulter-Counter (Plank et al., 1972).

The limits of the optical tracer method with regard to settling are especially important in the study of deep nepheloid layers because of their proximity to the bottom and because of their apparent importance in the sedimentation process.

Reported thicknesses of observed nepheloid layers vary widely. Some reported ranges are: 114 to 1090 m with a tendency for thickness to increase with depth (Hunkins et al., 1969), 300 to 2400 m (Eittreim *et al.*, 1969), 180 to 2700 m (Ewing and Connary, 1970) and 50 to 930 m (Plank *et al.*, 1972). In most cases, thickness has meant the distance from the bottom to the depth of minimum light scattering. It is questionable whether or not this distance is of much importance relative to the physical processes affecting the nepheloid material or to the processes of sedimentation. Most writers have also reported intensities of nepheloid layers by which they usually mean the ratio of maximum light scattering within the nepheloid layer to that at the depth of minimum scattering. This quantity may be relatively meaningless if the instrument used has not been calibrated in units of absolute light scattering. One reported range of intensities is from 1·0 to 7·8 (Plank *et al.*, 1972).

In examining the suspended matter which makes up the nepheloid layers, we might divide the areas of study into two general questions, which might equally apply to the study of any constituent of the sea. That is, where does it come from and where does it go?

In examining where it comes from, we first want to know the mineralogic and chemical composition of the material suspended in the layer. Until recently the only technique available to answer this question had been the filtering or centrifuging of large (200 litres) samples obtained by water sampling bottles. The material obtained by this method has been identified as lutite, a clay mineral (Ewing and Thorndike, 1965). However this technique is limited by the small sample size obtainable. In order to obtain sample sizes large enough to be used in X-ray diffraction and mass spectrometry studies, we have developed a deep-pumping apparatus which will allow the collection of large samples by filtering (Beer *et al.*, 1972). The advantage of this system is that no matter what the concentration of suspended matter is, pumping can be continued until clogging of the filter takes place. Surface monitoring of flow rate allows detection of filter clogging and a measure of the quantity of water filtered.

Along with more detailed chemical and mineralogical analaysis knowledge of the particle size distribution is necessary both as an aid in determining the source area of the material and for purposes of correlation with observed optical properties within the nepheloid layer.

Our measurements of particle size distributions (Plank *et al.*, 1972) showed no systematic differences between nepheloid and non-nepheloid waters which would indicate that the material above and within the nepheloid layer is of the same kind or from the same source area. This observation lends support to the hypothesis that the material within the nepheloid layer has settled from above, and is increasing in concentration near the bottom because of increased turbulent diffusion or slower settling velocity. To gain further insight into this problem it is necessary to obtain more detailed particle size distributions and for this purpose we are constructing a seawater particle analyzer which will provide increased resolution and range in measured size distributions.

The information so far obtained indicates several possible sources for the nepheloid material. Eittreim *et al.* (1969) suggest that the dominant mechanism which injects sediment into the bottom water of the North Atlantic is the action of turbidity currents which carry the material away from the continents down the slope to the North American basin. Our studies in the Panama Basin indicate that the continent is the primary source of the suspended matter in the bottom 100 m of the water column. Figure 7 shows a large negative gradient in the offshore direction in integrated particulate matter in the lower 100 m of the water column. It also seems that the Guayas River in southern Ecuador may be a major source area of suspended matter. In the nearshore region, particulate matter may be eroded by wave action down to depths of several hundred meters. Neudeck (1971)

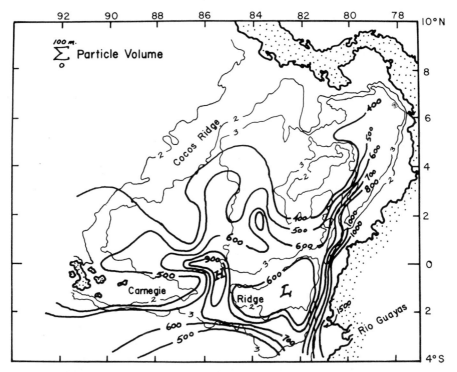

Fig. 7. Distribution of volume of suspended particulate matter integrated from the bottom to 100 m above the bottom ($2 \times 10^3 \ \mu m^3 \ ml^{-1}$).

observed ripples in bottom photographs taken in depths as great as 200 m on the shelf off the Oregon coast. Most of the ripples were symmetrical which suggests oscillatory currents, such as would be expected from wind waves or tides. Harlett (1971) used beam transmittance and bottom current measurements to demonstrate that the bottom nepheloid layer in the nearshore region thickens and thins in response to increases and decreases in current velocity. Several investigators (Joseph, 1955; Jerlov, 1958; Schubel, 1969; Schubel, 1971) have shown correlation between tidal currents and nepheloid distribution.

Organic matter and wind-blown or ice-rafted particles also reach the nepheloid layers by settling from the surface layers of the sea. However, it is not known to what extent they contribute to the total amount of material within the deep nepheloid layers. Zaneveld (1971) has discussed the changes in size distribution which might be expected when particles subject to oxidation settle through sea water, however further experimental work is necessary to determine the quantity and composition of particles which reach the deep sea by settling from the surface.

Another source of suspended matter in deep nepheloid layers is the deep sea sediments themselves which may be eroded by bottom currents. Questions still unanswered in this area concern the current velocities necessary to erode deep sea sediments and where and how often these velocities exist in the sea. Southard et al. (1971) have performed experiments on the erosion of calcareous ooze and found that velocities which caused erosion were not much greater than those commonly measured in the deep ocean, however Wimbush and Munk (1968) conclude that particle pick-up by hydrodynamic forces does not occur at the places where they made measurements of boundary layer velocities except perhaps during an occasional "benthic storm". At any rate, there seems to be little doubt that there are some areas of active erosion in the deep sea, but these may be of limited extent. Our studies in the Panama Basin (Fig. 7) indicate the likelihood of active erosion in the saddle in the Carnegie Ridge which forms the southern boundary of the basin. Geological studies also indicate active erosion in the saddle and deposition on the flanks of the ridge to the north and south (Van Andel et al., 1971).

The question of where the material in the nepheloid layer goes primarily involves an investigation of the factors which influence the distribution of particulate matter in the deep sea. This obviously indicates a need for greater knowledge of the abyssal circulation of the ocean. It is felt that once we have gained a more complete understanding of the response of suspended particulate matter to the processes of advection, turbulent diffusion, and settling, the distribution of deep

10*

nepheloid layers will become one of the most meaningful indicators of deep circulation. This is also dependent, of course, on our ability to make exact measurements of the kind and quantity of materials which make up the layers.

One factor which has already been recognized as important in the distribution, and even the existence, of deep nepheloid layers is bottom topography. Hunkins et al. (1969) found nepheloid layers in the Arctic Ocean primarily over ridges and rises while they were usually absent over abyssal plains. This agrees with our observations in the North Pacific and Panama Basin. On the other hand, workers at Lamont-Doherty have stated that nepheloid layers were observed at all stations in the North American Basin where the depth was greater than 3000 m (Eittreim et al., 1969) and at all stations in the Central, Northeast and Northwest Pacific Basins (Ewing and Connary, 1970). However, in most cases some correlation was observed between bottom topography and thickness and intensity of the layer. Our data from the Panama Basin (Fig. 7) shows that the amount of suspended matter in the lower 100 m of the water column clearly decreases over shallow portions of the Carnegie Ridge and over areas of subdued topography in the Northwestern portion of the basin.

The relationship between abyssal circulation and distribution of the deep nepheloid layers has been discussed qualitatively by several writers (Hunkins et al., 1969; Ewing et al., 1970; Ewing and Connary, 1970). Quantitative or theoretical results which are applicable to nepheloid distribution are contained in Jerlov's work on maxima in the vertical distribution of particles (Jerlov, 1958). Jerlov has also explained the presence of nepheloid layers at intermediate depths and the presence of thin layers of clear water beneath bottom nepheloid layers as being evidence of lateral spreading of particulate matter from source areas by horizontal advection and as evidence of decreased turbulence near the bottom (Jerlov, 1969).

Ichiye (1966) has attempted to relate observed light scattering profiles to profiles of vertical eddy diffusivity and vertical velocity. It has often been postulated that the exponential increase in light scattering which is commonly observed near the bottom is related to erosion and/or vertical transport of particulate matter due to the turbulent boundary layer at the benthic interface. It is felt that much work remains to be done in the areas of erosion, deposition, diffusion and particle dynamics within the benthic boundary layer before a complete understanding is gained of the relationship between observed distributions of nepheloid material and the processes which control the production and distribution of this material.

ACKNOWLEDGEMENT

This research was supported by the Office of Naval Research contract ONR N00014–67–A–0369–0007.

REFERENCES

Beer, R. M., Dauphin, P. and Sholes, T. (1972). Submitted to *Mar. Geol.*

Cromwell, T. (1953). *J. Mar. Res.*, **12**, 196–213.

Eittreim, S., Ewing, M. and Thorndike, E. M. (1969). *Deep-Sea Res.*, **16**, 613–624.

Ewing, M. and Connary, S. D. (1970). *Geol. Soc. Amer. Mem.* 126 (James D. Hays, ed.), pp. 4–82, Lamont-Doherty Geol. Observ. Palisades, New York.

Ewing, M., Eittreim, S. L., Ewing, J. I. and Le Pichon, X. (1970). *In* "Physics and Chemistry of the Earth", Vol. 8, Pergamon Press, London.

Ewing, M. and Thorndike, E. M. (1965). *Science N.Y.*, **147**, 1291–1294.

Harlett, J. C. (1971). Ph.D. thesis, Oregon State University, Corvallis, pp. 120.

Hunkins, K., Thorndike, E. M. and Mathieu, G. (1969). *J. Geophys. Res.*, **74**, 6995–7008.

Ichiye, T. (1966). *Deep-Sea Res.*, **13**, 679–685.

Ivanoff, A. (1959). *J. Opt. Soc. Amer.*, **69**, 103–104.

Jacobs, M. B. and Ewing, M. (1969). *Science N.Y.*, **163**, 380–383.

Jerlov, N. G. (1953). *Rep. Swedish Deep-Sea Exped.*, **3**, 73–97.

Jerlov, N. G. (1958). *Arch. Oceanogr. Limnolog.*, **11**, 227–250.

Jerlov, N. G. (1959). *Deep-Sea Res.*, **5**, 178–184.

Jerlov, N. G. (1964). *In* "Physical Aspects of Light in the Sea", pp. 45–49, Hawaii Press, Honolulu, Hawaii.

Jerlov, N. G. (1961). *Medd. Oceanogr. Inst. Göteborg, Ser. B*, **8**, 40 pp.

Jerlov, N. G. (1969). Københavns Universitet, *Rep. Inst. Fys. Oceanogr.*, **7**, 6 pp.

Joseph, J. (1955). *Proc. UNESCO Symp. Phys. Oceanogr., Tokyo*, 59–75.

Joseph, J. (1959). *Ber. Deut. Wiss. Komm. Meeresforsch.*, **XV**, 3, 238–259.

Ketchum, G. H. and Shonting, D. H. (1958). *Woods Hole Oceanogr. Inst. Ref.* No. 58–15, 28 pp.

Knauss, J. A. (1960). *Deep-Sea Res.*, **6**, 265–285.

Kozlyaninov, M. V. and Ovchinnikov, I. M. (1961). *Tr. Inst. Okeanol., Akad. Nauk SSSR*, **45**, 102–112.

Kullenberg, G. (1968). *Deep-Sea Res.*, **15**, 423–432.

Montgomery, R. B. and Stroup, E. D. (1962). *Johns Hopkins Oceanogr. Stud.*, **1**, pp. 68.

Morrison, R. E. (1970). *J. Geoph. Res.*, **75**, 612–628.

Neudeck, R. H. (1971). Master's thesis, Oregon State University Corvallis (in preparation).

Neuymin, G. G. and Paramonov, A. N. (1965). *Bull. Izv, Acad. Sci. USSR, Atmos. Oceanic Phys.*, **1**, (11), 64–67.

Pak, H., Beardsley, G. F., Jr. and Park, P. K. (1970). *J. Geophys, Res.*, **75**, 4570–4578.

Pak, H., Beardsley, G. F., Jr. and Smith, R. L. (1970). *J. Geophys. Res.*, **75**, 629–636

Paramonov, A. N. (1964). *Tr. Inst. Okeanol., Akad. Nauk SSSR*, **45**, 102–112.

Pickard, G. L. and Giovando, L. F. (1960). *Limnol. Oceanogr.*, **5**, 162–170,

Plank, W. S., Pak, H. and Zaneveld, J. R. V. (1972). *J. Geophys. Res.*, **77**, 1689–1694.

Schubel, J. R. (1969). *Neth. J. Sea Res.*, **4**, 283–309.

Schubel, J. R. (1971). *Neth. J. Sea Res.*, **5**, 252–266.

Southard, J. B., Young, R. D. and Hollister, C. D. (1971). *J. Geophys. Res.*, **76**, 5903–5909.

Spilhaus, A. F. (1965). Ph.D. thesis, Massachusetts Institute of Technology, Cambridge, pp. 242.

Timofeeva, V. A. and Neuymin, G. G. (1968). *Izv. Acad. Sci. USSR*, **4**, (12), 1305–1328.

Van Andel, T. H., Heath, G. R., Heinrichs, D. F. and Ewing, J. I. (1971). *Geol. Soc. Amer. Bull.*, **82**, 1489–1580.

White, W. B. (1969). *Tex. A & M Univ.*, Ref. 69-4-T, pp. 74.

Wimbush, M. and Munk, W. (1968). *In* "The Sea" (A. E. Maxwell, ed.), Vol. 4, pp. 731–758. Interscience Publishers, New York and London.

Wooster, W. S. and Cromwell, T. (1958). *Bull. Scripps. Inst. Oceanogr.*, **7**, 169–182.

Zaneveld, J. R. V. (1971). Ph.D. thesis, Oregon State University Corvallis, pp. 87.

Chapter 11

The Fluorescence of Dissolved Organic Matter in the Sea

EGBERT KLAAS DUURSMA

International Laboratory of Marine Radioactivity, IAEA Musée Océanographique, Principality of Monaco

I. INTRODUCTION

The fluorescence of sea water, when irradiated with ultraviolet light, was detected by Kalle (1949) after his earlier studies on the absorption of light (Kalle, 1937). For both studies he used the Pulfrich-Photometer of Zeiss (Fig. 1), which depends on visual detection of the light for its operation. It is mainly due to this investigator that knowledge on the origin, composition and distribution of fluorescent substances in the sea has been developed. The progress made in later studies has been due to the development of fluorimeters with instrumental detection of light. The optimal results can be obtained with instruments capable of investigating the absorption and emission spectra of fluorescent material in sea water.

II. METHODOLOGY

The method used by Kalle is given in Fig. 2A (Kalle, 1951). It is a requisite that the operator works in the dark or in very dim light, and

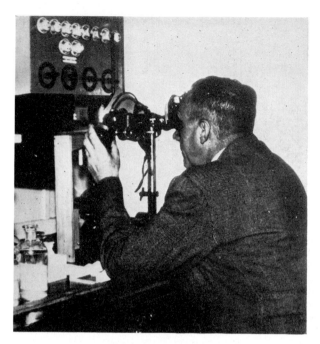

Fig. 1. Prof. Kurt Kalle behind the Pulfrich-Photometer on board the research vessel Gausz. (Photograph Prof. G. Dietrich.)

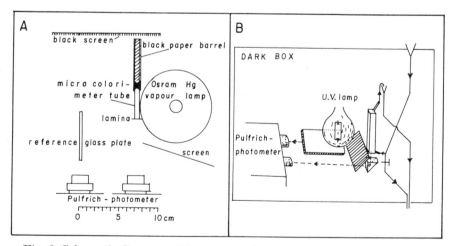

Fig. 2. Schematic diagram of fluorescence determination apparatus as used by (A) Kalle (1951) and (B) Duursma (1961). (A) and (B) are reproduced with permission of the Deutsche hydrographische Zeitschrift and Netherlands Journal of Sea Research, respectively.

acquaints himself to the conditions of dim light or dark for some time before he starts the measurements. In order to simplify the handling of samples, Duursma (1961) converted the system in such a way that the operator could work in a dark box, while the addition of samples was executed from the exterior of the dark box (Fig. 2B). The precision of the measurements has been improved by instrumental reading with help of spectrophotometers with fluorescence accessories and with fluorimeters specially constructed for fluorescence measurements.

Common to all these methods of detection of fluorescent substances is the fact that fluorescence emission is theoretically not proportional to the concentration of the fluorescent substances. This fact is inherent in any irradiation/emission system, because both types of radiation are absorbed in the medium. Although at very low concentrations the fluorescence seems to increase proportionally with increasing concentration, this increase becomes less and less at higher concentrations and might even invert to a decrease at very high concentration (Fig. 3B). In that case it is necessary to correct the measured fluorescence by a mathematical calculation which takes into account the absorption of both the exciting and the emitted light.

For an ultraviolet/fluorescence light system with a rectangular cell where the angle between the ultraviolet beam and the direction in which the fluorescence is measured is 90° (Fig. 3A), the fluorescence can be calculated with the formulae given by Duursma and Rommets (1961) (Fig. 4). For the most general case the corrected fluorescence, called absolute fluorescence ($F_{abs.}$), is given by the relation (1) which corresponds to the light system of Fig. 3A (1). This relation can be simplified when the irradiated portion of the cell extends to the bottom (Fig. 3A (2)), and when the reference solution is kept at such a low concentration that the absorption of light in this solution is negligible (μ_r and γ_r almost zero). Then the absolute fluorescence can be found by relation (2) of Fig. 4.

For the measurement of $F_{abs.}$ it is necessary to use ultraviolet fluorescence light beams of a selected part of the spectrum. The measurements of the light absorption coefficients μ and γ have to be made at exactly these same wavelengths in a spectrophotometer. $F_{abs.}$ can be given in the arbitrary unit defined by Kalle (1956, 1957). A solution of 1 mg quinine-bisulphate in 1 litre of 0·01 N H_2SO_4 has a fluorescence strength of 700 mFl. The mFl unit has the dimension (L^{-3}).

To avoid the problems connected with absolute measurements of fluorescence, it is always possible to work with diluted samples and use a calibration curve made up from the same material in solution. This curve can be standardized in absolute fluorescence units (mFl) by

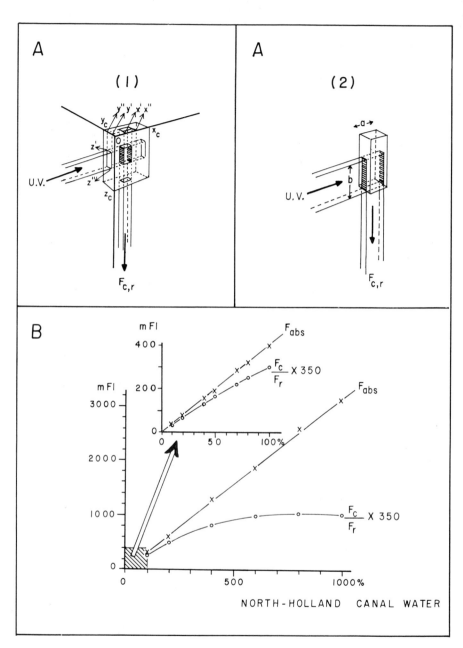

Fig. 3. A(1): Representation of ultraviolet irradiation and fluorescent emission in a rectangular cell. A(2): Irradiation and emission system of the Zeiss Fluorescence/spectrophotometer. B: Measured and corrected fluorescence of a series of dilutions of preconcentrated (1000%) North-Holland Canal water. Reproduced with permission of the Netherlands Journal of Sea Research. Key to symbols: A(1): x, y and z are the coordinates; $F_{c,r}$ is the fluorescence reading in relative units, where F_c is the reading for the solution and F_r that of the reference sample. A(2): a = path length of solution irradiated, b = depth of irradiation. B see caption to Fig. 4.

$$(1) \qquad F_{abs} = \frac{F_c}{F_r} 700 \left[q u.b.s. \right] \frac{\mu \, \gamma \left(e^{-\mu_r (x''-x')} - 1 \right) \left(e^{-\gamma_r (z''-z')} - 1 \right)}{\mu_r \gamma_r \left(e^{-\mu (x''-x')} - 1 \right) \left(e^{-\gamma (z''-z')} - 1 \right)} \, e^{x'(\mu-\mu_r)} \, e^{(z_c - z'')(\gamma - \gamma_r)}$$

$$(2) \qquad F_{abs} = \frac{F_c}{F_r} 700 \frac{a \, b \, \mu \, \gamma}{(e^{-a\mu} - 1)(e^{-b\gamma} - 1)}$$

Fig. 4. (1): Equation employed for the calculation of fluorescence in case A (1) of Fig. 3. (2): Equation employed for case A (2) of Fig. 3, with a [qu.b.s.] of 0·5 mg l⁻¹. Key to symbols: (see also Fig. 3). $F_{abs.}$ = the absolute fluorescence expressed in mFl units; F_c/F_r = the relative fluorescence (see Fig. 3); μ = the ultraviolet absorption coefficient of the sample; μ_r = the ultraviolet absorption coefficient of the reference sample; γ = the absorption coefficient of the sample at the wavelength of the fluorescence; γ_r = the absorption coefficient of the reference sample at the wavelength of the fluorescence; x', x'', z', z'', z_c, a and b = coordinates given in Fig. 3; [qu.b.s.] = the concentration of the quinine-bisulphate reference solution in mg l⁻¹ 0·01 N H_2SO_4.

measuring, for the most concentrated sample, the absolute fluorescence, the absorption of the ultraviolet light and the fluorescence light and applying the formula (2) given above. For oceanic conditions the correction is small, mostly less than 5%.

Self-made and commercial fluorometers have been applied in the last ten years. From these examples are: *in situ* fluorometer (Kullenberg and Nygård, 1971, Fig. 5, and Karabashev, 1970), the Turner fluorometer (Turner, 1968), the Baird-Atomic fluorometer (Traganza, 1969) and the Zeiss spectrophotometer with fluorometer accessory (Duursma and Rommets, 1961). The main difficulty connected with *in situ* fluorometers is that the fluorescence radiation in near surface waters at daytime is subject to interference from daylight. This can be overcome by having a completely shielded cell, but this is only possible when an *in situ* pump is used. Another method is to reduce the ambient light received by the detector, by letting the receiver face downwards and mounting an effective light trap in front of the receiver. However, the ambient light still reaching the detector in surface waters is two or three orders of magnitude greater than the fluorescence radiation. Kullenberg and Nygård (1971) have partly eliminated this interference by electronically discriminating the fluorescence signal from the ambient light signal. This can be done by the method where the ultraviolet light source is modulated, so that the equally modulated fluorescence light signal can be discriminated without interference from the essentially direct current signal from the ambient light.

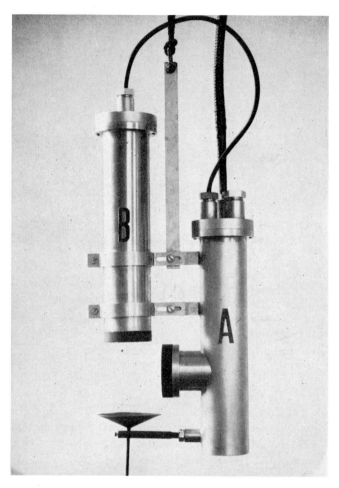

Fig. 5. *In situ* unit for measurement of the fluorescence of sea water. (Reproduced from Kullenberg and Nygård (1971) with permission of the authors.)

The most developed laboratory equipment is used by Traganza (1969) and is capable of investigating both the excitation spectra and the emission spectra of solutions. In the operation of this instrument, sea water is irradiated at selected wavelengths between 200 nm and 700 nm by manual adjustment of the excitation monochromator. The remainder of the wave band is scanned each time for emission with the emission monochromator. Once an emission peak is detected the emission monochromator is set at the peak wavelength and the search with the excitation monochromator is continued until the maximum excitation wavelength is determined. At this excitation wavelength the spectrum of the fluorescence light can be scanned automatically. An almost identical apparatus is used by Ivanoff and Morel (1971) but with a photomultiplier of high resolution. This photomultiplier has a very low dark current (type Lallemand) and is calibrated using a standard lamp of known colour temperature (Morel, unpublished).

Another technique, that has its advantages for remote fluorescence measurement of oil spills, fish oils and other fluorescent pollutants or natural substances at the surface is the "airborne fluorometer" of Perkin–Elmer (Hemphill *et al.*, 1969; Stoertz *et al.*, 1969). Here the solar-stimulated fluorescence is measured by equipment mounted on a helicopter against the background of reflected sunlight. The principle is

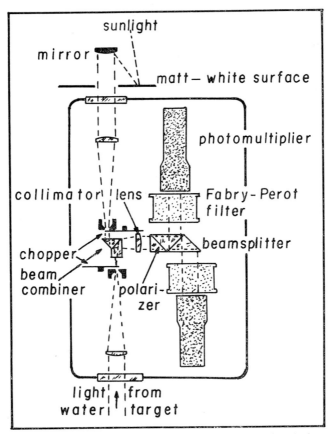

Fig. 6. Diagrammatic cross-section of the Fraunhofer line discriminator (Stoertz *et al.*, 1969). Reproduced from the Marine Technology Society Journal, Vol. 3, (1969) with permission of the editor.

based on the comparison of the Fraunhofer lines in sunlight with those in the light coming from the water target (Fig. 6). The presence of for example yellow fluorescent substances in the water has the result that the dark Fraunhofer lines in the yellow part of the spectrum will appear less pronounced than in the sunlight. By selecting the darkest Fraunhofer lines the equipment utilizes the above principle by viewing a single Fraunhofer line and comparing the relative darkness or depth of the line profile with the same line in the sun's spectrum.

For heavily polluted waters another type of fluorometer has been developed which is applied for the quantitative determination of oil in oil-spills from tankers (load on top) (Coomber, 1971). A critical point when measuring heavily polluted waters and especially oil-water emulsions, is the collection of material on the surface of the cells causing fouling of the optical path. This problem has been avoided by employing a system where the solution flows through a light-tight sampling chamber in the form of a free jet which passes through the ultraviolet beam and into the field of view of the emission detector without coming into contact with the cell windows. Condensation of water vapour is prevented by an air-purging curtain along the windows.

III. Specific Compounds

The analysis of specific organic compounds in sea water having fluorescent properties has been performed in three ways; firstly by direct analysis of separated compounds (Momzikoff, 1969a), secondly by analysis of the fluorescence spectrum (Traganza, 1969), and thirdly by synthetic experiments leading to the production of fluorescent substances (Kalle, 1963). Momzikoff extracted, with active carbon, 10·15 g of organic substances from 20 000 litres of prefiltered sea water (0·45 μm) from the Mediterranean, 2 miles off Monaco, at 5 m below the surface. This 10.15 g of viscous brown material contained fluorescent substances which were hydro-soluble. With chromatographic organic analysis three compounds could be isolated that showed fluorescent properties, namely (i) Isoxanthopterin, (ii) Riboflavin (Vitamin B2) and (iii) Lumichrome. The quantities were, however, very feeble, 200, 32 and 7 μg, respectively, in comparison to the quantity of original material of 10·15 g. This results in a concentration in sea water of 10^{-2} μg l^{-1} for isoxanthopterin, $1·6 \times 10^{-3}$ μg l^{-1} for riboflavin, and $0·35 \times 10^{-3}$ μg l^{-1} for lumichrome.

A comparison with the quantities of the same products in plankton can be found from other data of Momzikoff (1969b). He analyzed zoo-plankton caught with a 156 μm net from the same area where the sea

TABLE 1. Fluorescent substances detected in plankton and sea water of the Mediterranean off Monaco (Momzikoff, 1969).

	Sea water	Zooplankton (May/Sept.)		Ratio of substances in sea water as compared to plankton = (A)/(B)
	μg l^{-1}	Total plankton (wet)	2.5×10^{-5} g l^{-1} (May)* 3.0×10^{-5} g l^{-1} (Sept.)	
		μg g^{-1}†	μg g^{-1}‡	
Total dissolved organics extracted	500			
Isoxanthopterin:	10^{-2}	(May) 0.02 (Sept.) 0.7	4.6×10^{-7} 2.1×10^{-5}	22 000 480
Riboflavin:	1.6×10^{-3}	(May) 1.0 (Sept.) 0.8	2.3×10^{-5} 2.4×10^{-5}	70 67
Lumichrome:	0.35×10^{-3}	(May) 0.02 (Sept.) 0.02	4.6×10^{-7} 6.0×10^{-7}	760 580
	(A)		(B)	

* The zooplankton concentration in the same area are from Kane (1967).
† μg per gram of wet plankton.
‡ μg of substances in plankton in 1 litre sea water.

water samples were taken. Using the data on the concentration of plankton caught in an identical net in the same area of the Mediterranean (Kane, 1967), the quantities of the three fluorescent substances in the plankton present in one litre of sea water can be calculated. As can be seen from Table 1, these quantities are roughly 70 to 22 000 times lower than the quantities of the substances dissolved in sea water. This fact indicates that either the three mentioned compounds have a long residence time in sea water, or that other smaller plankton species are producing the substances in larger quantities than the investigated zooplankton.

FLUORESCENCE

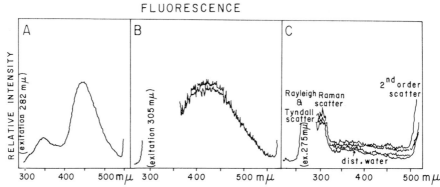

Fig. 7. A: Fluorescence spectrum of sea water collected in a surface concentration of *Trichodesmium sp.*, in the Sargasso Sea at 35° 26′ N, 67° 19′ W. B: Fluorescence spectrum of Atlantic Shelf water, collected at 30 m near Nantucket Shoals (40° 05′ N, 69° 37′ W). C: Fluorescence spectra of Sargasso Sea water of the same station as A (35° 26′ N, 67° 19′ W), collected at depths of 1 to 206 m. A, B and C are reproduced from Traganza (1969) with permission of the editor of the Bulletin of Marine Science.

Phytoplankton cultures and cultures of concentrated suspended matter incubated in sea water do indeed produce some fluorescent substances with a detectable characteristic spectrum (Traganza, 1969). Such a spectrum was, as an exceptional case, also found in sea water of the Sargasso Sea in a rich plankton bloom of *Trichodesmium sp.* in the surface water (Fig. 7A). This spectrum is different from that of the sea water of the Atlantic Shelf of the U.S.A., which contains only one broader peak in the same region of the spectrum (Fig. 7B). In the Sargasso Sea, below the plankton bloom in the surface water, the emission of fluorescence could hardly be detected in samples collected at depths of 1 m to about 200 m (Fig. 7C), compared with that from distilled water.

This study indicates that at high levels of plankton concentration (e.g. cultures or dense blooms in the sea) fluorescent substances can be detected in the water, but that a large production of measurable amounts in water with medium plankton concentration and in sub-surface waters is not expected. This fact is in agreement with the above results of Momzikoff for the medium or small plankton concentrations in the Mediterranean. Since also the spectrum of the cultures was

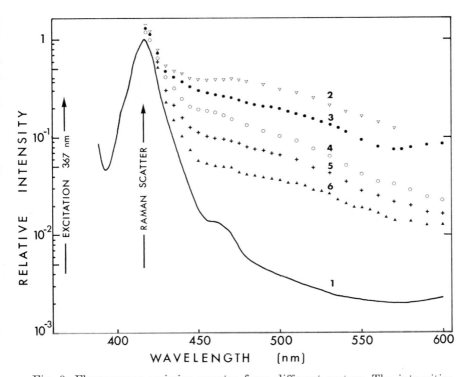

Fig. 8. Fluorescence emission spectra from different waters. The intensities are corrected by the spectral sensitivity factor of the equipment. They are divided by the intensity of the Raman scattering of pure water at 418 nm (excitation at 367 nm), and given on a logarithmic scale. (Morel, unpublished). 1 = Bidistilled water, passed through a column of active carbon. The small peak at 465 nm is Raman scattered light produced by parasitic excitation light of 405 nm; 2 = Sea water from the upwelling area of Mauritania; 18° 44′ N, 16° 35′ W; from 20 m below surface, 26th Apr. 1971; 3 = Water from the Var estuary, close to Nice, France; surface, 9th Sept. 1969; 4 = Sea water from the Villefranch-sur-mer bay; surface, 10th Sept. 1969; 5 = Sea water from the Mediterranean at 1000 m, 43° 29′ N, 7° 34′ E, 17th Sept. 1969; 6 = Sea water from the Mediter-ranean at 10 m, 43° 29′ N, 7° 34′ E, 17th Sept. 1969. Data reproduced with the permission of Dr. Morel.

different from that of the shelf-water, fluorescent substances of main-
land origin must be responsible for the amount found in the shelf-water.
Fluorescence spectra of sea water are mostly simple of shape, and the
differences for estuaries, upwelling areas and the open sea are in the
intensity and not the shape (Fig. 8).

Apart from such an input of fluorescent substances from the main-
land, Kalle (1963) has proved by carefully executed experiments that
the formation of fluorescent substances is possible by self-condensation
of dissociated constituents of carbohydrates. Especially compounds
like methylglyoxal, which are formed everywhere in nature, give in
the presence of nitrogen compounds (aminoacids) a kind of Maillard
reaction, a reaction well known in food products (formation of brown
coloured products) (Enders and Theis, 1938). This results in the formation of products with a melanin structure (Fig. 9). In order to give an
impression of the fluorescence intensity of different substances, Table 2
summarizes the findings of Kalle from determinations on natural
products and from the substances formed in his experiments.

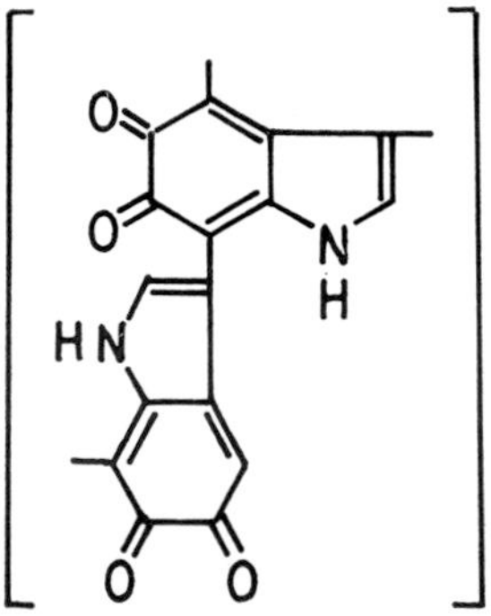

Melanin

Fig. 9. Basic structure of melanin (Duursma, 1965).

IV. DISTRIBUTION

The information on the distribution of fluorescent substances in the
open sea is limited to the data obtained by Kalle (1957), Ivanoff (1962),
and Traganza (1969). For coastal areas fluorescence determinations
have been made by Kalle (1949, 1956), Duursma (1961), Højerslev
(1971), Kullenberg and Nygård (1971), and Rommets and Postma
(1972).

TABLE 2. Fluorescence intensities of substances (Kalle, 1963).

a. *Solutions*	mFl
Atmospheric precipitation	1–85
Natural source water	0·2–100
Bordeaux wine	860
Light Italian vermouth	1250
Light export beer	2160
Dark malt beer	11 800

b. *Extracts from*	mFl ml mg^{-1}
Atmospheric air	0·0004–0·0027
Rocks	0·08–1·01
Bees' honey	5·9
House dust	30
Filter paper	51
Air 85 days in contact with dihydroxy acetone	56
Brown Lime-tree leaves, airdried	460

c. *Model experiments*   mFl ml mg^{-1}

(i) *Disc experiments*
(4 weeks contact; 25 mg of each substance, except
50 mg for potash; 3 drops of H_2O)

	mFl ml mg^{-1}
Ribose/glycine	10 000
Dihydroxy acetone/glycine	3000
Ribose/potash	200
Dihydroxy acetone/potash	200
Ribose/glycine/potash	10 000
Dihydroxy acetone/glycine/potash	6000

(ii) *Experiments with solutions*
(1 g carbohydrate l^{-1})

Dihydroxy acetone/potash	boiled for 10 min	700
Methylglyoxal/potash		1000
Dihydroxy acetone/glycine/ potash		2000
Methylglyoxal/glycine/potash		5000
Dihydroxy acetone/potash	4 days at 20°C	30
Methylglyoxal/potash		200
Methylglyoxal/potash (11 × concentrated)	55 days at 20°C	2000
Dihydroxy acetone/glycine/potash	4 days at 20°C	1000
Methylglyoxal/glycine/potash		3000

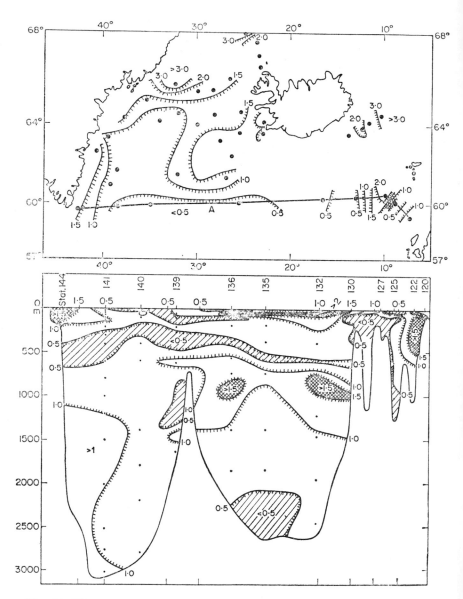

Fig. 10. Fluorescence (mFl) of sea water of the Irminger Sea between Green-land and Scotland. The upper figure gives the surface fluorescence values, the lower figure the fluorescence of deep waters of the section A. The figure is con-structed after Kalle (1957) by Duursma (1965). Reproduced with permission from Berichte der Deutschen Wissenschaftlichen Kommission für Meeresforschung.

For the open sea the fluorescence is very weak and this makes it very hard to determine accurately fluctuations in relation to other properties of the watermass. As Fig. 10 shows, the total variation of the fluorescence in the Irminger Sea is small, between about 0·5 and 3 mFl for surface waters and between 0·5 and 1·5 mFl for deeper waters. A different vertical distribution was found by Ivanoff (1962) in the Mediterranean (Fig. 11), by Karabashev (1970) in the Black Sea and by Ivanoff and Morel (unpublished) in the Atlantic. Here the fluorescence was low in the surface layer and increased with depth until a constant value. The constant value was observed in the Mediterranean from 75 m depth onwards, and in the Atlantic from 1000 m onwards.

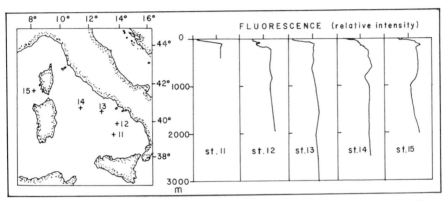

Fig. 11. Results of fluorescence measurements at five stations in the Mediterranean (Ivanoff, 1962). Reproduced with permission of the Académie des Sciences, Paris.

It is difficult to find a good explanation for the variations in the open sea fluorescence. In the Irminger Sea a relationship could not be established between the fluorescence and other parameters such as the phosphate and oxygen concentrations and measurements of the optical turbidity.

This is quite different for coastal waters, where the fluorescence is mostly related to the inflow of fresh waters with high fluorescence. In mixing areas of fresh and saline waters, and also further at sea, for example in the Gulf of Paria (Postma, 1954), the Baltic and the Kattegat (Kalle, 1949; Højerslev, 1971), in the Dutch Wadden Sea (Duursma, 1961; Rommets and Postma, 1972), in the North Sea (Kalle, 1956; Otto, 1967) and in the outflow area of the Amazon River (Yentsch, 1971), the fluorescence has a reverse proportionality to salinity (see as

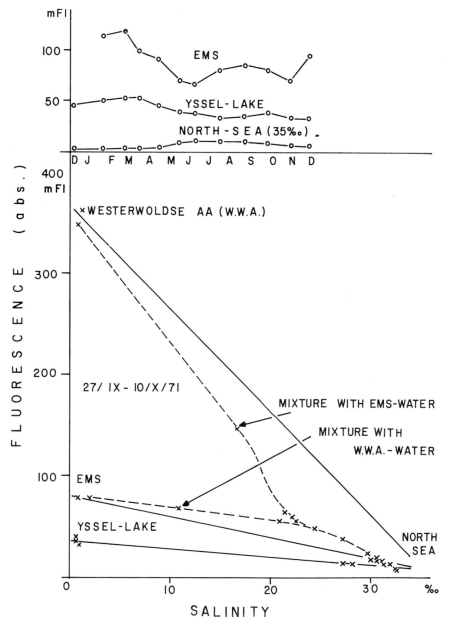

Fig. 12. Upper: Annual variation of fluorescence measured in the Ems, Yssel-lake and North Sea. Lower: Relationship of the fluorescence with the salinity of sea water from different parts of the Dutch Wadden Sea. Data reproduced from Rommets and Postma (1972) with permission of the authors.

an example Fig. 12). This indicates that the mainland-derived fluorescent substances have, in spite of some seasonal variations in the concentrations, conservative properties and are not rapidly decomposed in the sea. It is therefore possible to employ this conservative property of fluorescent substances originating from river inflow as an indicator.

When more than one type of freshwater is entering the same area, and when these different freshwaters have different fluorescent strengths, it is possible to trace their presence in sea water by the salinity/fluorescence relationship (Fig. 12). When the wavelength of fluorescence maximum is selected in the emitted spectrum, it is even easier to trace the fresh water in the sea water. Fluorescence monitoring has in this sense therefore an application to pollution studies.

V. BUDGET

Although not all investigators have expressed the fluorescence of sea water in representative numbers, it is possible to make a tentative budget of fluorescent substances in the sea. The different input pathways into the sea are rivers, rainfall, release from living and dead plankton and other organisms, and *in situ* formation of compounds of melanin structure from dissolved organic substances.

A. RIVERS

The fluorescence values, as determined by Postma and Kalle (1955), Kalle (1963) and Duursma (1961), for fresh waters, are of the magnitude of 30 to 150 mFl. Taking an arbitrary value of 50 mFl for the world-rivers, which discharge 3×10^{16} litres of water annually to the sea (calculated from Turekian (1968)), the total amount of fluorescence arriving in the oceans is $1·5 \times 10^{18}$ mFl litre. This amount is equal to the amount present in a 4·2 m thick surface layer of the oceans. For this calculation the mean ocean surface concentration is taken as 1 mFl (taken as average from Kalle's data) and the total ocean surface is taken as $360·8 \times 10^6$ km² (Bowden, 1965).

B. RAINFALL

Here again several assumptions have to be made. Gentle rainfall might have a higher fluorescence in the waterdrops than heavy rainfall, because the rain is essentially scavenging the fluorescent substances from the air. This might be concluded from Kalle's (1963) data on air bubbling through water and acetone and data on drops of condensation, fog, new and old snow and rain. In that case, a value of 12·5 mFl in the rainfall above the North Atlantic (Kalle, 1963) might be taken, and the input of fluorescent substances in this area with a rainfall of about 180 cm per annum will give a quantity per unit surface of 22·5 mFl m, which is equal to that found in a 22·5 m deep water layer (with 1 mFl fluorescence).

C. RELEASE FROM LIVING AND DEAD ORGANISMS

Several properties related to dissolved organic matter and plankton were followed over a period of one year at a station in the North Sea (Duursma, 1965). The fluorescence showed two peaks, one in spring and one in autumn in coincidence with the plankton blooms. However, these peaks disappeared within a few weeks. It was also found that the peaks did not synchronize with the peaks in dissolved organic carbon and nitrogen. Therefore the addition by plankton to the amount of fluorescent substances in the whole water column of the sea is probably negligible. This result is in agreement with the data given by Momzikoff (1969a,b) which show that the quantity of such compounds in plankton is only a small fraction of the total quantity dissolved in the water (Table 1).

D. *in situ* FORMATION

The experiments on the formation of fluorescent substances from different combinations of carbohydrates and aminoacids (Kalle, 1963), a summary of which is given in Table 2, resulted in fluorescent strength values of 30 to 10 000 mFl ml mg^{-1}. In other words the range of fluorescences obtainable from 1 mg ml^{-1} of material was 30 to 10 000 mFl. Clearly the quantity of substances in the sea that equally might react and produce fluorescent substances will be below the average amount of total organic matter, which is of the order of some mg per litre. The amounts of individual carbohydrates and aminoacids detected are generally below 10 μg l^{-1} (Duursma, 1965). All together a limit of 1 mg l^{-1} might be a reasonable value to be used in a budget calculation.

Taking the range mentioned above, 1 mg l^{-1} of material can produce 0·03–10 mFl. The lower value is below the fluorescence values found in the open sea, the highest value is about 3 to 10 times higher. Conclusively, it is possible that *in situ* formation of fluorescent substances can occur. Time does not play an important role, because a slow production rate is compensated by a slow decomposition rate.

E. SUMMARY

Summarizing the different possibilities, it seems obvious that there are three main sources of fluorescent substances. In coastal areas the dominating source will be run-off from land, while in the open ocean rainfall containing fluorescent substances from the atmosphere and *in situ* formation of fluorescent substances will add new material to the material already present. The addition from plankton is mostly immeasurable in the quantities produced, but their existence is shown by the detection of certain special fluorescent substances, that have identical structures in both plankton and sea water.

These budget calculations are, however, only tentative and much more information on the different facts of fluorescent substances is required to obtain a clear view on the production, residence time and distribution.

ACKNOWLEDGEMENT

My thanks are due to Kjell Nygård, Copenhagen, for the excellent presentation of the paper at the symposium, and for defending it during the discussion.

REFERENCES

Bowden, K. F. (1965). *In* "Chemical Oceanography" Vol. 1 (J. P. Riley and G. Skirrow, eds.), pp. 43–72. Academic Press, London.

Coomber, R. S. (1971). *Mar. Eng. Rev.*, 29–31.

Duursma, E. K. (1961). *Neth. J. Sea Res.*, 1, 1–148.

Duursma, E. K. (1965). *In* "Chemical Oceanography" Vol. 1 (J. P. Riley and G. Skirrow, eds.), pp. 433–475. Academic Press, London.

Duursma, E. K. and Rommets, J. W. (1961). *Neth. J. Sea Res.*, 1, 391–405.

Enders, C. and Theis, K. (1938). *Brennst.-Chem.*, 19, 360–365.

Hemphill, W. R., Stoertz, G. E. and Markle, D. A. (1969). *In* Proc. 6th Int. Symp. "Remote Sensing of Environment", pp. 565–585. Willow Run Laboratories, Ann Arbor, Michigan.

Højerslev, N. K. (1971). Københavns Universitet, *Rep. Inst. Fys. Oceanogr.*, 16, 12 pp.

Ivanoff, A. (1962). *C. R. Acad. Sci., Paris*, 254, 4190–4192.

Ivanoff, A. and Morel, A. (1971). Proc. Joint Oceanogr. Assembly (Tokyo, 1970), 178–179.

Kalle, K. (1937). *Ann. Hydrogr. Berlin*, 65, 276–282.

Kalle, K. (1949). *Deut. Hydrogr. Z.*, 2, 117–124.

Kalle, K. (1951). *Deut. Hydrogr. Z.*, 4, 92–96.

Kalle, K. (1956). *Deut. Hydrogr. Z.*, 9, 55–65.

Kalle, K. (1957). *Ber. Deut. Wiss. Komm. Meeresforsch.*, 14, 313–328.

Kalle, K. (1963). *Deut. Hydrogr. Z.*, 16, 153–166.

Kane, J. E. (1967). *Limnol. Oceanogr.*, 12, 287–294.

Karabashev, G. S. (1970). *Oceanology, Acad. Sci. USSR*, 10, 703–707. (Eng. Transl.).

Kullenberg, G. and Nygård, K. (1971). Københavns Universitet, *Rep. Inst. Fys. Oceanogr.*, 15, 10 pp.

Momzikoff, A. (1969a). *Cah. Biol. Mar.*, 10, 221–230.

Momzikoff, A. (1969b). *Cah. Biol. Mar.*, **10**, 429–437.

Otto, L. (1967). *Neth. J. Sea Res.*, **3**, 532–551.

Postma, H. (1954). *Verh. Kon. Ned. Akad. Wet.*, *Afd. Natuurkunde*, **20**, 28–64.

Postma, H. and Kalle, K. (1955). *Deut. Hydrogr. Z.*, **8**, 137–144.

Rommets, J. W. and Postma, H. (1972). Int. Rep. NIOZ, Texel, 1972–6, 9 pp.

Stoertz, G. E., Hemphill, W. R. and Markle, D. A. (1969). *Mar. Technol. Soc. J.*, **3**, 11–26.

Traganza, E. D. (1969). *Bull. Mar. Sci.*, *Univ. Miami*, **19**, 897–904.

Turekian, K. K. (1968). "Oceans", pp. 120. Prentice-Hall, Englewood Cliffs, New Jersey.

Turner, G. K. (1968). Fluoremetry in studies of pollution and movement of fluids. Fluoremetry Reviews. G. K. Turner Ass. Calif.

Yentsch, C. S. (1971). *In* Proc. Symp. "Remote Sensing in Marine Biology and Fishery Resources", pp. 75–97. Texas A & M Univ., College Station.

Chapter 12

The Use of Fluorescent Dyes
for Turbulence Studies in the Sea

H. WEIDEMANN

Deutsches Hydrographisches Institut, 2000 Hamburg 4, West Germany

I. INTRODUCTION

A. THE IMPORTANCE OF TURBULENCE IN OCEANOGRAPHY

Almost all the processes of movement in the atmosphere as well as in the ocean can be described, in the hydraulic sense, as being turbulent. This includes the established fact that an average current vector is superimposed by a complicated spectrum of statistical random motions. Hence, when observing the current passing by a certain point (Euler) fluctuations in direction as well as in velocity will be recorded; with Lagrange's method of consideration, a particle agitates accordingly in an irregular non-rectilinear track. The knowledge of the turbulence spectrum plays a major role in many oceanographic problems, for example in the important questions of the exchange and balance of

11

energy between ocean and atmosphere, especially however, for the understanding of all types of mixing and diffusion processes. These include both the eddy processes in boundary layers between oceanic watermasses of various origins as well as those which, within a watermass, give rise to dispersion and dilution of pollution materials which have been introduced into the medium.

B. TURBULENCE INVESTIGATIONS

(1) *Instrumental methods*

The quantitative investigation of turbulent motions, for physical reasons, is not easy. As the spectrum extends from smallest dimensions and periods up to the circulation systems of whole oceans, one is not in a position to record all these processes using one—and only one—method. In microstructures, in particular those of near molecular magnitude, the usual current measurement methods with mechanical sensors of high inertia are not applicable. Methods have been developed by which the cooling of heated wire (or similar sensors) is used as a measure for the intensity of turbulence; however, such methods are hardly suited for use in the sea.

For practical purposes, the action of the individual turbulent motions integrated over space and time are of primary interest. In order to investigate their action, it is possible to label certain small volumes of water, so that one can follow their total displacement as well as their relative movement. For this purpose, one may use small buoys which either float on the surface, or in a subsurface layer, determined by carefully adjusting their density, and simple devices such as drift-cards etc. The results of such investigations, however, do not provide enough information about the intensity of the turbulence in the smallest, and for the actual diffusion, especially important scales. Other methods are more suitable for this, which either use very small, fine dispersed particles (neutrally-buoyant suspended matter) or dissolved substances as tracers.

(2) *Tracer methods*

One must distinguish between natural and artificially introduced tracers. Substances naturally present in the sea belong to the former group, such as salts and trace elements as well as dissolved organic matter, dead particulate matter and biological objects without noticeable individual ability to move, such as fish eggs and fish larvae, bacteria, plankton etc. which behave similarly to inorganic and organic suspended matter. However, one can in this case also include among the natural tracers materials which reach the sea due to human activity, such as the pollutants transported by rivers or the radioactivity coming from nuclear weapon testing or nuclear energy stations, so long as they reach the sea without the explicit aim of a diffusion investigation.

However, to be able to undertake such investigations at a particular time and place, one must introduce tracers artificially. Certain substances are particularly suitable in so far as they can satisfy a series of conditions The most important are:

(1) Simple and accurate recording methods.
(2) Low costs.
(3) Harmless to human and marine organisms.

It could be preferable to use radioactive tracers, because they possess well defined properties and are easily detected even in extreme dilution, but the stipulation of harmlessness can in general only be satisfied by carrying out investigations on a small scale.

Therefore, as tracer one uses mainly watersoluble dyes. In the case of investigations with spatially and temporarily narrow limits, one is often satisfied to use the purely visual or photographic record which, in general however, permits only qualitative or limited quantitative results, e.g. Witting (1933) used rosaniline (Pulverfuchsin AB) for early experimental investigations of mixing in the Baltic Sea, by which he endeavoured to obtain quantitative results by photometrical comparison of the coloured water samples with standard control solutions.

However, for more accurate investigations, especially over a long period, quantitative recording is essential. Fluorescent dyes are the most suitable for this purpose, because they permit accurate measurements of the dilution to be recorded using physical methods with suitable equipment. The following sections are concerned with the properties of these dyes, the equipment which has been developed for their identification and the method of their application.

II. FLUORESCENT DYE TRACERS

A. LIST OF DYES USED

Apart from the three basic requirements for an ideal dye, as previously described, one must take further criteria into consideration, which have been compiled by various authors (e.g. Pritchard and Carpenter, 1960). These are:

(1) Good solubility, also no precipitation caused by sea salts.
(2) Good stability to the influence of daylight.
(3) Good optical colour contrast when compared with that of sea water.

(4) Fluorescence in a spectral range with low background noise in natural sea water.

After intensive investigations (quite independent of one another) both the Japanese (Japanese Government Agencies, 1958) and Pritchard and Carpenter (1960) arrived at the conclusion that the red organic dye, Rhodamine B, was the most suitable. Later, other authors advocated, apart from this colouring agent, other related dyes with the designation Rhodamine WT, Rhodamine 5G, Rhodamine BMG, Rhodamine C, as well as Pontacyl Brilliant Pink B.

In spite of several disadvantages, the yellow-green Fluorescein (Sodium fluoride, Uranine) has been used frequently as a colouring agent.

B. COMPARISON OF PROPERTIES

Very few authors give detailed specifications about the advantages and disadvantages of the various colouring agents based upon their own measurements and experience. The most detailed special investigation of this type has been made by Feuerstein and Selleck (1963); a few other specifications have also been given by Pritchard (1965).

The first mentioned work systematically investigated the properties of three of the dyes previously quoted (Rhodamine B, Pontacyl Brilliant Pink B, Fluorescein) to changes of temperature, salinity, pH value, background level, content of suspended matter as well as their photochemical decay rate.

The results obtained have been shown in various diagrams and tables, the most important of which are reproduced in this paper.

(1) *Intensity of fluorescence*

In Fig. 1 the relationship between the concentration of the dye and the intensity of fluorescence at constant temperature is shown for all three tracers. One recognizes that Rhodamine B and Fluorescein give about the same values, whereas the fluorescence of Pontacyl B.P.B. reaches only about 40% of these values. Accordingly, the limit of detection of the latter tracer in natural water is 2·5 times higher (15×10^{-11} instead of 6×10^{-11}).

(2) *Temperature*

Very detailed measurements of the influence of temperature lead to the result that the general decrease in fluorescence with increase of temperature can be described by

$$F_t = F_o \times \exp{(nt)}$$

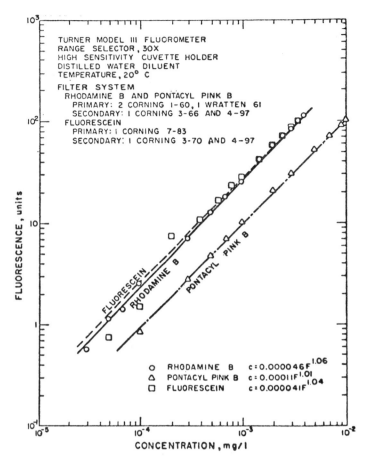

Fig. 1. Relationship between tracer concentration and fluorescence intensity (after Feuerstein and Selleck, 1963).

where F is the fluorescence in any unit (F_o by 0°C), t is the temperature in °C and n an individual constant for each tracer. Its value for Rhodamine B is: $n = -0.027/°C$; for Pontacyl B.P.B: $n = -0.029/°C$; however, for Fluorescein is only $n = -0.0036/°C$. A diagram derived from this (Fig. 2) permits the direct reading of the temperature dependent correction factors.

(3) Background fluorescence

The investigation of the background fluorescence present in natural water samples from coastal waters have, as expected, given locally very differing data, except that those of Rhodamine B and Pontacyl B.P.B. are about 10^2–10^3 lower than that of Fluorescein.

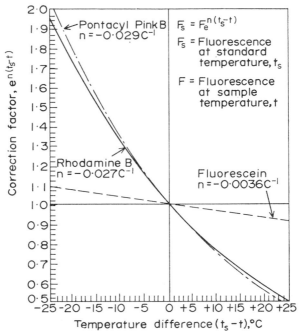

Fig. 2. Temperature correction curves for fluorescence analyses (after Feuerstein and Selleck, 1963).

(4) *Suspended matter*

The natural content of suspended solids present in such samples, however, can in two ways lead to changes in the tracer fluorescence: first, through the attenuation effect, that is the absorption and scattering of the excitation beam as well as of the fluorescent light by the suspended particles; the second through the adsorption of colouring agent molecules on the particles.

(a) *Attenuation.* The first effect (attenuation) appears to play a significant role only when the concentration of suspended solids is relatively high; one can eliminate this effect by appropriate changes in the method of measurement (by centrifugation or sedimentation of the samples before measurement).

(b) *Adsorption.* The second effect (adsorption) can, on the contrary, lead to considerable uncertainty when measuring in coastal waters. As adsorption can produce very varying values according to the type of suspended material, its particle size etc. and the tracers used, as well as its relationship to the chlorinity, it is hardly possible to obtain universally valid quantitative evidence. Nevertheless, in individual cases one can contrive quite useful formulae or diagrams for reduction of the measurements, as for example, carried out by Talbot and Henry (1968). Their diagram (Fig. 4) based upon theoretical considerations and measurements (Fig. 3) allows, at least, the order of magnitude of the adsorption effect to be correctly estimated. Similar results for suspended sediments were obtained by Kremser (1972),

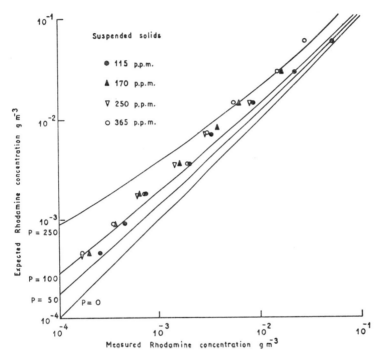

Fig. 3. Relation between measured and expected Rhodamine B concentrations obtained by serial dilution, using natural river water samples (after Talbot and Henry, 1968).

who investigated the influence of phytoplankton, too, without finding a measurable effect. He suggests that a reduction of the concentration of 2% should be considered as the limit beyond which the effect of adsorption can be neglected; an effect of this magnitude may already be observed with the sediment content of 5 mg l^{-1}. Feuerstein and Selleck (1963), when comparing the three chosen tracers, found that contrary to Rhodamine B the other two dyes showed no measurable adsorption when the coloured water was enriched with natural sediments. Some of the other recommended types of Rhodamine, especially

Rhodamine WT (Dupont) were preferred by some authors because of their smaller adsorption tendency (Pritchard, 1965; Carter and Okubo, 1965).

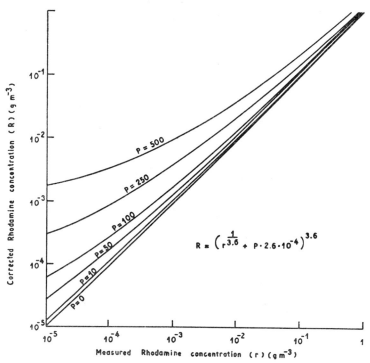

Fig. 4. Calculated correction curves for Rhodamine B adsorption, for various values of the suspended load (after Talbot and Henry, 1968).

(5) pH

Figure 5 illustrates the influence of the pH value, which was investigated in all three tracers by Feuerstein and Selleck. Obviously, one can neglect this influence within the pH range normally met in natural waters; it merely appears that care is advisable with extremely low values (pH < 5) when using Rhodamine B and especially with Fluorescein.

(6) *Photochemical decay*

Photochemical processes can also cause the impairment of the colouring agent's concentration in a solution as well as biological and chemical actions. Quantitative specifications, however, are, in this case, of very limited value because they are only obtainable under laboratory

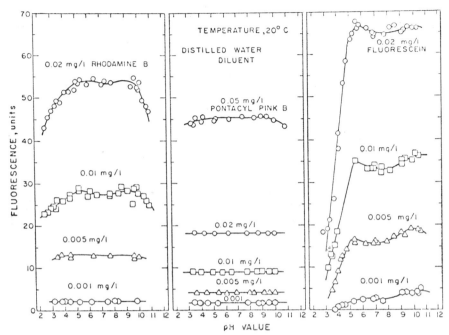

Fig. 5. Effect of pH value on fluorescence intensity of three tracers (after Feuerstein and Selleck, 1963).

conditions, which have little similarity to conditions occurring in nature. In the natural environment, not only is the daylight changeable but individual parts of the spectrum penetrate to different depths depending on the amount of suspended and dissolved matter in the water. Unfortunately, none of the authors give a clear indication which part of the spectrum is responsible for the decay. Kremser (1972) estimates from considerations on the molecular structure of Rhodamine that light quanta with a wavelength of about 400 nm should be necessary. His experiments showed no restriction to this particular wavelength; the influence of the shorter wavelengths, however, appeared to be much stronger than that of wavelengths longer than 400 nm.

The values published by Feuerstein and Selleck (1963) can be accepted as being relative values only. During their laboratory experiments they observed a half-life of 3 h for Fluorescein, 31 h for Rhodamine B and 68 h for Pontacyl B.P.B. during radiation by full sunlight, whereas with a cloudy sky the decay took about 5 times longer. Joseph et al. (1964) describe an experiment lasting several weeks on the North Sea island of Heligoland, during which a heavily diluted solution of Rhodamine B was placed in a shallow, open dish (which was covered by a polyethylene film only) and placed in the open in daylight. Average half-life values of 3 to 6 days were observed, that is about the same order of magnitude as observed by Feuerstein and Selleck. However, during an experiment at the sea surface (owing to vertical turbulence) after a short time the colouring agent usually begins to occupy a layer many metres thick, from which only that part which is in the immediate proximity of the surface is affected by photochemical decay. If one compares the loss in this layer with the total amount used, then it appears to be hardly substantiated within the general accuracy of measurement limits during an experimental period of several weeks, as is confirmed by the results of the large-scale experiment Rheno (1965) using Rhodamine B (Weidemann, 1973).

Kremser (1972) recommends, as a result of his very detailed investigations, that the integrated ultraviolet radiation should be measured

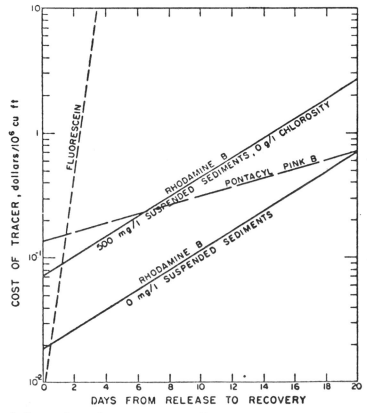

Fig. 6. Comparison of tracer costs, considering length of residence and water quality (after Feuerstein and Selleck, 1963).

during an experiment and, if an error of more than 2% is to be expected a correction should be applied to the total dye quantity.

In contrast to Rhodamine B, the photodecay rate of Fluorescein is obviously so large that the application of this dye is greatly restricted; the exceptions are its use during nocturnal investigations or by sub-surface injection at great depths, or if merely optical (photographic) evaluations of short-term surface investigations are planned.

C. COMPARISON OF ECONOMY

As an interesting conclusion to their investigations, Feuerstein and Selleck (1963) give a comparison of the costs involved between the three tracers on the basis of the attributes or measurement values examined (Fig. 6). One recognizes that Fluorescein for experiments which last not more than 1 day, and Rhodamine B in sediment-free water up to 20 days (in muddy water up to 6 days) are more economical than Pontacyl B.P.B. when one compares the cost of marking a definite volume of water (in the diagram, 10^6 ft^3).

III. FLUOROMETRIC INSTRUMENTS

A. GENERAL REQUIREMENTS

All fluorometers used in practice are based upon the physical property of the colouring agent solution to absorb radiation in a particular spectral range, thereby stimulating the molecules to emit a longer

TABLE 1. Absorption and emission of dyes.

Author	Colouring agent	Spectra (wavelength nm)	
		Absorption	Emission
Pritchard and Carpenter (1960) (Fig. 7)	Rhodamine B	ca. 450–600 max. 550	ca. 560–610 max. 575
Feuerstein and Selleck (1963)	Rhodamine B	550	570
	Pontacyl	560	578
	Fluorescein	480	510
Pritchard (1965)	Rhodamine 5G	535 (narrow)	590–600
Karabashev and Ozmidov (1965)	Rhodamine B	550	max. 590
	Fluorescein	480	max. 530
Fründel et al. (1971)	Rhodamine B	—	510 .. > 700 max. 590
Karabashev (1972)	Rhodamine S	530–580 max. 550	max. 610
	Fluorescein	max. 480	max. 540

wave secondary radiation. Very little (and partly divergent) information can be found in the literature about the aforementioned recommended colouring agents which concerns the specific absorption and emission spectra (Table 1).

Naturally, most information concerns Rhodamine B which is the tracer most frequently used, the emission spectrum of which lies in the yellow-orange part of the spectrum in which a natural background worth mentioning is seldom to be found.

According to Pritchard (1965) there exist algae which have the same spectral properties as those of Rhodamine B, and when they are present in the background, readings can rise up to about 100 times the normal value.

The equipment which is described below is designed and used for Rhodamine B measurements. However, certain items have also been used, employing the appropriate modified sets of filters, for the measurement of Fluorescein or natural chlorophyll.

B. PHYSICAL PRINCIPLES

The basic principle for all the fluorometers used is identical: emanating from a source of light, which usually radiates in a wide part of the spectrum, the light is directed through a primary filter. Two properties are required for this filter: (1) highest possible transmittance in those parts of the spectrum in which the dye used has its absorption maximum; and (2) as complete blocking as possible in those parts of the spectrum in which the fluorescence is recorded on the secondary side. After passing through the primary filter, the light falls (sometimes focused) upon the water sample to be measured, which is either in sample cuvettes, continuous flow cuvettes or behind a pressure-proof window directly *in situ*. The emitted fluorescence is received through a secondary filter (again, partially with optical bundling) and focused on to a highly sensitive light detector. The secondary filter must also fulfil two requirements: (1) highest possible transmittance in the spectral range of highest fluorescent intensity, (2) no fluorescence of its own in this range, and (3) as complete as possible elimination of the excitation radiation.

The basic sensitivity of the equipment is critically dependent upon the extent to which these requirements are fulfilled but will, incidentally, be limited by the natural background. As none of the known filter sets can positively guarantee 100% absorption in the region of the spectrum to be blocked, transmitters and receivers of light are arranged, as a safeguard, at right-angles to one another so that no direct, but only scattered primary light can reach the receiver.

The light-sensitive element within the receiver is normally a photo-multiplier or a photodiode. Contrary to the narrow band width of the filter, broad band sensors are preferred, so that with a change of dye it is sufficient merely to change the filter set.

C. LABORATORY FLUOROMETER

(1) *Description of Turner Mod. 111*

The fluorometer first used and described by Pritchard and Carpenter (1960) was the Turner Mod. 111. It is a laboratory instrument the exchangeable doors of which are designed as cuvette holders, and thereby permit very simple change of samples and cuvettes. It can

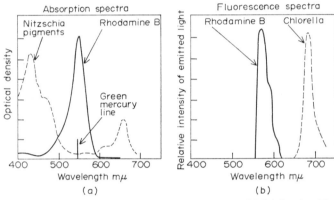

Fig. 7. Absorption (*a*) and fluorescence spectra (*b*) of Rhodamine B, compared with some biological background pigments (after Pritchard and Carpenter, 1960).

use sample cuvettes of various sizes as well as continuous flow cuvettes of 5 and 20 cm³ content. This model, up till now, has been used by numerous authors all over the world for diffusion investigations carried out in the sea, lakes and rivers. Carpenter recommends that it be fitted with the following filter combinations for experiments using Rhoda-mine:

> Primary: Corning 1–60, Kodak Wratten 61,
>   Corning 1–60 (green)
> Secondary: Corning 3–66, Corning 4–97 (orange).

The equipment operates according to the scheme reproduced in Fig. 8. It consists of an optical bridge, by which a part of the primary light from a mercury ultraviolet lamp and the secondary light are converted by a chopper into an 180° phase-shifted alternating light beam. The intensity of the reference primary light beam is regulated by a servo-system until equal amplitudes are attained. The fluorescence intensity can then be read off from a scale divided into 100 units, or can be transmitted as an electrical signal by a potentiometer coupled with the indicator to any suitable recorder.

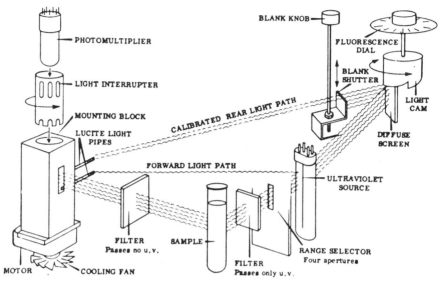

PHOTOMULTIPLIER

LIGHT INTERRUPTER

MOUNTING BLOCK

LUCITE LIGHT PIPES

CALIBRATED REAR LIGHT PATH

FORWARD LIGHT PATH

FILTER
Passes no u.v.

SAMPLE

MOTOR   COOLING FAN

FILTER
Passes only u.v.

BLANK KNOB

FLUORESCENCE DIAL

BLANK SHUTTER

LIGHT CAM

DIFFUSE SCREEN

ULTRAVIOLET SOURCE

RANGE SELECTOR
Four apertures

Fig. 8. Turner Mod. 111 fluorometer, schematic of optical bridge system (after Pritchard and Carpenter, 1960).

The sensitivity of this equipment, limited by background, as with all instruments, lies in the range of 1×10^{-11} g cm^{-3}. The amount of primary light can be adjusted by changeable diaphragms (aperture ratio 1–3–10–30-times). Therefore, the equipment covers about $3\frac{1}{2}$ decades for the measurement of fluorescent intensity; namely 3000 . . .30, 1000 . . . 10, 300 . . . 3, and 100 . . . 1 scale units. The range of measurable concentration values is, however, larger because the fluorescence intensity does not rise linearly with the dyes' concentration (Fig. 9). One can recognize in the log–log diagram the inversion effect which occurs especially with larger concentrations (roughly between 10^{-6} and 10^{-5}); the increasing absorption of the exciting and fluorescent light by the highly coloured water overwhelms the increase in fluorescence. For practical application however, this effect is unimportant because sailing through such high concentrations must be avoided for other reasons (additional turbulence caused by the ship's propeller, contamination of the ship, of pumps, pump-pipes, and measuring cuvettes).

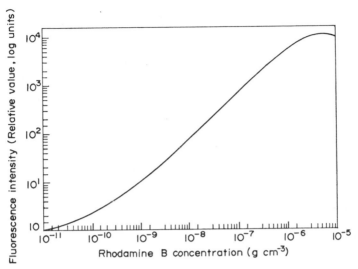

Fig. 9. Relationship between fluorescence intensity and Rhodamine B concentration (after Joseph *et al.*, 1964).

(2) *Calibration and accessories*

The calibration is carried out in a suitable manner with graduated standard solutions of the colouring agent concerned. It is recommended to begin with the weakest solution, especially as Rhodamine B adheres strongly to the walls of the glass cuvette. (When using continuous flow cuvettes, rinsing can be necessary for more than half an hour after a highly concentrated solution has been passed through, before the indication finally drops to the original basic value.)

To operate this equipment continuously on board ship necessitates the employment of a pump to take in sea water with the help of pipes, mounted over the ship's side or under the ship's bottom, then through hoses into the continuous flow cuvette. The time lag which occurs can be ascertained by experiment and must be considered when evaluation is made. This is particularly important when a longer hose with a depressor is used (e.g. Schuert, 1970) to sample water from greater depths.

D. *in situ* FLUOROMETERS

Owing to the uncertainty of this method, several attempts were made to design *in situ* equipment by which it is possible to bring the complete measuring equipment to the site where the fluorescence is to be measured.

(1) *Modified Turner (DHI)*

Certainly, the simplest method is to build the laboratory fluorometer into a watertight container, together with an underwater pump, and then lower it on a cable. This method was applied by the Hamburg Group (Weidemann, 1973) (Fig. 10), who, moreover, put their surface

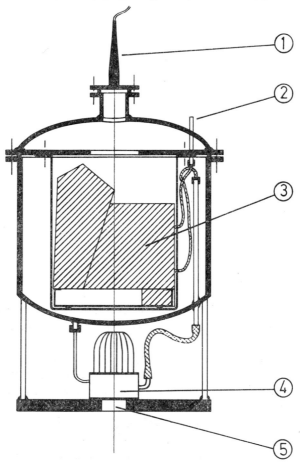

Fig. 10. Scheme of *in situ* fluorometer of the German Hydrographic Institute. 1 = multi-core cable; 2 = water outlet; 3 = fluorometer Turner Mod. 111; 4 = pump, 5 = water inlet.

fluorometer into a similar container and placed it in a well directly above the ship's bottom. By these means the length of the pipe and hose system between water inlet and measuring cuvette could with both types of installation be greatly reduced. As a result, it was no longer necessary to take the time lag into consideration for the surface measurements, yet during vertical measurement it was noticeable that a slight "hysteresis" between the lowering and hoisting registrations occurred. The problem of the long rinsing out after the passing of water of high concentration can likewise be considerably reduced.

As a result of building-in to a container, it was no longer possible to change by hand the diaphragms for alteration of the range. Therefore, a mechanical, continually changeable diaphragm was coupled to the indicator, which at a very low signal opened the diaphragm fully so that complete sensitivity was maintained, which however, with increasing intensity gradually reduced the primary aperture down to $\frac{1}{10}$ (full scale). By this feedback no doubt the calibration curve is distorted more than it was originally, but the range is increased to 3 (instead of, as formerly, 2) decades. There now remains the choice of two measurement ranges only: 1 . . . 1000 (with basic diaphragm 30) or 3 . . . 3000 units (with basic diaphragm 10). The last mentioned setting is practically only necessary during the early stages of an experiment, whereas, for the later stages the range 1 . . . 1000 is sufficient.

(2) *Zone Research Mod. 701 (Costin, 1965)*

This equipment (Fig. 11) which as an *in situ* fluorometer has been especially developed for tow operation, also works according to the

Fig. 11. Towed *in situ* fluorometer ZRI Mod. 701 (photograph: Zone Research Inc.).

12

same principle, as the laboratory equipment mentioned above. The
optical bridge system is shown in Fig. 12. The essential difference to
that of the Turner equipment is, apart from the arrangement of the
components, the replacement of the mechanical diaphragms to balance
the optical bridge, by a circular neutral density wedge filter, the
variation of which covers 2 or 3 decades in logarithmical graduation,
equivalent to about 3–5 decades of concentration variation.

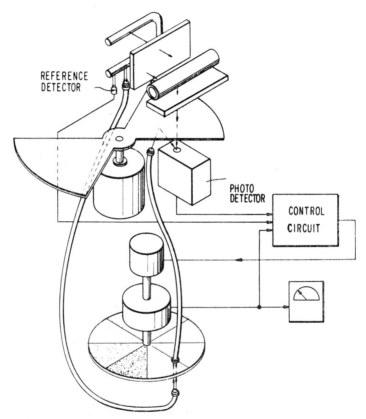

Fig. 12. ZRI Mod. 701 fluorometer; scheme of optical bridge system (after
Zone Research Inc., Instruction Manual).

The filters used for Rhodamine B differ only on the primary side of
the Turner equipment: Schott BG-36 (didymium) and Wratten 58 are
used; on the other hand, the same Corning 3–66 and 4–97 are used on
the secondary side. Figure 13 shows the spectral range obtained in this
manner.

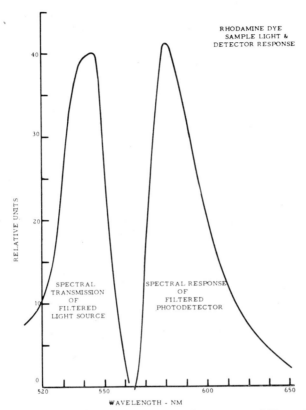

Fig. 13. ZRI Mod. 701 fluorometer, spectral response of filters for use with Rhodamine dye (after Zone Research Inc., Instruction Manual).

A pump is not used with this equipment; it is necessary to tow with sufficient speed in order to guarantee a flow through the cuvette, which for the purpose of daylight screening, is placed within a repeatedly curved inner canal. The indication threshold of this apparatus also lies between 10^{-11} and 10^{-10} g cm^{-3}. In order to be able to supervise the towing depth a pressure sensor and a temperature sensor are incorporated.

(3) *Karabashev/Ozmidov*

Karabashev and Ozmidov (1965) developed and used the first real *in situ* equipment without cuvette for their experiments with Fluorescein in the Black Sea. It is constructed in principle as a conventional fluorometer, however, the primary and secondary side components are contained in separate, pressure-proof cases arranged in parallel, and the path of light is directed at right-angles by a prism. Owing to the restriction to Fluorescein it is sufficient to use an incandescent lamp for excitation. The range of measurement is given in 10^{-6} to 10^{-10} g cm^{-3}; that of the active water volume seen by the optical system is ca. 100 cm^3. Nevertheless, it is only possible to use this apparatus at night or in sufficiently great depths as no protection against daylight has been included.

A newer version of this fluorometer has been described by Karabashev and Solovev (1968); its main improvements are the usability for both Fluorescein and Rhodamine, and the adaptation to near-surface measurements by a daylight protection system. Karabashev (1972) reports on comparison experiments between his *in situ* and a continuous-flow-cuvette fluorometer, in which the time lag, the averaging effect and the smoothing of the naturally occurring gradients by the flow system became obvious, particularly during the first few hours of an experiment.

(4) *Kullenberg*

In order to eliminate the influence of daylight, G. Kullenberg (1968) constructed and used an *in situ* apparatus for his Rhodamine B experiments in which four pressure-proof boxes are fastened to a frame (Fig. 14). The special feature of this is that two separate multipliers

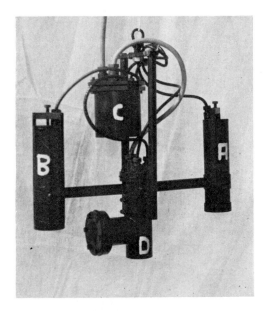

Fig. 14. *In situ* fluorometer after Kullenberg (1968). A, B = photomultipliers; C = pressure sensor; D = light source.

(A and B) are used, of which one (A) merely measures the daylight intensity from below; meanwhile the other (B) in addition receives the fluorescent radiation excited by the primary light (D). In this manner the daylight influence will be eliminated by connecting the two multipliers in a compensation circuit. The sensitivity range switching is operated electrically; altogether, the range covers 4–5 concentration decades between about $5 \cdot 10^{-11}$ and $5 \cdot 10^{-7}$ g cm^{-3}. The filters employed are: primary (to filter out the green mercury line): Schott OG 2/1 mm, BG 18/3 mm and BG 20/6 mm; secondary: Schott VG 9/2 mm and Kodak Wratten 23A.

The equipment is intended exclusively for vertical measurements from a stationary or slowly moving ship; the depth is measured by a built-in pressure sensor (C).

(5) *Variosens (Früngel)*

A new *in situ* fluorometer, which differs in some details from those previously discussed, is described by Früngel *et al.* (1971). The essential

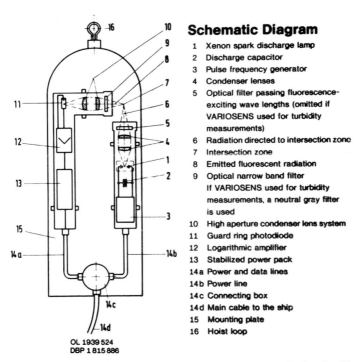

Schematic Diagram

1 Xenon spark discharge lamp
2 Discharge capacitor
3 Pulse frequency generator
4 Condenser lenses
5 Optical filter passing fluorescence-exciting wave lengths (omitted if VARIOSENS used for turbidity measurements)
6 Radiation directed to intersection zone
7 Intersection zone
8 Emitted fluorescent radiation
9 Optical narrow band filter
   If VARIOSENS used for turbidity measurements, a neutral gray filter is used
10 High aperture condenser lens system
11 Guard ring photodiode
12 Logarithmic amplifier
13 Stabilized power pack
14a Power and data lines
14b Power line
14c Connecting box
14d Main cable to the ship
15 Mounting plate
16 Hoist loop

OL 1939 524
DBP 1 815 886

Fig. 15. Variosens *in situ* fluorometer, scheme (after Impulsphysik, Hamburg).

features of this equipment, which is produced under the name "Vario-sens", are the replacement of the continuous light source by an impulse light source (xenon spark discharge lamp) as well as the automatic compensation of the influence of daylight. The latter is achieved by a continuous illumination of the receiver side silicium photodiode with the help of a luminescence diode, which directs an adjustable basic current into the photodiode (Fig. 15).

The use of an impulse light source, which is operated with an adjust-able frequency of maximum 10 Hz, has the advantage of considerable saving of energy when compared with the conventional incandescent or mercury-vapour lamps (total consumption 6 W) so that this equip-ment can be used without difficulty with longer cables for work in greater depths or in towing operations.

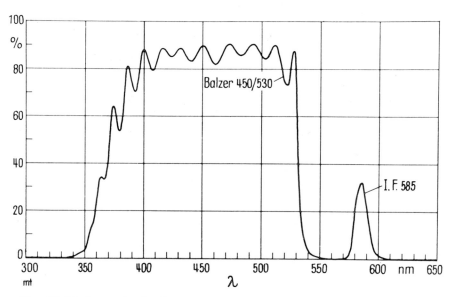

Fig. 16. Variosens *in situ* fluorometer; spectral response of filters for use with Rhodamine (after Früngel *et al.*, 1971).

For tracer measurements using Rhodamine B, the apparatus was equipped with a primary filter, Balzer 450/530 and a secondary filter, Schott I.F. 585. Figure 16 shows the spectra transmitted by these filters. The basic sensitivity lies at ca. $5 \cdot 10^{-11}$ g cm^{-3}.

The equipment, which apart from the type illustrated in Fig. 17, can also be produced in other versions with variations in the arrange-ment of transmitter and receiver as well as for different maximum pressures. The instrument can also be adapted to *in situ* measurements of chlorophyll, light scattering or attenuation. In this case, the spectral filter can be interchanged or, alternatively, replaced by a grey filter.

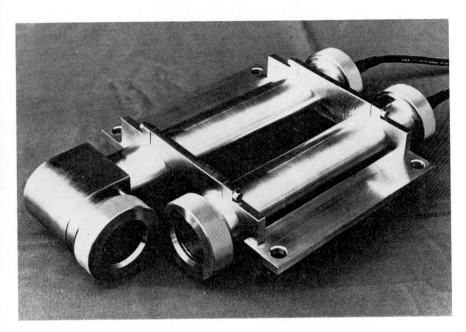

Fig. 17. Variosens *in situ* fluorometer; mounted in parallel mode (photograph: Impulsphysik, Hamburg).

III. THE FLUOROMETRIC TRACER METHOD IN PRACTICE

A. TYPE OF EXPERIMENTS

(1) *Depth*

One can differentiate between experiments in the surface layer and in deeper layers, according to the depth at which the dye is released as well as to the required techniques which necessarily diverge from one another in many respects.

(*a*) The colouring agent is usually delivered as a highly concentrated solution. Therefore, it is convenient to prepare it on land before the experiment by adding methanol or ethanol (for Rhodamine B about 40%) and adjust it to a density of about 1·0. The adjustment is not critical in a surface experiment, as long as the density of the solution remains lower than that of the water; in any case, wave turbulence at the surface will quickly reduce any density differences. On the other hand, in experiments in a deep layer it is essential to adjust the density of the solution to be close to the density of the water in that layer. Even when the solution has already been prepared on land according to the expected water density, it will be advisable to measure temperature and salinity *in situ* before the experiment, and to correct the density of the solution by the addition of an appropriate quantity of alcohol, water or concentrated solution. It may also be useful to adjust the temperature of the solution before release, as Karabashev and Ozmidov (1965) tried with an apparatus specially constructed for this purpose (Fig. 18). If such precautions are omitted, there is a risk that considerable parts of the dye solution injected under water do not stay at the intended depth, but either sink to the bottom, or rise to the thermocline or even to the surface.

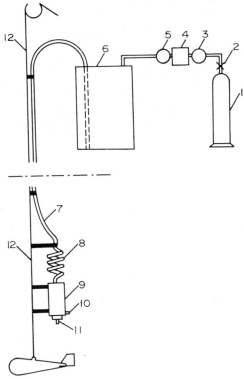

Fig. 18. Scheme of release system for subsurface tracer experiments (after Karabashev and Ozmidov, 1965). 6 = pressurized dye container; 7 = rubber hose; 8 = copper coil; 9 = 2 litre vessel; 10 = outlet; 11 = resistance thermometer, 12 = hydrographic wire.

(b) The method of measurement also differs. In surface experiments it is sufficient to use, as main equipment, a ship-board laboratory instrument with a suction mechanism (see above) which is intended principally to measure horizontal spread of the colour patch. Additionally, the vertical distribution of the colouring agent can be determined by vertical profiles (if an *in situ* instrument is available) or also from individual water samples only.

(2) *Release mode*

In both types of experiment described, surface as well as subsurface, there are two ways possible to release the colouring agent: (1) the whole amount is released in the shortest possible time (instantaneous release), or (2) the dye is let out at a constant rate over a period of time which is long compared with the scale of the experiment (continuous release).

(a) *Selection of release mode.* Both types of experiments can produce equal information about the magnitude of the diffusion parameter: the results of the one type can be compared with those of the other. The decision rests with the conditions in question: when this concerns the simulation of accidents, dumping action etc. then the instantaneous release method is preferable; whereas the conditions pertaining to sewage outlets etc. can be investigated far better by the continuous release method. The latter is especially true when water with strongly changing currents (e.g. tides) is concerned, in which the dye distribution becomes quasi-stationary only after a considerable period of time.

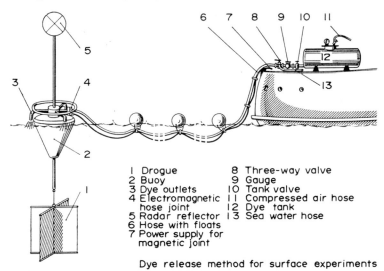

1 Drogue	8 Three-way valve
2 Buoy	9 Gauge
3 Dye outlets	10 Tank valve
4 Electromagnetic	11 Compressed air hose
hose joint	12 Dye tank
5 Radar reflector	13 Sea water hose
6 Hose with floats	
7 Power supply for	
magnetic joint	

Dye release method for surface experiments

Fig. 19. Scheme of surface dye release system (after Joseph *et al.*, 1964).

12*

(b) *Methods and devices.* For instantaneous release, the total amount of colouring agent determines if it is necessary to use special equipment to spread the colour. Small amounts (a few kilograms of colouring agent) can be poured direct into the water without difficulty; larger amounts, on the other hand, are better spread by special equipment as has been described, for example, by Joseph *et al.* (1964) (Fig. 19) in order to achieve a circular initial patch as symmetrical as possible.

For continuous release, especially in subsurface experiments, equipment with special diffusers are the best; such as the Russian equipment shown in Fig. 19 or the Canadian procedure (Fig. 20) described by Murthy and Csanady (1971).

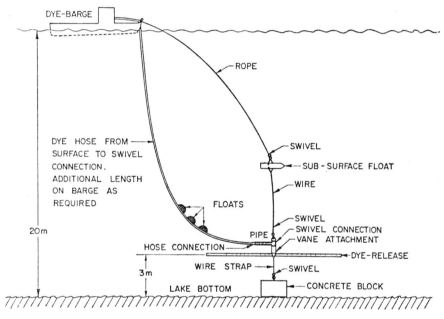

Fig. 20. Scheme of continuous bottom release system (after Murthy and Csanady 1971).

(3) *Positioning problems*

In experiments of the instantaneous type, the most ideal way of determining the ship's position is to relate it to the centre or concentration maximum of the patch which is shifting with relation to the sea bottom due to the mean current. In surface experiments this centre can be marked by using a suitable radar buoy with a drogue, in subsurface experiments a similar method is possible whereby the drogue must be suspended from the surface buoy at the depth of the colouring agent injection. In any case, it is recommended for navigation that dead reckoning procedure from course and speed is used for plotting the measurements, and not according to location-bound navigation systems (e.g. Decca, landmarks etc.).

In continuous release experiments, on the other hand, one is interested in general in the geographical distribution and the dilution as functions of distance from inlet, so that here the use of the above mentioned location-bound navigation procedure suggests itself. This type of experiment has the advantage that one can wait until a stationary distribution has appeared, and then one can measure the distribution in all three dimensions without undue haste. On the other hand experiments of the instant release type demand that the measurements take place as quickly as possible and quasi-synoptic in order to record the state of mixing at defined times after the beginning of the experiment. In large scale experiments (e.g. Rheno, 1965) these conditions cannot be strictly fulfilled in spite of the participation of several ships (Weidemann, 1973).

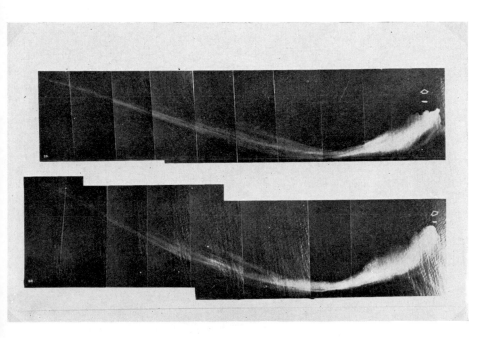

Fig. 21. Aerial photographs of a surface dye experiment near Cape Kennedy (after Carter and Okubo, 1965). Upper: 2 h 4 min after release; lower: 2 h 50 min after release.

B. SOME TYPICAL RESULTS

It is not possible to discuss here the results of the very numerous investigations which have been published to date; a collection of many important results of the past ten years has been made (amongst others) by Okubo (1970). One or two particularly interesting points or results of general interest, however, should be recognized here as examples for the use of these methods.

(1) *Ekman shear*

The first example (Fig. 21) comes from a publication by Carter and Okubo (1965); it shows two aerial photographs which were taken at short intervals from each other during a tracer experiment. The clearly visible curvature of the dye patch is certain evidence of the existence of the Ekman current. In a work by Katz *et al.* (1965) there is a drawing (Fig. 22) which explains clearly the principle for this phenomenon;

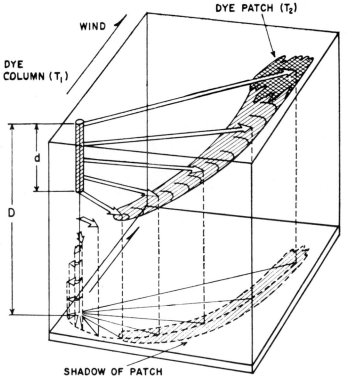

Fig. 22. Scheme of Ekman effect on surface dye experiments (after Katz *et al.*, 1965).

the colouring agent which is diffusing from the surface fleck downwards gets into the shear of the Ekman spiral. In clear water this coloured deep water can be observed from the air.

(2) Concentration decrease with time

The second diagram (Fig. 23) is reproduced in a paper by Abraham and Van Dam (1970), after Van Dam (1968). It contains a synopsis of the reduction observed in the maximum concentration as a function of time from a very large number of North Sea experiments using Rhodamine B (1 kg colouring agent each time). In spite of all the scattering, there is some regularity to be found with an asymptote of about

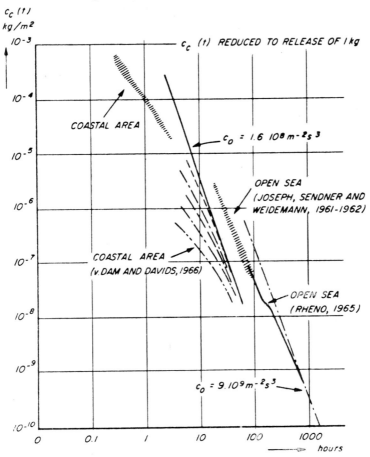

Fig. 23. Relationship between maximal tracer concentration, reduced to 1 kg, and time after release, from various North Sea experiments (after Abraham and van Dam, 1970).

t^{-3}, and this was reached during the Rheno (1965) large scale investigations mentioned above.

(3) *Subsurface survey example*

Schuert (1970) described an example of an extremely deep subsurface experiment. Figure 24 shows a resulting registration of this experiment (which took place near Hawaii) and an attempt to interpret this is also given in this figure. The difficulty of this technique was quite obviously the "blind" work, and additionally the particularly great time lag. Nevertheless, successful three dimensional surveys of the colouring agent distribution were obtained at about 300 m depth.

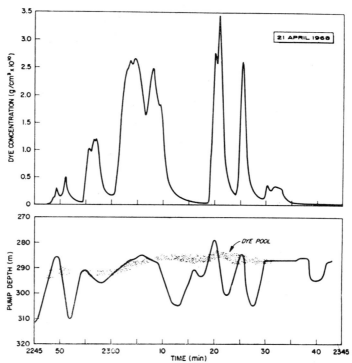

Fig. 24. Subsurface dye experiment near Hawaii (after Schuert, 1970). Upper: record of concentration; lower: record of pumping depth, with dye pool as determined from both plots after correction for time lag.

(4) *Multi-ship survey example*

The last example in Fig. 25 shows the results of a survey which was undertaken during the Rheno large scale experiment in 1965. So far, the difficulty of sufficiently accurate recording of the very small scale details of the partial concentration distribution over such large areas is obvious. The accuracy of results derived from such experiments, therefore, should not be over-estimated; in any case they are not sufficient, to date, to provide a clear decision for or against the one or the other theoretical interpretation of turbulent diffusion. On the other hand, there exists, or can be quickly obtained (thanks to these experimental methods), sufficient empirical data that one can, possibly together with numerical model calculations, arrive at reliable forecasts.

Fig. 25. Large-scale experiment, Rheno 1965, Survey No. 7 (14 days after release) with tracks of four vessels (after Weidemann, 1971).

V. Summary

The importance of oceanographic turbulence research justifies the development of special tracing methods on the basis of fluorescent colouring agents. The results of detailed investigations into their properties, their advantages and disadvantages for particular tasks as well as the fluorometers developed and used for laboratory and *in situ* operation for measurement of their concentration have been summarized in this chapter. Supplementary observations have dealt with the most important methods of application and several typical examples of the results obtained have been described.

ACKNOWLEDGEMENTS

The author wishes to thank particularly his colleague H. Sendner, who assisted him in the perusal of the extensive literature, and Mrs. Petersitzke who translated this manuscript into English.

REFERENCES

Abraham, G. and Van Dam, G. C. (1970). *FAO Techn. Conf. Mar. Pollut. (Rome)*, MP/70/E-1.

Carter, H. H. and Okubo, A. (1965). *Chesapeake Bay Inst., Johns Hopkins Univ., Rep.*, NYO–2973–1.

Costin, J. M. (1965). *In* "Symposium on Diffusion in Oceans and Fresh Waters" (T. Ichiye, ed.), pp. 68–70. Lamont Geological Observatory Columbia University, Palisades, New York.

Feuerstein, D. L. and Selleck, R. E. (1963). *J. Sanit. Eng. Div., Proc. Amer. Soc. Civ. Eng.*, **89**, 1–21.

Früngel, F., Knütel, W. and Suarez, J. F. (1971). *Meerestechnik/Mar. Technol.*, **2**, 241–247.

Japanese Government Agencies (1958). *In* "Peaceful Uses of Atomic Energy", *Proc. U.N. Int. Conf. 2nd*, 1958 *(Geneva)*, **18**, 404–409.

Joseph, J., Sendner, H. and Weidemann, H. (1964). *Deut. Hydrogr. Z.*, **17**, 57–75.

Karabashev, G. S. (1972). *Beitr. Meeresk.*, **30/31**, 67–80.

Karabashev, G. S. and Ozmidov, R. V. (1965). *Izv. Acad. Sci. USSR Atmosph. Oceanic Phys. Ser.*, **1**, 1178–1189.

Karabashev, G. S. and Solovev, A. N. (1968). *Izv. Akad. Nauk SSSR., Ser. Fiz. Atmos. Okeana*, **4**.

Katz, B., Gerard, R. and Costin, M. (1965). *J. Mar. Res.*, **70**, 5505–5513.

Kremser, U. (1972). *Beitr. Meeresk.*, **30/31**, 101–125.

Kullenberg, G. (1968). *Københavns Univ., Rep. Inst. Fys. Oceanogr.*, **3**, 1–50.

Murthy, C. R. and Csanady, G. T. (1971). *Water Res.*, **5**, 813–822.

Okubo, A. (1970). *Chesapeake Bay Inst., Johns Hopkins Univ., Techn. Rep.*, **62**.

Pritchard, D. W. (1965). *In* "Symposium on Diffusion in Oceans and Fresh Waters" (T. Ichiye, ed.), pp. 146–147. Lamont Geological Observatory Columbia University, Palisades, New York.

Pritchard, D. W. and Carpenter, J. H. (1960). *Bull. Int. Assoc. Sci. Hydrol.*, **20**, 37–50.

Schuert, E. A. (1970). *J. Geophys. Res.*, **75**, 673–682.

Talbot, J. W. and Henry, J. L. (1968). *J. Cons. Int. Explor. Mer*, **32**, 7–16.

Van Dam, G. C. (1968). *Rijkswaterstaat, Math.-Fys. Afd.*, Nota MFA 6812.

Weidemann, H. (1973). Ed. "The I.C.E.S. Diffusion Experiment RHENO 1965", *Rapp. Proc.-Verb. Reun. Cons. Int. Explor. Mer*, **163**, 111 pp.

Witting, R. (1933). *Soc. Sci. Fenn., Comment. Phys.-Math.*, **7**.

Chapter 13

Remote Optical Sensing in Oceanography Utilizing Satellite Sensors

PAUL E. LA VIOLETTE

U.S. Naval Oceanographic Office, Washington D.C. 20390, U.S.A.

I. INTRODUCTION

One of the basic problems of physical oceanography is the synoptic and continuous examination of an oceanic region's physical properties. At sea, an oceanographer may study intensely the small area of ocean about his ship, but be unable to compare it to critical conditions simultaneously occurring 5 or 100 km away.

Instrumented satellites have extended the oceanographers' small study area and have furnished him with the ability to view repeatedly vast regions of the ocean. These views, however, are at present restricted by satellite weight limitations, cloud cover, sensor resolution, orbit orientation and altitude, and data handling technology. In addition, because the manner of data collection is limited necessarily to electromagnetic radiation, the oceanographic information directly obtainable is restricted to the immediate surface layers of the ocean.

Despite these restrictions, oceanographers experienced in remote sensing are trying to capitalize on the advantages of satellites. First, they are refining present satellite sensors and exploring the use of new ones. Secondly, they are examining closely the satellite information now available with the hope that they may infer physical relationships

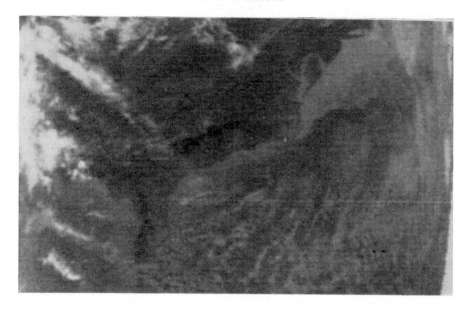

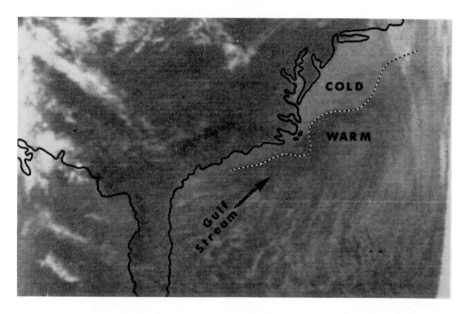

Fig. 1. Infrared Imagery of East Coast of United States. Daytime infrared imagery of the east coast of the United States showed the thermal gradients of the North wall of the Gulf Stream as it appeared on 8th April 1970.

not immediately apparent. Finally, they are using satellites to collect telemetered data from time-proven, conventional instruments aboard unmanned buoys.

II. THE NOAA SATELLITE, ITS PRIME RADIOMETER AND GROUND STATIONS

There has been no satellite which could be called an "oceanographic" satellite. Although much satellite-gathered oceanographic information is available, it is obtained from satellite sensors whose main purpose is not oceanography. These sensors that have provided the most information have been: infrared radiometers, visual sensors, and in the case of the manned spacecraft missions, photographic cameras. Some of the methods with which the information from these sensors is being employed oceanographically are discussed in the sections which follow. This section will describe the operational NOAA (née ITOS) satellite, one of its prime sensors, and the ground stations associated with it. NOAA's orbit, data acquisition, and transmission are typical of United States meteorological satellites now in use that are capable of furnishing oceanographic data.

The NOAA satellites are similar to the research NIMBUS and earlier operational TIROS satellites in that they circle the globe in polar orbit, with these orbits timed to occur at approximately the same time each day on the earth's lighted and dark sides. Their prime source of data are provided by a scanning radiometer (SR), whose scanning component is a rapidly rotating mirror. As the mirror rotates, it sweeps the earth from horizon to horizon at right angles to the nearly north-south path of the spacecraft. The single scan spot—the elemental field of view at any instant—passes through a beam splitter and spectral filters to provide the 10·5 to 12·5 μm infrared channel and an 0·5 to 0·7 μm visible channel.

The optics of the SR infrared channel provide a 5·3 mrad field of view, or approximately 8 km scan spot of the earth directly below the satellite. As the mirror rotates, the scan spot moves toward the horizon and the area viewed increases in size. In equatorial regions, data from adjacent orbits overlap at about 1670 km from a point directly below the satellite. The nadir angle at the overlap is approximately 60°; and the earth area viewed by a single scan spot, about 15 by 22 km in size. As the satellite continues toward the poles the amount of overlap between contiguous passes increases.

The 0·5 to 0·7 μm visible channel is an interesting adjunct to the thermal channel. The visible channel optics give a 2·7 mrad field of

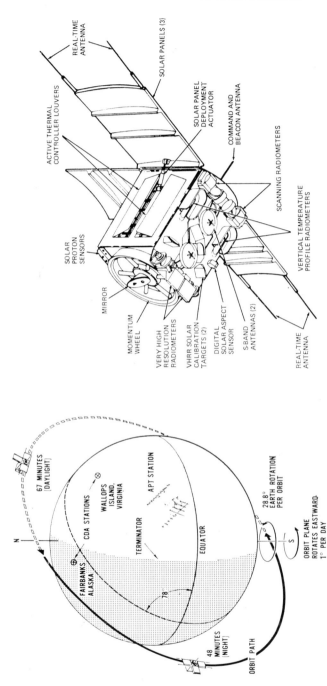

Fig. 2. The NOAA Satellite and its Orbit. The present NOAA operational satellites consist of three-axis-stabilized earth-oriented spacecraft designed to provide full day and night coverage of the entire surface of the earth on a daily basis. Their orbits are circular at approximately 1500 km altitude, and near-polar with a 78° angle of inclination retrograde to the equator. Total orbital period is approximately 115 min (67 min in sunlight and 48 min in the earth's shadow). The earth rotates beneath the orbit 28·8° during this period, allowing the satellite to observe a different portion of the earth's surface with sufficient overlap from orbit to orbit.

The orbit is sun-synchronous and precesses (rotates) eastward about the earth's polar axis approximately 1° per day, or at the same rate and direction as the earth's average annual revolution about the sun. The satellite will always cross the equator at the same time in the afternoon northbound and the same time in the early morning southbound (local time). Sun-synchronous orbits compensate for seasonal variations by keeping the orbital plane of the satellite in a constant position with reference to the sun for consistent illumination throughout the year.

view or slightly less than a 4-km scan spot at directly below the satellite. The visible channel is extremely useful during the daylight passes in distinguishing clouds from anomalies in the thermal imagery data collected simultaneously in the infrared channel. (A more detailed description of the NOAA satellite, its orbit and its sensors may be found

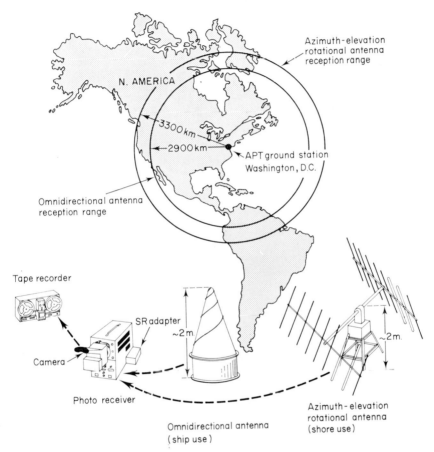

Fig. 3. The Field Automatic Picture Transmission (APT) Satellite Station. The portable field Automatic Picture Transmission (APT) stations mentioned in the experiments are small and relatively unsophisticated in comparison to the large permanent APT stations operated by the weather service. Each field APT station consisted of an EMR Model III A Weather Satellite Photo Receiver with Camera Pack and SR Adapter, a tape recorder for later imagery playback or for conversion of the incoming analog signal to digital data, and either an azimuth-elevation rotational antenna or an omnidirectional antenna. Although the rotational antenna gives the best reception on land, the omnidirectional antennas are used to compensate for ship or aircraft movement.

in "Modified Version of the Improved TIROS Operational Satellite (ITOS D–G)", by Schwalb (1972)).

The early NOAA satellites have been equipped to transmit slow-scan television pictures of the sunlit portion of the earth. Since these video pictures are transmitted on the same frequencies as the scanning radiometer, video transmissions from these early satellites were normally programmed to occur during the day and SR transmission at night. The SR on NOAA 2 will completely replace the video camera system as the primary source of direct readout data. Thus, this satellite and those in the series that follow will give daytime as well as nighttime thermal data.

The SR data are broadcast immediately upon collection by the satellite at a frequency of 137·50 or 137·62 MHz. Any Automatic Picture Transmission (APT) station within the satellite's line of sight can receive the data via this transmitted signal. Simultaneously to the broadcast, the SR data are stored on board the spacecraft for later transmission to earth upon command of either of two Command Data Acquisition (CDA) stations.

III. OCEANOGRAPHIC USE OF SATELLITE INFRARED DATA

The NOAA and NIMBUS satellites have provided extensive infrared data for a number of oceanographic studies. An example of such a study was make by Szekielda (1970) in a clear-sky ocean area with strong seasonal upwelling—the eastern coast of the Somali Republic. The emphasis in Szekielda's study was on the synoptic sea surface temperature patterns disclosed by the NIMBUS 2 satellite in its momentary overflight of the region. No atmospheric corrections were made to the satellite data, as actual temperature values were not as important to the study as the temperature patterns and the daily and seasonal changes in these patterns.

During June, July and August the stress of the monsoonal winds creates strong thermal gradients along the Somali coast. Analyses of NIMBUS 2 infrared data for these months in 1966 show clearly the shifting patterns of these thermal gradients as well as the day to day variation in their strength. In general, the analyses show a pronounced anticyclonic gyre and indicate that two distinct areas of upwelling occurred off the Somali coast during June and July but not August 1966.

In addition to quantitative data such as were used in the above study, thermal imagery photographs may be made of the satellite infrared data. The variation in the gray tones of these photographs depicts

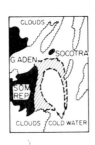

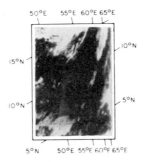

NIMBUS 3-direct readout infrared imagery, August 1970

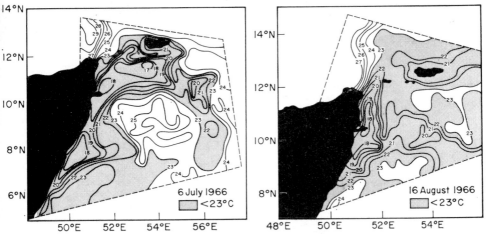

NIMBUS 2-stored infrared data (equivalent blackbody temperatures)

Fig. 4. Infrared Imagery and Data Analyses. Satellite infrared data may be used for a variety of oceanographic purposes. The analyses of NIMBUS 2 infrared stored data presented above were used in a general study of monsoonal upwelling along the Somali coast. The NIMBUS 3 infrared imagery was received by an APT station aboard the R/V *Chain* while off the Somali coast and was of immediate use to an on-going oceanographic survey of the upwelling.

In comparing the upwelling conditions for the two years, the analysis of NIMBUS 2 data showed the July sea surface temperature pattern was remarkably similar to the August 1970 pattern. The August 1966 and the August 1970 patterns were dissimilar, however, with the 1966 analysis showing a general reduction in the surface temperature gradients and a related earlier cessation.

equivalent variations of the earth's blackbody radiations as seen by the satellite's infrared sensor.

Properly utilized, infrared imagery provides a rapid method of identifying and monitoring strong thermal gradients in clear-sky regions. Such imagery can be a highly useful survey tool to the scientist

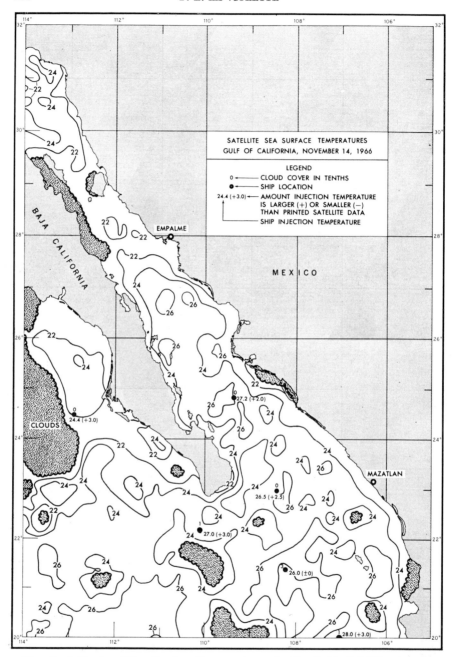

Fig. 5. A Study of Sea Surface Temperatures in the Gulf of California using
Satellite Infrared Data—NIMBUS 2. Five days of NIMBUS 2 infrared radiometer data, 11th to 15th November 1966, were examined individually and then
combined in a selective composite designed to arrive at sea surface temperatures

[continued opposite

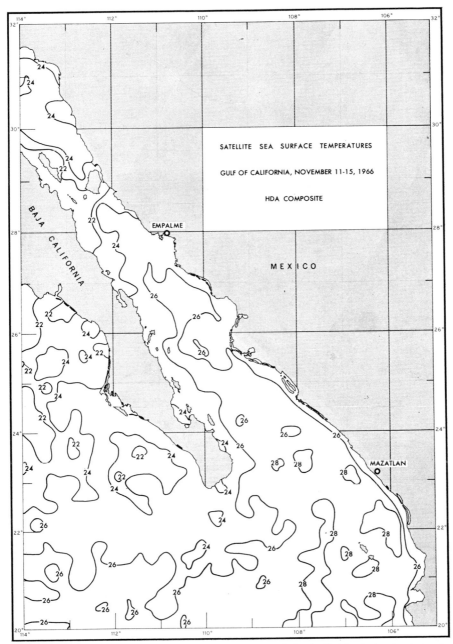

SATELLITE SEA SURFACE TEMPERATURES

GULF OF CALIFORNIA, NOVEMBER 11-15, 1966

HDA COMPOSITE

in a region of transitory clouds. An examination of the individual charts showed a small migration of warm water northward into the Gulf during the five-day period. The variation was small however, and the five-day High Daily Average (HDA) composite may be taken as representing the sea surface temperature conditions of the region with the transitory clouds removed.

at sea. It could provide an oceanographer with answers to questions such as: Does the study region's dynamic structure change as the survey ship makes it traverse from one day's station to the next; and is the orientation of the ship's survey track such that critical dynamic features are not missed?

In an effort to test the usefulness of satellite infrared imagery to such a survey, an APT satellite station was established aboard the Woods Hole Oceanographic Institute's R/V *Chain* during an August 1970 cruise off the Somali coast (La Violette and Stuart, 1972). The APT station received infrared data directly from the NIMBUS 2 and ITOS 1 satellites. The imagery obtained from the data compared favorably with the thermal patterns derived by the ship's conventional instruments and showed that the surface temperature field was essentially stationary for the two-week period 14 to 30 August. In addition, when the imagery was compared to Szekielda's analyses, the August 1970 surface temperature patterns were found to be similar to the July rather than the August 1966 satellite data.

The infrared sensors utilized in these experiments were designed primarily to furnish meteorological information. Naturally, in trying to use meteorological sensors for oceanographic measurements extensive problems arise.

One of the most difficult of these problems is the interference of clouds and atmospheric moisture with the infrared sensor recording of the ocean's equivalent blackbody radiation. When clouds are present between the sensor and the ocean, the sensor records cloud rather than ocean radiation temperatures. The surface thermal features of the ocean in these cases are effectively blocked from the satellite's view. Thin clouds and water vapor affect sensor data by making the ocean's surface appear cooler than it actually is. The amount of apparent cooling is directly proportional to the amount of water vapor in the air.

A number of methods have been formulated whereby transitory clouds over an ocean or smaller region may be artificially removed and the surface temperature structure examined. Some methods use the maximum temperatures which occur over a number of days; others examine the satellite signal and establish a signature curve of temperature (Smith *et al.*, 1970). The method described in the following example utilizes a selective composite of several days' data (La Violette and Chabot, 1969).

The Gulf of California was chosen to test the compositing method, because the Gulf has relatively simple sea surface temperature structure and distinctive land boundaries easily defined in the infrared sensor data. Although one of the most cloud-free oceanic areas of the world, enough

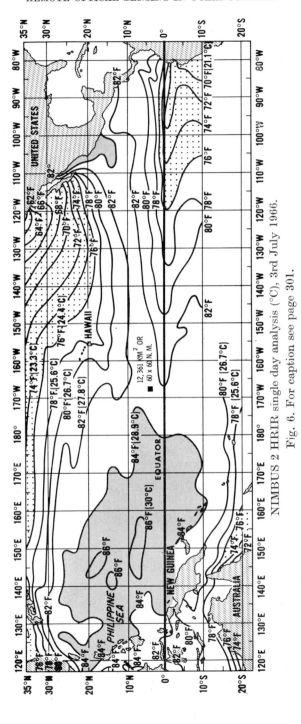

NIMBUS 2 HRIR single day analysis (°C), 3rd July 1966.
Fig. 6. For caption see page 301.

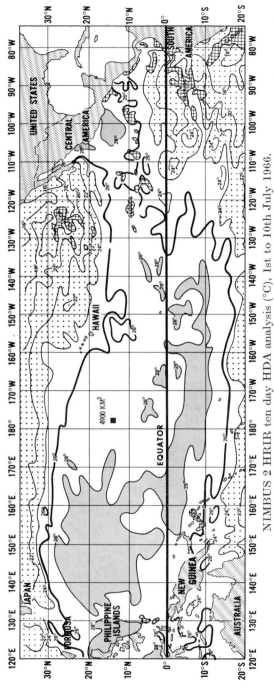

NIMBUS 2 HRIR ten day HDA analysis (°C), 1st to 10th July 1966.
Fig. 6. For caption see page 301.

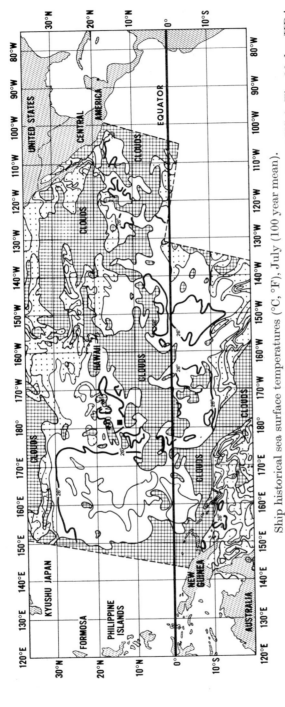

Ship historical sea surface temperatures (°C, °F), July (100 year mean).

Fig. 6. Infrared Data Analysis of Pacific Ocean vs. Historical Sea Surface Temperature—NIMBUS 2. The 10-day HDA Pacific Ocean chart was constructed from NIMBUS 2 infrared radiometer data in a similar fashion to the Gulf of California Study. The historical sea surface temperature chart is the July average of 100 years of ship injection temperatures.

clouds are normally present to cause some interference with a single day's examination by satellite. The region, therefore, seemed ideal to test the compositing method. The region was first examined in detail for each of five consecutive days—11th to 15th November 1966—and the results then combined in a High Daily Average (HDA) composite.

The averaged satellite temperature value for each basic resolution point in the study represented an approximate area of 22 km square. These temperature values were "corrected averages" arrived at by excluding from the data all readings lying outside boundary temperatures of 6° and 31°C. These boundary limits were derived from historical extremes of sea surface temperature data for November in the Gulf of California.

The five-day HDA composite was made by examining the daily temperature average at a point for each of the five days and printing the highest daily average which occurred during the period. This process was repeated for the entire HDA chart. The resulting printed values were designed to reflect the smallest concentration of clouds or minimum amount of moisture over each point for the five-day period.

The five-day HDA composite was compared to sixteen sea surface temperature stations taken by the R/V *Yolanda* during the period 10th to 16th November 1966. These station data were lower than the HDA composite data by an average of 1·3°C. The −1°C difference is consistent with other sea surface temperature studies made using NIMBUS 2 infrared data and may be a correction value needed for the satellite radiometer.

In an attempt to test the HDA composite method on an oceanic scale, ten days of NIMBUS 2 infrared data—1st to 10th July 1966—were combined in a composite of the Pacific Ocean (La Violette and Allison, 1969). An examination of multiple-day composits of cloud pictures, showed that ten days was the minimum period necessary to remove most clouds and still retain a synoptic picture of the ocean's temperature structure.

The ten-day HDA Pacific Ocean composite chart was printed in the same fashion as the five-day HDA Gulf of California study. In this case the basic resolution point was a square 70 km on a side. The HDA chart is a grid of those resolution points containing the highest daily averages which occurred over the ten days and representing the smallest concentration of clouds and the minimum amount of moisture over each point for the ten days.

The resulting temperature chart was then compared to the historical mean sea surface temperature chart in July. Examination of the two charts in Fig. 6 shows that while much of the general structure of the two charts are the same, the HDA composite chart shows more details than the smoothed averages of the historical chart.

The most important eventual application of the method will be to observe the variation of an ocean's temperature field from one ten-day period to the next. Furthermore, it should be possible to replace the artificially removed cloud cover and to examine the causal effect the water temperature had on cloudiness. This type of study, coupled with an examination of the variation in layer depth, may be used to give a dynamic picture of the energy exchange of the air-sea regime.

Another problem in the utilization of satellite infrared sensors for oceanographic purposes, is the type of energy being measured. The "sea surface temperature" collected by infrared sensors is the equivalent blackbody radiation of the first millimeter of the surface layer. Is this radiation temperature equal to the energy layer, sometimes more than a meter thick, which oceanographers call the sea's surface? Sufficient radiation loss occurs at night, so that the surface radiation temperature in the early morning hours can easily represent a dynamic mixing to a depth of several meters. During the daytime, however, insolation offsets the radiation loss by a variable amount. If there is no mechanical mixing, such as when wind speeds are less than 10 knots, then infrared daytime measurements may be the radiation of a surface film quite different in temperature than the water a few centimeters deeper. Since the basic resolution of the infrared sensors presently in use is approximately 8 km, it is felt that most daytime data is an aggregate of different mixing conditions. The resulting temperature may be considered an average of these conditions and, therefore, representative of the region's general surface temperature.

As a test of the accuracy with which daytime and nighttime infrared satellite data can depict actual ocean conditions, an experiment was conducted in the Gulf of California during May 1971 by the United States and Mexican Governments with the author as Principal Investigator and Coordinator. The experiment, called LITTLE WINDOW 2, involved data collection by satellite, aircraft and ship sensors within a 200-km square or window in the Gulf. Final results of this experiment have not been derived. Preliminary results, however, show good correlation between the satellite and surface truth data.

Two field APT satellite receivers, one at La Paz and one aboard the R/V *David Starr Jordan* were uniquely used in this experiment to collect infrared data for use in the postsurvey analysis. These field data augmented those SR data stored aboard the satellite for later transmission to the National Environmental Satellite Service (NESS) at Suitland, Maryland. In addition to collecting data, the field station's satellite infrared and visual photographs provided LITTLE WINDOW 2 scientists with a constant visual representation of the region's sea surface temperature field and cloud cover.

Upon completion of the survey phase of LITTLE WINDOW 2, the infrared data collected by the APT stations were digitized and contoured. These analyses were then compared to the analyses of SR (stored) data. Early comparisons show that the APT station SR data generally contain the least system noise of the two methods of data collection. Extensive comparison work must still be done, but these early results indicate that field APT stations are capable of high quality infrared data collection, and that these data may be used in oceanographic surveys and in their postsurvey analysis.

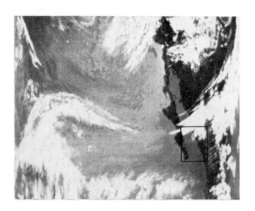

DRSR imagery.   DRSR data analysis.

Fig. 7. Infrared Imagery and Preliminary Analysis of LITTLE WINDOW 2 DRSR data—NOAA 1. Little Window 2 was an ocean-truth experiment comparing satellite derived data to aircraft and ship derived data. The daytime infrared imagery and sea surface temperature analysis shown above are from Direct-Readout Scanning Radiometer data received from NOAA 1 on 5th May 1971. Simultaneously to the overflight of the Gulf by the satellite, a research aircraft flew a track over the region collecting airborne radiation temperature in a square 200 km on a side. Five research vessels of the United States and Mexico were positioned at strategic points in the square to provide oceanographic data to aid in the comparison study.

IV. OCEANOGRAPHIC USE OF SATELLITE VISUAL SENSOR DATA

Pictures of the earth's surface from space have been available from visual range sensors aboard a number of satellites. However, only those sensors aboard the polar orbited NOAA (ITOS), NIMBUS, and ESSA, and the equatorial geosynchronous ATS satellites have produced pictures with the detail necessary for oceanographic work.

One of the most interesting oceanographic features apparent in satellite visual sensor pictures is the delineation, made by clouds, of strong thermal current or upwelling regions. For example, pictures taken by ATS 2 on 10th April 1967 apparently define the boundaries of the cold Peru Current as well as its westward extension, the Pacific South Equatorial current (La Violette and Chabot, 1968). These and other satellite pictures seem to indicate that a regular relationship exists between the water temperature of the major ocean currents and the regional cloud conditions.

The inference that variation in the strength and direction of the currents both seasonally and from year to year may be diagnosed by examination of these cloud patterns is unavoidable, and studies in this direction have been made (La Violette and Seim, 1969).

Further examinations of cloud patterns indicate they can be of more oceanographic use than the mere delineation of gross current features. For example, the actual time of a seasonal change in the atmospheric circulation of a region—and therefore the time of change in oceanic conditions affected by this atmospheric circulation—may be dated by satellite pictures showing changes in the region's daily cloud patterns.

Other indications may include unusual storms in a region, which would indicate strong vertical mixing and, thus, a deeper thermocline. Conversely, the lack of cloud cover and long periods of intense insolation would indicate a strongly stratified water column with a shallower than normal thermocline.

All cloud patterns peculiar to a region are not necessarily associated with the conditions of the water below. However, by combining the knowledge of a region's cloud patterns as shown in satellite pictures with a good comprehension of its oceanographic conditions, the variation in the vertical and horizontal oceanic temperature structure of the region may be determined.

The pictures taken by the NOAA (ITOS) and odd-numbered ESSA satellites have been archived in a format suitable for computer manipulation. The National Environmental Satellite Service has formulated several methods of working with the data including global mosaics, brightness composites and gray-tone enhancements (Booth and Taylor, 1968). The immediate use of these manipulations is for meteorological purposes, but they can be used for oceanography as well.

The brightness composites may be used to remove transitory clouds over ice fields artificially and thus define the areal coverage of the ice during the compositing period. The method is similar to the example

13

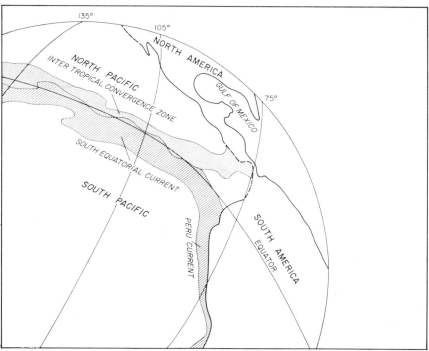

Fig. 8.—see caption opposite.

given of infrared data compositing. Within the time period of the composite, each day's satellite picture is examined, resolution point by resolution point, and the darkest tone for each resolution point printed. If the proper number of days is chosen (minimum number so as not to hide ice movement and maximum number to discard transitory clouds), the only portion of the print that remains white is the ice field and any persistent cloud cover.

Ice information also can be furnished to scientists in the field using APT satellite receivers. In April 1972, an aerial investigation of the ice movement off the east coast of Greenland was guided by APT satellite pictures received aboard the aircraft during its flight over the ice. These visual-range pictures from ESSA 8 and NIMBUS 4 provided the U.S. Navy scientists aboard the aircraft with vital data on the distribution and structure of the ice as well as associated weather conditions. Because of the northern latitude of the survey approximately 50% overlap occurred during consecutive passes, and several passes from each satellite were usable each day. Thus, pictures of the study area were received almost continuously between 08.00 and 14.00 G.M.T.

V. OCEANOGRAPHIC USE OF MANNED SPACECRAFT PHOTOGRAPHY

Photographs of the ocean from space have proved to be a unique, tantalizing tool for oceanic investigations. The high resolution of these color photographs has shown a wealth of detail impossible to duplicate by other methods now available. Surface and near-surface conditions appear in the photographs as sea scars, rips and long period waves. Many of these features have been shown to exist over large areas on a scale previously unimagined.

Probably the most interesting surface phenomena visible on the photographs are the effects made noticeable by the sun's reflection off the water's surface. This reflection, called "sun glitter" (Chapter 3 in

---

Fig. 8. Cloud Patterns associated with the Peru and Pacific South Equatorial Currents—ATS 2. This unusual vidicon picture, transmitted from ATS 2 at 1400 EST 10th April 1967, shows the region of upwelling associated with the Peru Current, and its westward extension, the Pacific South Equatorial Current. These two current systems are outlined by the low surface clouds shown as light gray off the South American coast and are clearly defined as the dark region immediately west of South America and adjacent to the Equator. The picture also shows Baja California, Yucatan Peninsula, Gulf of Mexico and the coastline from Colombia to Chile. Visible to the north of the Pacific South Equatorial Current are the cloud patterns associated with the Intertropical Convergence Zone.

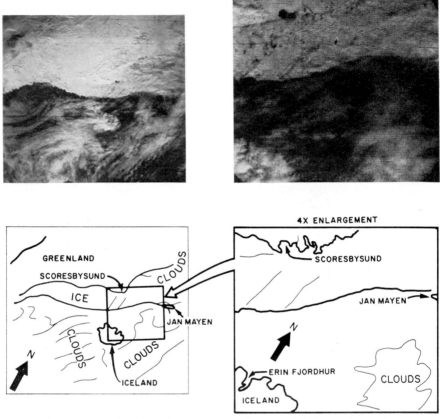

Fig. 9. Airborne APT Receiver Satellite Imagery Photograph of the Survey Area, East Coast of Greenland—NIMBUS 4 Pass 9966, 19th April 1972. Visual as well as infrared imagery may be used for operational purposes during an oceanographic survey. The NIMBUS 4 vidicon pictures shown above were received by an APT station aboard the U.S. Navy research aircraft, *Arctic Fox*, while in flight off the coast of Greenland. This was part of a series of flights made during April 1972 that collected data on the region's ice-sea relationship using airborne remote sensing instrumentation. The satellite's pictures provided the aircraft's scientists with continual information on the ice distribution and condition as well as the region's cloud cover. The enlargement was made by the aircraft's APT receiver using a tape recording of the satellite's original analog signal.

this book), usually appears as a bright irregular zone on the space pictures. The strength of the surface winds and the sea conditions seem to be the main factors influencing the pattern and brightness of the sun glitter. The unusual surface structures which are revealed in the areas of sun glitter—long lines, striations, discontinuities, and gyres—cannot

Fig. 10. The Islands of Socotra and The Brothers—APOLLO 7. As seen by the crew of APOLLO 7 on 15th October 1968 from 100 miles in space, the waters around the islands of Socotra and The Brothers, caught in the sun's reflection, reveal complex surface phenomenon impossible to view by ordinary means. Situated just off the East African coast and north of one of the strongest upwelling areas in the world, the islands form a deflective barrier to the northeastward movement of the cold, upwelled water. The vortices, slicks, swells and other lines which are visible reveal current direction, internal waves, and regions of convergence and divergence. Such information is invaluable in understanding the oceanography of one of the world's largest undeveloped fishing regions.

easily be explained. Of all the features visible in space photography, these are probably the most in need of surface investigation.

The fact that all of the pictures have been in color has aided visual examination. The contrast of color enhances the differences between oceanic structures of interest, delineating features that seem continuous when seen in black and white.

Broadband color film has its limitations, however, in differentiating phenomena close together on the spectral scale. Multispectral photography taken during the flight of APOLLO 9 was an attempt at expanding the definition characteristics of the presently used color system. The spectral ranges of the films used during this mission are similar to the ranges planned for the proposed Earth Resources Technology Satellites.

The GEMINI color pictures have been subjected to a series of experiments to extract spectral information for depth determination. One experiment conducted by Ross (1968) is based on the idea that in shallow, relatively clear water, the amount of light reflected from the bottom is inversely proportional to depth. Consequently, areas of equal color intensity may be supposed to indicate areas of equal depth. Ross assigned false colors to certain densities and from these was able to construct bathymetric charts, which are compatible with available hydrographic charts.

Photographs taken by astronauts aboard MERCURY, GEMINI, and APOLLO spacecraft have been confined to obvious, unusual oceanic conditions and to flight paths. With the exception of those of APOLLO 6 and 8 and the full earth pictures taken on the way to the moon, the manned spacecraft photographs of the earth have been limited to those taken from an equatorial band stretching from approximately 35° N to 35° S and from a height of approximately 180 km. Thus, photographs of the Gulf of California, northern Gulf of Mexico, Florida and the Red Sea abound, while the fishing banks of Newfoundland, Iceland and the North Sea have not been photographed.

APOLLO 9 was the most recent manned spaceflight with earth photography as part of its prime mission. Not until late 1973, when SKYLAB will be orbited, will new earth-oriented photographs be available. The photographs that have been taken to date, however, provide much material for study and must be examined carefully to extract total information.

VI. Future Satellites and Their Sensors

A host of new satellites are now being planned or prepared for launch. These satellites and their oceanographically applicable instruments are listed in Table 1. Of particular interest are the Very High Resolution Radiometer (VHRR) to be placed on NOAA 2 (née ITOS D) and the Visual and Infrared Spin Scan Radiometer (VISSR) planned for GOES 1 (née SMS A). As its name implies, the VHRR is a two-channel (infrared and visible) radiometer with a surface resolution of approximately one kilometer at the sub-satellite point for both channels. The VISSR is a nine-channel (one infrared and eight visible) radiometer, with a surface resolution of approximately 10 km for the infrared channel and approximately one kilometer for each of the visible channels. A full disc, infrared and visual picture of the earth from each channel of the radiometer will be collected every twenty minutes. The data from the VHRR and VISSR will be transmitted by their satellites to any properly equipped ground station within the satellites' line of sight. The high resolution and multichannel configuration of the radiometers, however, will require ground readout facilities with far greater flexibility and image quality than even the most sophisticated of present APT facilities. Because these exceptional quality sensors will be orbited aboard operational satellites, and therefore, will be aboard each succeeding satellite in the series, it behooves interested oceanographers to investigate the possibility of constructing these special APT ground stations. Information on the requirements of such a station may be obtained from the Director of NASA's Goddard Space Flight Center, Greenbelt, Maryland.

As the data from the elaborate sensors aboard this new suite of satellites become increasingly available, proven as well as new data handling techniques will produce an increasing source of oceanographically applicable data. The limit of their application will depend on the ingenuity of the members of the oceanographic community and the utility the community makes of this ingenuity.

VII. A Lexicon on "Oceanographic" Satellite Names

The names of satellites and their operational systems are a confusing conglomerate of acronyms. It is hoped that the following discussion and Table 2 will arrange them in some logical order.

The Television Infrared Observational Satellites (TIROS) were weather research satellites operated by the National Aeronautics and Space Administration (NASA). There were ten of these research satellites launched between April 1960 and July 1965.

The successful development of TIROS led to the TIROS Operational System or TOS. As the name implies, the satellites used in the TOS system are operational rather than research satellites and are managed by the National Oceanographic and Atmospheric Administration's (NOAA) National Environmental Satellite Service (NESS) after being initially launched and checked out by NASA's Goddard Space Flight Center (GSFC). In its early stages this system required two TOS

TABLE 1. Proposed NASA-NOAA satellites.

Satellite designation (pre-launch/ post-launch)	Sensor mode	Sensor	Sensor resolution (km)	Data transmission mode	Approximate launch date	Approximate altitude (km)	Orbit	Operating agency
ITOS D/NOAA 2	IR/visible	SR	8/4	DR and stored	October 1972	1450	Polar	NOAA
	IR/visible	VHRR	1·0	DR and stored				
Nimbus D/Nimbus 5	IR/visible	SCMR	0·5	DR and stored	January 1973	1110	Polar	NASA
	Microwave	ESMR	25	Stored				
	IR	TWIR	10	Stored				
ERTS A/ERTS 1	Near IR/visible	RBV	0·1	DR and stored	June 1972	910	Polar	NASA
	Near IR/visible	MS	0·1	DR and stored				
Skylab	Multispectral	Camera	0·1 varied	Manned mission	Mid to late 1973	460	50° Inclined Equatorial	
	IR	IR spectrometer	0·5 varied					
	Multispectral	IOCH radiometer	0·1 varied					
	Microwave	Microwave scatterometer	10					
ITOS E/NOAA 3	IR/visible	SR	8/4	DR and stored	Late 1973	1450	Polar	NOAA
	IR/visible	VHRR	1·0	DR and stored				
SMS A/GOES 1	IR/near IR/visible	VISSR	1·0	DR	Oct. 1973	20 225	Geosyn. 100° W at Equator	NOAA
ERTS B/ERTS 2	Near IR/visible	RBV	0·1	DR and stored	Late 1973	910	Polar	NASA
	Near IR/visible	MS	0·1	DR and stored				

satellites to be in orbit at all times. One satellite broadcasted visible-range sensor data to any Automatic Picture Transmission (APT) ground station within the satellite's line of sight. The second satellite limited its transmission of visible-range data to Command Data Acquisition (CDA) stations at Chincoteague, Virginia, and Gilmore Creek, Fairbanks, Alaska.

The first satellites of the TOS system were the Environmental Survey Satellites (ESSA). Nine of these satellites were launched, the even numbered ones equipped to transmit data to APT stations, the odd numbered ones programmed to transmit data to CDA stations.

With the launching of TIROS M in December 1969, the second generation of the TOS system was inaugurated. Once in orbit, TIROS M was redesignated Improved TIROS Operational Satellite (ITOS). ITOS performed the work of two ESSA satellites in that it transmitted data to both APT and CDA stations. These satellites had the additional capacity to collect infrared data. With the establishment of the U.S. Government Agency, NOAA, the second ITOS in the series was named: NOAA 1.

The numerical and alphabetical designation of the satellites within a series denotes their launch status. Prior to its launch, a satellite is given a letter designation for planning purposes. After launch and once it has attained orbit, it is reclassified with a numeral to designate its sequence in the satellite series. For example, ITOS D when launched will be redesignated NOAA 2. When NASA's Synchronous Meteorological Satellite (SMS) A is placed in orbit, it will become part of NOAA's new Geosynchronous Meteorological Environmental Satellite Series. It's name, thereafter, will be GOES 1.

The Application Technology Satellite (ATS) series includes research satellites in geosynchronous and equatorial orbits. There are two ATS satellites currently in operation: ATS 1 and 3. ATS 1 holds a position over the Pacific, while ATS 3 is in orbit over South America.

NIMBUS is pleasantly different in being named after the Latin word for rain clouds rather than being an acronym. It is aptly named as it was developed to study clouds with visual range sensor systems during the day and with infrared sensors at night.

NIMBUS was originally designed to overcome basic limitations in the TIROS satellites and to become the first in a series of operational satellites. However, for various reasons, NIMBUS was redesignated a research and development satellite and ESSA 2 replaced it as the first operational satellite. It has remained a research satellite and has been the work platform for many of the radiometers which were placed on or planned for placement on operational satellites.

TABLE 2. Satellites capable of oceanographic data acquisition (as at 1st June 1972).

Satellite	Data mode	Sensor	Launch date	End operations	Operational period	Orbit	Inclination (°)	Average altitude km(nm)	Picture resolution[1] km(nm)
Nimbus 1	Infrared	HRIR	28 Aug. 64	23 Sept. 64	26 days	Polar	98·7	678 (366)	8(4·3) at 925(499)
	TV	AVCS		23 Sept. 64	26 days				0·9(0·5) at 925(499)
Nimbus 2	Infrared	HRIR	15 May 66	15 Nov. 66	174 days	Polar	100·3	1137 (614)	9(5) at 1100(595)
	Infrared	MRIR		28 Jul. 66	75 days				56(30) at 1110(600)
	TV	AVCS		31 Aug. 66	109 days				0·9(0·5) at 1100(595)
Nimbus 3	Infrared	HRIR	14 May 69	3 Jan. 70	256 days	Polar	81·0	1071 (571)	9(5) at 1100(595)
	Infrared	MRIR		4 Feb. 70	288 days				56(30) at 1100(595)
	TV	IDCS		25 Jan. 70	278 days				0·9(0·5) at 1100(595)
Nimbus 4	Infrared	THIR	8 Apr. 70	8 Apr. 71	12 months	Polar	99·8	1093 (589)	9(5) at 1100(595)
	TV	IDCS		Operational	Continuous				0·9(0·5) at 1100(595)
ITOS 1	Infrared	THIR	23 Jan. 70	18 Jun. 71	511 days	Polar	102·1	1446 (780)	9(5) at 1100(595)
	TV	AVCS		19 Aug. 71	573 days				0·9(0·5) at 1100(595)
NOAA 1	Infrared	HRIR	11 Dec. 70	19 Aug. 71	251 days	Polar	102·0	1447 (781)	9(5) at 1100(595)
	TV	AVCS		Operational	Continuous				0·9(0·5) at 1100(595)
ESSA 1	TV	AVCS	3 Feb. 66	12 June 68	861 days	Polar	97·8	769 (415)	3·7(2·0) at 1435(775)
ESSA 3	TV	AVCS	2 Oct. 66	9 Oct. 68	738 days	Polar	101·0	1434 (774)	3·7(2·0) at 1435(775)
ESSA 5	TV	AVCS	20 Apr. 67	20 Feb. 70	1037 days	Polar	101·9	1386 (748)	3·7(2·0) at 1435(775)
ESSA 7	TV	AVCS	16 Aug. 68	19 Jul. 69	338 days	Polar	101·7	1452 (784)	3·7(2·0) at 1435(775)
ESSA 8	TV	AVCS	15 Dec. 68	Operational	Continuous	Polar	102·0	1653 (892)	3·7(2·0) at 1435(775)
ESSA 9	TV	AVCS	26 Feb. 69	Operational	Continuous	Polar	102·0	1686 (910)	3·7(2·0) at 1435(775)
ATS 1	TV	SSCC	7 Dec. 66	Operational	Continuous	Geo-synchronous (151° W)	0·2	36 614 (19 757)	4·6(2·5) at 36 600(19 750)
ATS 2	TV	AVCS-camera 1	6 Apr. 67	19 Jul. 67	106 days	Equatorial	28·3	N.A.	0·9(0·5) at 11 100(5990)

ATS 3	TV	AVCS-camera 2	5 Nov. 67[3]	19 Jul. 67	106 days	Geo-synchronous (73° W)	0·4	35 518 (19 166)	19(10) at 11 100(5990) at
	TV	MSSCC[2]		Operational	Continuous				3·7(2·0) at 35 500(19 155)
	TV	IDCS		Operational	Continuous				7·8(4·2) at 35 500(19 155)
Mercury 6 (MA-6)	Color[4] photogr.	35 mm camera system	20 Feb. 62	20 Feb. 62	4 h 55 min 23 s	Varied orbits	32·5	211 (114)	Variable depending on camera angle
Mercury 7 (MA-7)	Color photogr.	35 mm camera system	24 May 62	24 May 62	4 h 56 min 05 s	Varied orbits	32·6	215 (116)	Variable depending on camera angle
Mercury 8 (MA-8)	Color photogr.	70 mm camera system	10 Mar. 62	3 Oct. 62	9 h 13 min 11 s	Varied orbits	32·6	222 (120)	Variable depending on camera angle
Mercury 9 (MA-9)	Color photogr.	70 mm camera system	15 May 63	16 May 63	34 h 19 min 49 s	Varied orbits	32·5	215 (116)	Variable depending on camera angle
Gemini 3	Color photogr.	70 mm camera system	23 Mar. 65	23 Mar. 65	4 h 52 min 31 s	Varied orbits	32·6	193 (104)	Variable depending on camera angle
Gemini 4	Color photogr.	70 mm camera system	3 June 65	7 June 65	97 h 56 min 22 s	Varied orbits	32·6	222 (120)	Variable depending on camera angle
Gemini 5	Color photogr.	70 mm camera system	21 Aug. 65	29 Aug. 65	190 h 55 min 14 s	Varied orbits	32·6	258 (139)	Variable depending on camera angle
Gemini 7	Color photogr.	70 mm camera system	4 Dec. 65	18 Dec. 65	330 h 35 min 01 s	Varied orbits	28·9	298 (161)	Variable depending on camera angle
Gemini 6	Color photogr.	70 mm camera system	15 Dec. 65	16 Dec. 65	25 h 51 min 24 s	Varied orbits	28·9	298 (161)	Variable depending on camera angle
Gemini 8	Color photogr.	70 mm camera system	16 Mar. 66	17 Mar. 66	10 h 41 min 26 s	Varied orbits	28·9	298 (161)	Variable depending on camera angle
Gemini 9	Color photogr.	70 mm camera system	3 June 66	6 June 66	72 h 20 min 55 s	Varied orbits	28·8	280 (151)	Variable depending on camera angle
Gemini 10	Color photogr.	70 mm camera system	18 Jul. 66	21 Jul. 60	70 h 46 min 45 s	Varied orbits	28·8	295 (159)	Variable depending on camera angle
Gemini 11	Color photogr.	70 mm camera system	12 Sept. 66	15 Sept. 66	71 h 15 min 45 s	Varied orbits	28·8	297 (160)	Variable depending on camera angle
Gemini 12	Color photogr.	70 mm camera system	11 Nov. 66	15 Nov. 66	94 h 32 min 33 s	Varied orbits	28·9	300 (162)	Variable depending on camera angle
Apollo 4[5]	Color photogr.	70 mm camera system	9 Nov. 67	9 Nov. 67	8 h 23 min	N.A.	32·7	N.A.	Variable depending on camera angle
Apollo 6[5]	Color photogr.	70 mm camera system	4 Apr. 68	4 Apr. 68	10 h	Varied orbits	32·5	178 (96)	Variable depending on camera angle
Apollo 7	Color photogr.	70 mm camera system	11 Oct. 68	22 Oct. 68	260 h 09 min 45 s	Varied orbits	31·6	274 (148)	Variable depending on camera angle
Apollo 9	Color photogr.	70 mm camera system	3 Apr. 69	13 Apr. 69	241 h	Varied orbits	32·0	Varied	Variable depending on camera angle

[1] Where values are applicable, resolution is measured at the subsatellite point. For example, the Nimbus 2 HRIR at an altitude of 1100 km(595 nm) has a subsatellite ground resolution of 9 km(5 nm).

[2] Color pictures from ATS 3 are available only through January 1968. Black and white pictures are at present received using the green signal.

[3] Although launch date for ATS 3 was 5 Nov. 67, the MSSCC system did not commence operations until 8 Nov. 67; the IDCS became operational on 7 Nov. 67.

[4] The Oceanographic value of mercury photographs is difficult to determine; these flights are included for completeness only.

[5] Apollo 4 and 6 were unmanned flights; pictures were taken automatically every 10 s.

N.A. = Not applicable.

REFERENCES

Booth, A. L. and Taylor, V. R. (1968). *Nat. Env. Sat. Serv.*, TM 9, Washington, D.C.

La Violette, P. E. and Allison, L. (1969). *Trans. Amer. Geo. Union*, EOS, **50**, 213. Washington, D.C.

La Violette, P. E. and Chabot, P. L. (1969). *Deep-Sea Res.*, **16**, 539–547.

La Violette, P. E. and Chabot, P. L. (1968). *Deep-Sea Res.*, **14**, 123–124.

La Violette, P. E. and Seim, S. E. (1970). Naval Oceanographic Office TR-215. Washington, D.C.

La Violette, P. E. and Stuart, L. Jr. (1972). *Trans. Amer. Geo. Union*, EOS **53**, 213. Washington, D.C.

Ross, D. S. (1968). *In* Technical Papers of the "Ninth Meeting of *ad hoc* Spacecraft Oceanography Advisory Group", Naval Oceanographic Office, Washington, D.C.

Smith, W. L., Rao, P. K., Kottler, R. and Curtis, W. R. (1970). *Mon. Wea. Rev.*, **48**, 604–611.

Szekielda, K. H. (1970). NASA preprint X-651-70-415, G.S.F.C. Greenbelt, Md.

Schwalb, A. (1972). NOAA Technical Memorandum NESS 35.

Vermillion, C. H. (1969). NOAA SP-5080.

Chapter 14

The Remote Sensing of Spectral Radiance from below the Ocean Surface*

ROSWELL W. AUSTIN

*Visibility Laboratory, Scripps Institution of Oceanography,
University of California, San Diego, La Jolla, California, U.S.A.*

I. INTRODUCTION

Since the advent of manned spaceflight, there have been many interesting examples of photography obtained from space over ocean and coastal regions. Some of these photographs show ocean bottom features with remarkable clarity and detail. Some show turbid plumes of sediment-bearing water in the littoral zone caused by the action of tides or longshore currents, or of effluent extending seaward many kilometers from river mouths. Still others show changes in water coloration which are apparently associated with changes in the living or detrital organic material in the water. The potential applications of such space photography to the solution of oceanographic problems and the even greater potential of directly-obtained spectroradiometric data are manifestly obvious.

The visible region of the electromagnetic spectrum is unique in that it allows information to be carried from below the ocean surface to a

* This work was supported in part by the Spacecraft Oceanography Project Office of the U.S. Naval Research Laboratory under Contract N00014-69-A-0200-6033.

remote sensor. More specifically, it is only in the narrower spectral region from about 400 to 600 nanometers that sensible data can be obtained from water depths greater than a few meters. At wavelengths longer than 600 nm, water absorption quickly attenuates the signal generated by bottom or water reflectance. In fact, at some wavelengths between 600 and 700 nm, the component of the remotely sensed signal which is due to the radiance of the sky reflected by the ocean surface becomes larger than the component which originated from below the surface. Beyond this wavelength, the remote sensing technique becomes more useful for the appraisal of surface than of subsurface phenomena. At the shorter wavelengths, that is below about 400 nm, the water absorption also increases rapidly in all but the clearest of oceanic waters. Perhaps even more significant at the shorter wavelengths is the rapid increase of the veiling effects of the atmosphere. As the wavelength decreases that portion of the radiation emanating from the ocean which reaches the remote sensor becomes insignificant when compared to the radiance of the scattered daylight in the path of sight.

The remote optical sensor utilizes the spectral, spatial and temporal information encoded in the radiance signal which it receives. Inferences about such variables as the water depth, turbidity, or chlorophyll level are drawn after this encoded information has been processed in some manner. Unfortunately, the apparent signal with which the remote sensor must work usually differs markedly from the inherent signal which exists at or just below the water surface. As it is this latter signal which contains the desired information about the water or the ocean bottom, it is important to understand how it is attenuated and modified in travelling upward through the atmosphere to the sensor.

The apparent radiance of the ocean surface when viewed from altitude z will be denoted L_z. This radiance may be broken down into two components. Thus,

$$L_z = L_o T_a + L^*, \tag{1}$$

where L_o is the inherent radiance of the sea if viewed just above the surface, T_a is the atmospheric transmittance for image-forming light over the path of sight, and L^* is the path radiance, i.e. the contribution to the apparent radiance caused by the scattering of ambient daylight into the path of sight.

The inherent radiance, L_o, in turn may be further broken down into two components: the underwater radiance transmitted through the water/air interface, L_w, and the radiance due to the Fresnel reflection of sky light by the water surface. Thus,

$$L_o = L_w + L_r \tag{2}$$

and combining eq. (1) and (2) we obtain,

$$L_z = (L_w + L_r)T_a + L^*. \tag{3}$$

In the following sections we will examine in greater detail the magnitude and spectral dependence of the various factors which contribute to or affect in some way the radiance which is available to a remote sensor above the atmosphere. In Section V we will present examples of the application of the concepts and techniques we have described. Examples of apparent radiances computed from field measurements of the component parameters will be given.

II. THE WATER SURFACE EFFECTS

The presence of the water/air boundary has a significant effect on the amount of energy which enters the ocean and on the magnitude and directional properties of the radiance signal which leaves the water surface on its upward journey to the remote sensor. Fortunately, the phenomena of refraction, reflection, and transmission at the interface can be calculated from a knowledge of the indices of refraction and the local surface windspeed. These phenomena, as they affect the remote sensing problem, will be reviewed briefly in the next paragraphs. Figure 1(a) and 1(b) establish the nomenclature.

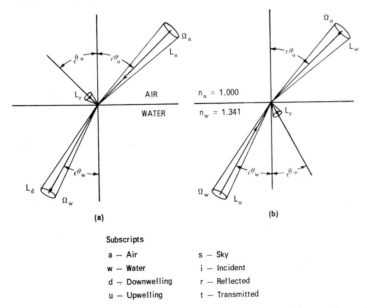

Subscripts			
a — Air		s — Sky	
w — Water		i — Incident	
d — Downwelling		r — Reflected	
u — Upwelling		t — Transmitted	

Fig. 1. Geometry and nomenclature for radiance at the air/water boundary.

A. REFRACTION

The refraction at the calm surface may be computed from Snell's law, i.e. $n_a \sin \theta_a = n_w \sin \theta_w$. The index of refraction for air, n_a, will be taken as 1·000. Tables of refractive index for sea water, n_w, are given in Dorsey (1940). It should be noted that the index of refraction for water varies slightly with wavelength, temperature, and salinity. However, small variations in this index will not affect the computation of the radiance components in the remote sensing problem to any significant degree. We will assume in what follows, therefore, a mean refractive index for sea water of 1·341. Figure 2 shows the relationship between the angle θ_a in air and θ_w in water.

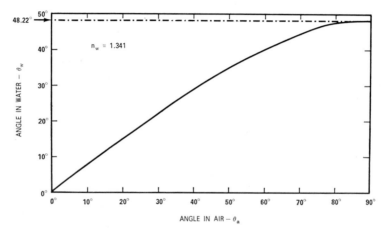

Fig. 2. Curve for refraction at the air/water boundary. Upwelling radiance having angles greater than 48·22° suffers total reflection when incident on a flat calm surface.

The first important implication of the refraction at the surface for the remote sensing problem is that the entire sky dome is compressed underwater into a cone subtending a half angle of 48·22° and, conversely, that all radiance observed by a sensor above the surface originates from within this same underwater cone. Wind-roughening of the surface, of course, will extend and diffuse the edge of this cone somewhat. The second important implication is that flux contained within a small solid angle, Ω_w, below the surface will be spread into a larger solid angle, $\Omega_a = n^2\Omega_w$, above the surface. Therefore, a radiance emerging from the water surface will be decreased by the factor $1/n^2$ or 0·555 by virtue of refraction alone.

B. REFLECTION

The reflectance of the water/air boundary may be computed from the Fresnel reflectance formulae. For polarized light, these are

$$r_\perp = \frac{\sin^2 (\theta_a - \theta_w)}{\sin^2 (\theta_a + \theta_w)} \tag{4}$$

and

$$r_\| = \frac{\tan^2 (\theta_a - \theta_w)}{\tan^2 (\theta_a + \theta_w)}. \tag{5}$$

The combined form for unpolarized light is

$$r = \frac{1}{2}\frac{\sin^2 (\theta_a - \theta_w)}{\sin^2 (\theta_a + \theta_w)} + \frac{1}{2}\frac{\tan^2 (\theta_a - \theta_w)}{\tan^2 (\theta_a + \theta_w)}. \tag{6}$$

Figure 3 shows $r_\perp$, $r_\|$, and r plotted against both θ_a and θ_w. It will be noted that $r_\|$ goes to zero when $\theta_a + \theta_w = 90°$, i.e. when $\tan (\theta_a + \theta_w) = \infty$. The particular angle which meets the condition (commonly

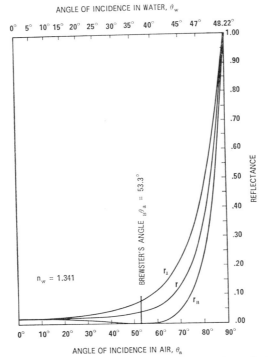

Fig. 3. Reflectance of undisturbed air/water boundary for polarized light ($r_\perp$ and $r_\|$) and for unpolarized light (r).

14

referred to as Brewster's angle) may be computed from the relationship $\tan {}_{B}\theta_{a} = 53\cdot3°$. A remote sensor viewing the ocean surface at this angle may be rendered insensitive to the sky light reflected by the surface if the camera or radiometer is fitted with a properly-oriented polarizer. Unfortunately, at this viewing angle, the air mass is increased to 1·67, which may severely increase problems associated with the atmospheric transmittance and path radiance in a significant number of instances. Additionally, the introduction of the polarizer must reduce the desired subsurface water signal by at least a factor of two and this may prove detrimental in some sensing systems having high information rates.

Thus far, the assumption has been that the ocean surface is smooth. However, the surface of the sea is seldom calm. By the action of the wind, for instance, the tipping of the surface causes the effective reflectance of the surface to change significantly near grazing incidence for light entering the water from above and near the critical angle for light emerging from below. In this connection, Duntley (1952), (1954) and Cox and Munk (1954) studied the manner in which the statistical distribution of wave slopes is dependent upon surface windspeed. Gordon (1969), using the simplified assumption that the crosswind and upwind slope statistics may be represented by a single circular Gaussian distribution of slopes, has generated time-averaged reflectances for various windspeeds from 0 to 19 m s⁻¹. Figure 4 shows the manner in which the time-averaged reflectances vary with windspeeds of 0, 4, 10,

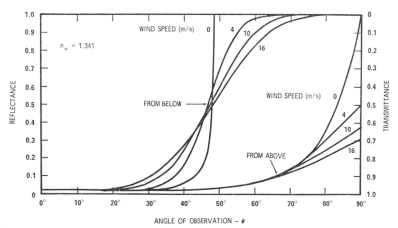

Fig. 4. Reflectance and transmittance of water/air boundary as a function of the angle of observation for several windspeeds. Curves at left are for radiance coming up from below the surface; those on the right are for radiance incident from above.

and 16 m s^{-1}. Tables 1 and 2 present these reflectances for selected angles of incidence for the same windspeeds. In addition, Table 2 lists the effective emergent reflectance for diffuse upwelling irradiance. The computation of the emergent reflectances was made assuming that the upwelling radiance just below the surface is constant with angle.

TABLE 1. Time-averaged air/water surface reflectance from above

Observation angle θ_a	Windspeed m s^{-1}			
	0	4	10	16
0°	0·0211	0·0211	0·0212	0·0212
10°	0·0211	0·0212	0·0213	0·0214
20°	0·0213	0·0214	0·0217	0·0220
30°	0·0222	0·0226	0·0232	0·0239
40°	0·0253	0·0262	0·0276	0·0291
50°	0·0346	0·0366	0·0394	0·0420
60°	0·0610	0·0646	0·0686	0·0709
70°	0·1354	0·1365	0·1316	0·1247
80°	0·3502	0·2919	0·2371	0·2046
90°	1·0000	0·4934	0·3642	0·3002

For sea water, refractive index $n_w = 1\cdot341$.

TABLE 2. Time-averaged water/air surface reflectance from below

Observation angle θ_w	Windspeed m s^{-1}			
	0	4	10	16
0°	0·0211	0·0211	0·0213	0·0217
10°	0·0211	0·0213	0·0218	0·0228
20°	0·0218	0·0227	0·0255	0·0334
30°	0·0265	0·0325	0·0613	0·0961
35°	0·0350	0·0602	0·1234	0·1686
40°	0·0588	0·1559	0·2367	0·2741
45°	0·1529	0·3801	0·4065	0·4131
50°	1·0000	0·6718	0·5988	0·5629
55°	1·0000	0·8905	0·7715	0·7055
60°	1·0000	0·9807	0·8967	0·8277
r_e	0·485	0·478	0·470	0·463

For sea water, refractive index $n_w = 1\cdot341$.
r_e = emergent reflectance of water/air boundary for diffuse upwelling irradiance.

Whereas this is not precisely correct, the computations do serve to show the magnitude of this reflectance and the manner in which it is effected by wind-roughening of the surface.

For most windspeeds and viewing geometries of interest in remote sensing, the effect of wind on the magnitude of the time-averaged reflectance values as indicated in Fig. 4 and Tables 1 and 2 is not a major one. However, the roughening of the surface does have a major effect in determining the portion of the sky that is reflected upward toward the sensor. Thus, for example, the angular size of the sun's glitter path will be determined by the windspeed. And as a consequence, the direction and field of view of the sensor, as well as the optimization of the orbit of the spacecraft (or time of observation from an aircraft), will be determined largely by the necessity for keeping the glitter pattern out of the field of view of the sensor for those systems designed to work with radiances originating from below the surface.

C. TRANSMISSION

The radiance transmitted through the water/air boundary is the residue of the incident radiance after the reflected component discussed previously has been subtracted. Thus, radiances having near-vertical angles of incidence suffer a reflectance of only 0·021 and, therefore, have a transmittance of 0·979. Figure 4 shows how the transmittance varies with windspeed and angle of observation for both entering ("from above") and emerging ("from below") radiances. The effect of the surface transmittance or of changes in it has little significance to the remote sensing problem except when the solar zenith angle is large (low sun) or when the angle of observation is large.

In the case of the low sun, the amount of flux transmitted through the surface into the water decreases rapidly with increasing angle of incidence; hence, the amount of upwelling radiance resulting from the interaction of this input flux with the water or the bottom is similarly reduced. The roughening of the ocean surface by the wind will produce a significant increase in the time-averaged transmittance through the surface for both the sunlight, when the sun is low, and the radiance of the sky immediately above the horizon.

For observation angles in air greater than about 50° (about 35° from the vertical underwater), the time-averaged transmittance of the surface for upwelling radiance from below becomes affected in a major way by windspeed. Thus this transmittance is decreased for angles almost out to the critical angle (48°) and beyond this angle, radiance that would have suffered total internal reflectance at a calm surface will be transmitted. It should be noted that the effect of the wind-related change in transmitted radiance from below is to cause the radiance at a point on the surface to be the sum of radiances from a time-varying set of directions below the surface. The resulting degradation in resolution will not be significant if the spatial resolution of the sensor is larger than the attenuation length for the radiance in the water.

III. ATMOSPHERIC DEGRADATION OF THE SIGNAL

Thus far we have seen that the apparent radiance which is available to the remote sensor is composed of various components operated on by the water/air interface and the atmosphere. Having described the nature of the reflectance and transmittance at the interface as it affects the remote sensing of subsurface phenomena, we will now proceed to describe the manner in which the atmosphere acts to degrade the information contained in the inherent radiance of the water, L_u.

The radiance at the surface, L_o, as given in eq. (2) may be rewritten to take into account the effects of refraction, transmittance, t, and reflectance, r, of the water surface as follows:

$$L_o = \frac{t}{n_w^2} L_u + r L_s. \tag{2a}$$

It is to be understood that all radiances are in general a function of both wavelength and angle, while the transmittance and reflectance of the water surface are a function of angle and windspeed but within our assumptions are not dependent on wavelength. L_s is the radiance of the sky in the set of directions which, when reflected by wind-roughened water surface, will fall within the field of view of the sensor. Combining eq. (2a) with eq. (1), we obtain,

$$L_z = \left[\frac{t}{n_w^2} L_u + r L_s \right] T_a + L^*. \tag{3a}$$

Here T_a and L^* are functions of wavelength and of the direction and length of the path of sight.

Only the radiance L_u contains useful information about water or the ocean bottom. All other terms in eq. (3a) act to decrease or obscure this inherent signal in its upward passage to the sensor. We will next provide some examples of the magnitude and spectral dependence of these degradation factors of atmospheric origin.

The application of visible region remote sensing to oceanography assumes first, of course, that there are no clouds in the path of sight. Unfortunately, there are regions of the oceans that have a high probability of cloud cover as, for example, in the northern latitudes in winter. In these areas frequent repetitive coverage is required if the remote sensor is to obtain a clear view of the surface. Other areas enjoy a high probability of cloud-free skies and are therefore more susceptible to the application of these techniques. Excellent cloud atlases are available from which one can determine these probabilities, such as Miller and Feddes (1971) and Greaves *et al.* (1971).

In apparently cloud-free areas there still exists the possibility of scattering from tenuous, thin clouds or from a turbid aerosol. Often it is difficult or impossible without *a priori* information to determine if a change in radiance is caused by atmospheric, surface, water, or bottom phenomena. In these ambiguous situations, sophisticated processing or careful analysis and interpretation by specialists is required. There are, however, numerous examples of photography for which there is little difficulty in correctly attributing the origin of the signal.

A. ATMOSPHERIC TRANSMITTANCE

The transmission of image-bearing radiance by the atmosphere, even in presumably clear areas, will vary with time. Numerous models have been prepared and techniques for the computation of atmospheric transmittance using such models have been devised by others. An excellent and readable review is provided by Moon (1940), while more recent effort is presented by Rozenberg (1966). McClatchey *et al.* (1971), provide a set of five atmospheric models and comprehensive supporting information from which transmittances may be calculated for various paths of sight.

Figure 5 presents representative curves for the transmittance of a vertical path of sight (air mass, $m = 1$) as a function of wavelength. The solid curves show the transmittance for four models. Curve A is

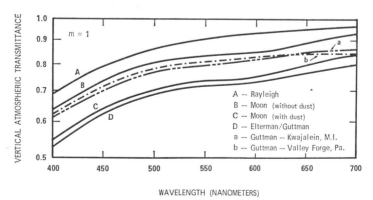

Fig. 5. Vertical beam transmittance of the atmosphere for various atmospheric models (curves A, B, C, D) and as measured by Guttman (1968) (curves a and b).

for an atmosphere having Rayleigh (molecular) scattering only. Curve B was obtained using Moon's model, but does not show the losses he computed for scattering by dust particles. Curve C was also obtained using Moon's model, but includes dust. Curve D is based on the model of Elterman. The dashed curves are for data obtained by Guttman (1968) through a maritime atmosphere in the mid-Pacific at Kwajalein in the Marianas Islands, and through continental air at Valley Forge, Pennsylvania. The data selected for plotting were for the clearest days at these two sites. However, all of Guttman's data with one exception fell within the bounds of curves A and D. The selected clear-day transmittances were higher than is predicted by Elterman's model (which Guttman used as a guide), or than is obtained using Moon's atmosphere containing dust. The shape of all the curves is much the same except for A, which does not show the effect of the Chappuis ozone absorption band.

A considerable body of data obtained in the southern California area by the Visibility Laboratory shows variation from 0·62 to 0·84 in the vertical transmittance for "clear" days. These data were obtained with a photopically-corrected sensor (wavelength of maximum transmission 555 nm). The data show some seasonal trends and a dependence upon the type of meteorological air mass within the seasons, but these effects may be peculiar to the local meteorological situation and not of general applicability. Of the types of air masses observed, the stable continental tropical air was usually the clearest and accounted for most of the observed transmittances above 0·75. Stable maritime polar air was next in clarity, with most of the observations ranging from 0·8 to 0·65. The remaining measurements were made in unstable air or in transitional periods between two types of air mass. The transmittances for these situations usually fell below 0·7, with the unstable air somewhat lower than the transitional air masses. All of these days were described as clear, with cloud cover usually less than one octa. The changes in transmittance were due primarily to changes in the scattering along the path of sight caused by water vapor and haze in the atmosphere. The spectral dependence of the transmittance under these conditions would be less pronounced than for the clearer atmospheres, but the general shape of the transmittance curve would not be grossly different from those shown in Fig. 5.

B. SKY RADIANCE AND SUN GLITTER

We have seen that the remote sensor signal contains a component which is proportional to the radiance of the sky in the directions reflected by the wind-roughened ocean surface. It was also shown in Fig. 4 and in Table 1 that the water surface reflectance varies with the angle of observation and windspeed. However, the variation in reflectance with observation angle is slight for steeply downward paths of sight, increasing from 2·1 to 2·5% for angles 0° to 40° from the vertical. As for windspeed, its effect on the value of space or time-averaged reflectance, i.e. within the field of view and integration time of the sensor, is not significant until observation angles in excess of 70° from the vertical are used. Such large angles are not generally useful for systems intended for the assessment of water color because of the long paths of sight through the atmosphere.

The direction of view determines not only the path length but also whether or not specularly-reflected sunlight is included within the field of view. Rozenberg and Mullamaa (1965) and Cox and Munk (1955) have shown the distribution of radiance in the solar glitter pattern for various solar zenith distances. According to their data a system operating with windspeeds up to 10 m s^{-1} would not be significantly affected by specularly-reflected solar radiance if the observation angles were greater than 30° to 40° from the solar specular point. Thus the radiance generated at the surface at greater angular distances is essentially that of the reflected sky light.

The angular distribution of the sky radiance is affected primarily by the solar zenith angle and atmospheric scattering. Figure 6 shows some distributions in the sun-zenith plane for selected solar positions, using data obtained on clear days in San Diego, Bocaiuva, Brazil (Richardson and Hulburt, 1949) and Stockholm (Hopkinson, 1954). All the curves are based on photopic measurements and, as might be expected, show similar shapes and magnitudes. Their usefulness lies in predicting, from them, the manner in which the sky radiance varies with solar zenith angle and angle of observation. Although the shape may vary some with wavelength for a given condition, the variations with time and place probably exceed those due only to wavelength.

Just as the specular reflection of a single point in the sky (viz. the sun) is spread out over a large angular region on the wind-roughened ocean surface, so a sensor viewing a particular spot on the surface away from the sun will see a large angular region of the sky reflected by the randomly distributed wave facets within the field of view. Therefore, the radiance seen reflected by the ocean surface is the convolution of the sky radiance surrounding the specularly reflected point in the sky and the statistical distribution of wave slopes at the point of observation on the surface. If the angle of observation is in the region of the minima of the curves in Fig. 6, the reflected sky radiance will be minimized and will not be greatly affected by windspeed.

θ_s	DATE	PLACE	REFERENCE
24°	6-2-67	San Diego	Vis Lab
39°	6-2-67	San Diego	Vis Lab
43°	10-7-69	San Diego	Vis Lab
50°	5-1947	Bocaiuva	Richardson & Hulburt (1949)
52°	9-2-64	San Diego	Vis Lab
67°	10-2-53	Stockholm	Hopkinson (1954)

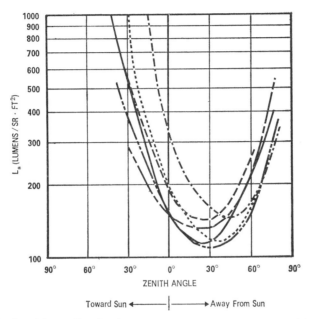

Fig. 6. Sky luminance distributions in sun-zenith plane for several locations and solar zenith angles, θ_s.

The spectral distribution of the radiance of the sky will depend upon the amount of scattering matter in the atmosphere and its size distribution, and the angular position of the point in the sky under consideration with respect to the sun and the horizon. In the darker portions of the sky, as for example near the minima in Fig. 6, the sky is the bluest and the correlated color temperatures for its spectral radiance may be 40 000 to 60 000°K. As the point of observation approaches the sun or the horizon, the color temperature decreases, trending toward that of sunlight, i.e. 5500 to 6500°K. Figure 7 shows several spectral distributions for daylight, all normalized at a wavelength of 560 nm. The curves are from the data of Henderson and Hodgkiss (1963) and from

14*

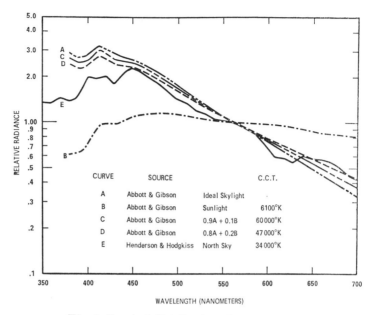

Fig. 7. Spectral distributions for sky radiance.

Nickerson (1960), the latter being a compilation of the earlier work of Abbott and Gibson. The Abbott and Gibson work provides a convenient means for combining various proportions of direct sunlight (above the atmosphere) and idealized blue sky light (Rayleigh atmosphere illuminated by sunlight) to obtain the spectral distribution of natural daylight having correlated color temperatures from 6100°K to 60 000°K. Their work used the measurements of the spectral distribution of sunlight by Abbott in 1923 which do not have very high spectral resolution. The measurements agree generally, however, with more recent measurements having higher resolution, but may show too much radiance in the blue end of the spectrum. Originally, the Abbott and Gibson results were intended to show the relative spectral irradiance of daylight, but the curve shapes should be equally applicable to sky radiances. The curve from Henderson and Hodgkiss, on the other hand, was obtained by viewing a north sky at an inclination of 45° with an acceptance angle of 6°. It presents relative data from the mean of 10 normalized sky radiance measurements.

The curves of Fig. 7 are relative and provide only information about the shape of the spectral distribution of sky radiance. If we inspect Fig. 6 we find that the absolute values of sky luminance at the minima of the clear-day curves vary only about 16% from the mean value of

$128\ell \times sr^{-1} \times ft^{-2}$ and the absolute values of the zenith sky (if we exclude the case for the sun only 24° from the zenith) are within 18% of the mean value of $170\ell \times sr^{-1} \times ft^{-2}$.

While the number of examples used to obtain these ranges may be small, the fact remains that remote sensing is limited to clear days. Also, the direction of observation is restricted to the region from the nadir to no further than about 60° from the nadir, usually not including the region within 40° of the solar specular point in order to exclude sun glitter. These restrictions work to limit the range of sky luminance or radiance values that will be of concern to the problem. Some examples of absolute values of spectral sky radiances and their application will be given in Section V.

C. PATH RADIANCE

The radiance due to sunlight and sky light scattered into the path of sight between the ocean surface and a sensor above the atmosphere is the largest single component of the apparent radiance signal at the remote sensor. It has been observed to be four to five times larger than the inherent radiance from the ocean on clear days over southern California coastal waters. Occasionally, this path radiance may be relatively small, at times and places where the atmosphere is particularly dry and free of scatterers. Such conditions may occur frequently, for example, over waters in the vicinity of the large desert areas of Mexico, North Africa and the Saudi Arabian peninsula. Yet, even though these may be important special cases, they do not represent the situation over most of the world's oceans, nor is the path radiance likely to be appreciably less even under these best of circumstances.

Because the several components of the apparent signal cannot be separated, it is not possible to measure the path radiance directly from space. Duntley *et al.* (1957) describe the elements of the problem in considerable detail and have discussed methods of determining the path radiance by computation from data derived from aircraft measurements of the path function, which is the path radiance per unit length, and the path transmittance. Such data for performing these computations are not generally available for a specific condition, however, and other techniques have been developed which allow the determination of the earth-to-space path radiance (and beam transmittance of the path of sight) from surface-based measurements (Gordon *et al.*, 1963; Duntley *et al.*, 1964; Duntley *et al.*, 1970).

Briefly, the radiance of the sky in a particular direction is the path radiance for that path of sight. Duntley and his colleagues at the Visibility Laboratory have demonstrated through the analysis of a body of data collected contemporaneously on the ground, in the air, and from spacecraft that the path radiance from orbital altitude $L^*(\theta, \phi)$ for downward-looking paths of sight making an angle θ with the zenith direction and ϕ with the sun-zenith plane may be predicted satisfactorily from ground-based measurements of the sky radiance, i.e. $L_s(\theta', \phi') = L^*(\theta', \phi')$, and the beam transmittance of the total atmosphere as determined by ground measurements of the apparent radiance of the solar disk, $T_a(\theta_s)$. The upward path direction, $\theta', \phi',$ is such that the scattering angle, β', i.e. the angle between the sun and the path of sight, is the same as the scattering angle β that applies to the downward path, θ, ϕ. Figure 8 illustrates the geometry of this measurement in the sun-zenith plane. For solar zenith angles less than $45°$, the $L_s(\theta',\phi')$ measurement is made with the scattering angle fixed at $90°$, as measurements at the proper angle, β', would require looking down below the horizon. Fortunately the scattering functions for natural aerosols with heterochromatic light do not vary greatly in the backward direction, and it has been shown that the error introduced by this compromise is not of significant proportions (Gordon *et al.*, 1963).

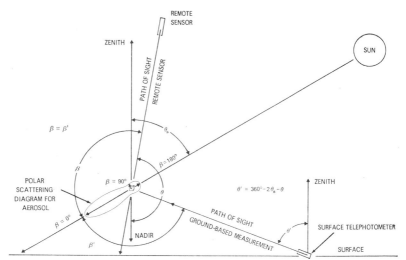

Fig. 8. Remote sensor path radiance determination, angular relationships.

The relationship equating $L^*(\theta, \phi)$ for the required path of sight to the variables observable from the surface is

$$L^*(\theta, \phi) = L_s(\theta', \phi')\left[\frac{1 - T_a(\theta_s)^{\cos\theta_s/\cos(\pi-\theta)}}{1 - T_a(\theta_s)^{\cos\theta_s/\cos\theta'}}\right], \qquad (6)$$

where $T_a(\theta_s)$ is the atmospheric beam transmittance in the direction of the sun. In Section III.A we found that the atmospheric transmittance does not vary markedly with wavelength. Thus the path radiance $L^*(\theta, \phi)$ resembles the sky radiance $L_s(\theta', \phi')$ in its spectral dependence and the curves in Fig. 7 serve to show the relative shape of curves for path radiance as a function of wavelength. In the event that complete spectral data are not available, the determination of the absolute values for atmospheric beam transmittance and sky radiance at two or three wavelengths allows one to obtain an approximation of the spectral path radiance.

IV. The Inherent Radiance Emitted by the Water

The radiance which emerges from below the water surface results from the interaction of the directional light field with the water and the ocean bottom. The details of this interaction and the associated equations required for determining the upwelling radiance are rather complex and the computation of the upwelling radiance by means of these interaction equations requires knowledge of the directional nature of the downwelling light field above the water and a complete description of the scattering and absorbing properties of the water. Instead, for hueristic development, we will present the much simpler equations relating the upwelling irradiance below the water surface, E_u, and that above the water surface, E_w, to the downwelling irradiance above the surface, E_s. From these we can obtain a measure of the effect which the various reflectances and transmittances have on the emerging irradiance and, with the aid of some simplifying assumptions, can gain an insight into the radiance leaving the water.

Figure 9 presents the elements of the problem. The ocean is represented as having a surface with a reflectance, r_d, for downwelling irradiance which is different from the reflectance, r_e, for the upwelling or emergent irradiance by virtue of the change in the angular distributions of the upwelling and downwelling light fields. As the transmittances through the surface are $t_d = 1 - r_d$ and $t_e = 1 - r_e$, they, too, will be different. Thus, the directional nature of the light fields is captured in the reflectances and transmittances. The reflectance and transmittance of the water, R_w and T_w, are dependent upon depth because the angular distribution of the light field changes with depth. However, the resulting changes in R_w and T_w are not great after the first few meters. As a consequence, we will assume they are constant with depth and the same for the upward and downward direction.

In the following development the water reflectance, R_w, is the ratio

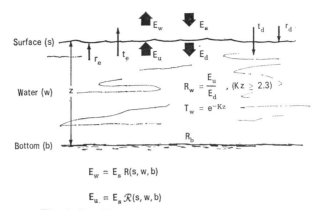

$$E_w = E_s R(s, w, b)$$

$$E_u = E_s \mathscr{R}(s, w, b)$$

Fig. 9. Irradiance interaction with the ocean.

of the upwelling to the downwelling irradiances which would obtain in the absence of any return from the bottom. In most cases, this condition could be assumed to exist when the round trip transmission of the total water column, T_w^2, is less than one per cent. Under these circumstances the upwelling spectral irradiance, E_u, is uniquely determined from a knowledge of R_w and the downwelling spectral irradiance, E_d, below the surface. The total irradiance transmittance of the water column, T_w, is simply

$$T_w = \exp(-Kz),$$

where K is the irradiance attenuation coefficient and z is the depth of water over the bottom.

The upwelling irradiance just below the surface, E_u, is given by

$$E_u = E_s \mathscr{R}(s, w, b)$$

where the interior reflectance $\mathscr{R}(s, w, b)$ is given by

$$\mathscr{R}(s, w, b) = \frac{t_d \left[R_w + \dfrac{T_w^2 R_b}{1 - R_w R_b} \right]}{1 - r_e \left[R_w + \dfrac{T_w^2 R_b}{1 - R_w R_b} \right]} \tag{8}$$

The total externally-reflected irradiance, E_w, may be found by the equation

$$E_w = E_s[r_d + t_e \mathscr{R}(s, w, b)] = E_s R(s, w, b)$$

where

$$R(s, w, b) = r_d + \frac{t_d t_e \left[R_w + \dfrac{T_w^2 R_b}{1 - R_w R_b} \right]}{1 - r_e \left[R_w + \dfrac{T_w^2 R_b}{1 - R_w R_b} \right]} \tag{9}$$

is the total effective exterior reflectance for the combination of the surface, water and bottom.

Tables 1 and 2 provide information from which the sizes of r_d, r_e, t_d, and t_e may be estimated. The R_w, water reflectance, will vary with the wavelength and type of water but will generally range from 0·01 to 0·10. The water transmittance, T_w, will vary markedly with wavelength and depth and must be determined for each situation. The bottom reflectance, R_b, will vary widely as the bottom composition changes from silts and dark mud to white coral sand. We may expect it to fall in the range of 0·05 to perhaps 0·35. When these ranges of values are applicable, we can make some further simplifications as follows:

$$R_w R_b \ll 1, \quad \therefore (1 - R_w R_b) \approx 1$$

$$r_d = 0\cdot02, \quad t_d \approx 1$$

$$r_e \approx 0\cdot48, \quad t_e \approx 0\cdot052.$$

Then eqs (8) and (9) become,

$$\mathcal{R}(s, w, b) \approx \frac{R_w + T_w^2 R_b}{1 - 0\cdot48[R_w + T_w^2 R_b]} \tag{8a}$$

and

$$R(s, w, b) \approx 0\cdot02 + \frac{0\cdot52[R_w + T_w^2 R_b]}{1 - 0\cdot48[R_w + T_w^2 R_b]} \tag{9a}$$

Obviously, the R_w term accounts for the return signal containing information about the water, and the $T_w^2 R_b$ term accounts for transmitting the input irradiance to the bottom, reflecting it at the bottom, and transmitting it back to the surface. In circumstances where $T_w^2 R_b \ll R_w$, we may further simplify the expressions with the approximation that $0\cdot48[R_w + T_w^2 R_b] \ll 1$ making the denominators in both equations approach unity and we then obtain

$$\mathcal{R}(s, w, b) \approx R_w + T_w^2 R_b. \tag{8b}$$

and

$$R(s, w, b) \approx 0\cdot02 + 0\cdot52[R_w + T_w^2 R_b] \tag{9b}$$

Equations (8b) and (9b) are simple approximate statements for $\mathcal{R}(s, w, b)$ and $R(s, w, b)$ that are not only useful for computation under the restrictions given but are also helpful in intuiting the nature and magnitudes of these reflectances.

Whereas the solution of actual remote sensing problems would require knowledge of the spectral values of R_b, R_w, and T_w for the particular bottom and water involved, it is often useful for the purpose of estimating the magnitude of the available signal in a particular situation, to have information about the magnitude of these variables and the way in which the R_w and K values found in nature may be correlated. Unfortunately, data on the spectral reflectance of ocean and coastal bottoms are sparse indeed. However, some examples of values of R_w and K may be obtained from various tabulations in the literature, e.g. Jerlov (1968) and Tyler and Smith (1970). In the latter, Tyler and Smith have published spectral irradiance data obtained with a submersible spectroradiometer in nine widely differing bodies of water. They also present tables of spectral attenuation coefficient (irradiance K) computed from this irradiance data for all nine locations and spectral reflectance functions, R_w, for four of the cases.

The terms in the irradiance interaction equations have implicit dependence on the angular distribution of the radiance both above and below the surface. We may use this fact to obtain some simple approximations for the upwelling radiance emerging from the water. It will be recalled that in Section II.B the upwelling irradiance was assumed to be completely diffuse, i.e. the upwelling radiance from the lower hemisphere was constant with angle. It follows from this assumption that upwelling radiance below the surface is

$$L_u = \frac{E_u}{\pi} = \frac{E_s \mathscr{R}(s, w, b)}{\pi}. \tag{10}$$

Furthermore, only the central 48·2° cone of radiances from below the surface will be present above and the angular distribution of this component of the above-water radiance field will be similarly uniform. To the extent that this holds true, we can write a similar expression for the upward flowing radiance above the water:

$$L_w = \frac{E_s R(s, w, b)}{\pi}. \tag{11}$$

Now by comparison of eqs. (8) and (9), we find that

$$R(s, w, b) = r_d + t_e \mathscr{R}(s, w, b). \tag{12}$$

It will be noted that r_d is the Fresnel reflectance of the surface, and as the input irradiance, E_s, is preponderantly due to direct sunlight which is, by design, being specularly reflected away from the remote sensor, the contribution to L_w from the r_d term will be small in the direction of the sensor. With this approximation we can combine eqs. (10), (11), and (12) and obtain the simple approximate relationship that

$$L_w \approx t_e L_u. \tag{13}$$

We found in Section II.B that the emergent reflectance, r_e, was on the order of 0·48. Thus the transmittance for emerging irradiance is $t_e = 0.52$. It is interesting to compare this with the more precise relationship obtained in Section II.A, viz.

$$L_w = \frac{t}{n_w^2} L_u, \tag{14}$$

where $t/n_w^2 = 0.543$. The closeness of these two factors is not of course accidental, but results from the dependence of the emergent reflectance on the index of refraction. Note that the t in eq. (14) is the radiance transmittance for near normal incidence from below, and $t = 1-r$ where r is given in Table 2.

More precise values of the radiance L_u can, of course, be obtained by direct measurement. In order to obtain surface truth data for documenting remote sensing experiments, the Scripps submersible spectroradiometer was reconfigured and calibrated to measure spectral radiance. At sea the instrument was suspended vertically downward beneath a small buoy well away from the attending surface vessel. Upwelling spectral radiance measurements were then obtained at two predetermined depths, 2·2 m, and 7·7 m. These data were then used to calculate an upwelling radiance K, and this in turn was used to calculate the radiance which would be found just below the surface, i.e. L_u. Examples of measurements obtained in this manner are given in the next section.

V. Samples of Apparent Spectral Radiance over the Ocean

A series of coordinated measurements were performed in the summer of 1971 off the coast of southern California. Measurements of the upwelling spectral radiance were obtained by the method just described in waters that were subjectively determined to be blue (4 August), blue-green (3 August), and green (5 August). The subjective descriptions were further documented by careful determinations of hue match with Munsell color samples. The results of the measurements and subjective color determinations are given in Fig. 10.

Using equipment and techniques described by Duntley et al. (1970), abridged measurements of the pertinent optical properties of the atmosphere were made from a nearby island. On 3 August atmospheric measurements were made at four wavelengths. Because of equipment difficulties, measurements at only a single wavelength were obtained on 4 and 5 August. The techniques suggested in Section III were applied to obtained the spectral shape of the atmospheric transmittance,

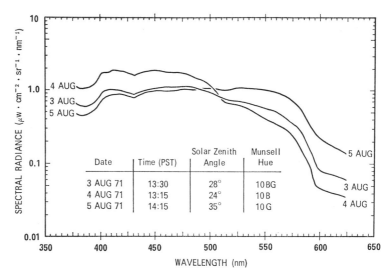

Fig. 10. Upwelling spectral radiance just below the surface computed from measurements at depths of 2·2 nd 7·7 m.

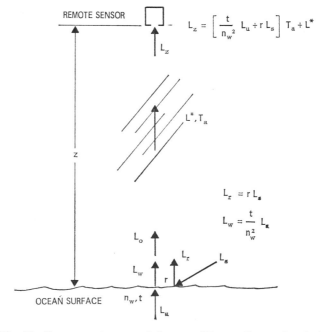

Fig. 11. Component parts of the upwelling radiance signal, L_z.

T_a, the sky radiance, L_s, and the path radiance, L^*. The absolute values of these components were adjusted to agree with the values obtained at the wavelengths of the measurements. Figure 11 summarizes some of the concepts of the earlier sections and will be helpful in visualizing

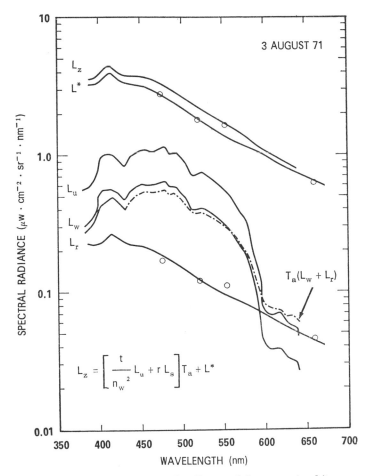

Fig. 12. Computed apparent spectral radiance of the ocean (and its components) as observed above the atmosphere. Blue-green water.

the meaning of the curves in Figs. 12, 13 and 14. The latter show the results of the computations for the three days. The four circled points on the L^* and L_r curves for 3 August are the values computed from the shore station measurements. Curve C in Fig. 7 (Abbott and Gibson's 0·9 ideal sky light plus 0·1 sunlight) was selected as having the best fit to the measured points. As the atmospheric conditions were judged to be similar for the three days, the same curve shape was assumed for the other two cases on 4 and 5 August in the absence of data to the contrary.

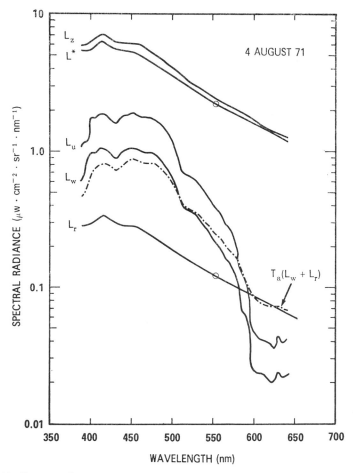

Fig. 13. Computed apparent spectral radiance of the ocean (and its components) as observed above the atmosphere. Blue water.

The measured upwelling spectral radiances, L_u, were multiplied by t/n_w^2 to obtain the corresponding above-the-surface radiances, L_w. The zenith sky radiances, L_s, were multiplied by the Fresnel reflectance $r = 0.021$ to obtain the water surface radiances due to reflected sky light, L_r. The sum of L_w and L_r, the total inherent radiance leaving the surface, was multiplied by the computed atmospheric transmittance, T_a, and summed with the computed L^* to obtain the resultant apparent radiance at the remote sensor, L_z.

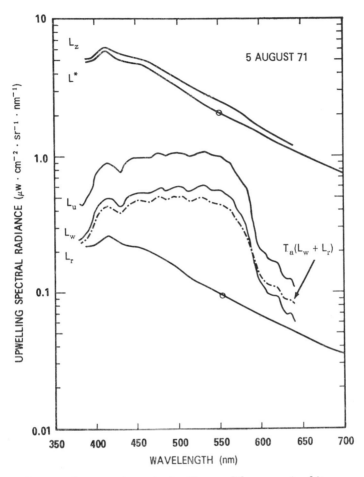

Fig. 14. Computed apparent spectral radiance of the ocean (and its components) as observed above the atmosphere. Green water.

Several facts can be readily seen by inspection of the three sets of curves. First, the differences in water color encountered on the three days is easily seen in the spectroradiometric curves L_u and L_w for the upwelling radiances from the water. Second, the addition of the reflected zenith sky radiance, L_r, to the inherent above-surface water radiance, L_w, has only a minor effect on the shape of the curves. Third, the spectral radiance transmitted through the atmosphere to the remote sensor (that is, the component $T_a(L_w + L_r)$ before the addition of the path radiance) still contains the major portion of the information contained in the original spectral water signature. And fourth, the addition of the desaturating radiance caused by ambient daylight scattered into the path of sight by the intervening atmosphere (i.e. the path radiance, L^*) is by far the most significant single problem in visible region remote sensing. In the three examples the information containing transmitted signal was never more than about 20% of the path radiance. The path radiance will vary with time and place and may be significantly smaller on some occasions. We may greatly improve the potential of optical remote sensing, however, by using techniques which emphasize changes in the spectral signatures obtained from immediately adjacent areas or which look for subtleties in the shape of the signature. Various derivative spectroradiometry and radiance ratio techniques have been tried and show promise for extracting information from data that is seemingly dominated by path radiance. These techniques have the advantage that they can be optimized to provide the maximum sensitivity to a phenomena of particular interest such as the concentration of chlorophyll or other pigments in the water.

The three examples which have been presented show a technique for computing the apparent radiance of the ocean arriving at a sensor outside the atmosphere. The results are probably representative of the situation found over silt-free deep water where the path of sight is steeply downward and the sun glitter is outside the field of view. However, where the bottom-reflected component of the underwater signal is apparent, the L_u and its corresponding L_w term may increase appreciably. The reflectance of the bottom, R_b, may differ spectrally from the water reflectance, R_w, in a dramatic way, and the radiance at the longer wavelengths (but still below the water cutoff around 575 nm) may increase several-fold in clear waters where the T_w^2 term does not dominate. This has the effect of making the $L_w T_a$ component of the apparent signal at the remote sensor much closer to, or even larger than, the path radiance term.

A similar situation can be found when silt-laden waters having high reflectance in the red spectral region are carried near the surface. Here the path radiance contribution is small and the atmospheric transmittance is high, which again tends to make the $L_w T_a$ term dominant.

In summary it has been shown that over water where bottom and turbidity effects are not prominent, the signal from the water available to the remote sensor may be only 20 to 25% of the signal contributed by path radiance. The component of the signal from the sky light reflected by the ocean surface is usually trivial compared to the signal from the water over most of the blue and green spectral region on clear days with steeply downward paths of sight. The problem of detecting changes in water depth, chlorophyll content, or tracing coastal or ocean currents, depends on the ability of the sensor to detect either changes in shape of the spectral signature of the apparent radiance or changes in the absolute values from point to point in the field of view. Over most of the spectral region where the water transmittance is the highest, the path radiance has the major degrading effect on the water-generated radiance signal. Small changes in the path radiance can and often do cause changes in the apparent radiance which may be easily confused with changes associated with phenomena below the ocean surface. Prominent changes in water color may be rendered subtle by the atmosphere. The remote optical sensing of subsurface ocean phenomena, while having exciting and valuable potential for synoptic ocean exploration, is nonetheless a procedure which is highly dependent on the ability of the atmosphere at a particular time and location to transmit the signal to the sensor.

REFERENCES

Clarke, G. L. and Ewing, G. C. (1971). WHOI Ref. No. 71–74, Woods Hole Oceanographic Institution, Woods Hole, Massachusetts, U.S.A.

Cox, C. (1956). In "Contributions", University of California, San Diego, Scripps Institution of Oceanography, San Diego, California, U.S.A., 193–207.

Cox, C. and Munk, W. (1954). *J. Mar. Res.*, **13**, 198 227.

Cox, C. and Munk, W. (1955). *J. Mar. Res.*, **14**, 63–78.

Dorsey, N. E. (1940). "Properties of Ordinary Water-Substance." Reinhold Publishing Company, New York, U.S.A.

Duntley, S. Q. (1952). "The Visibility of Submerged Objects." Visibility Laboratory, Massachusetts Institute of Technology, Cambridge, Mass., U.S.A.

Duntley, S. Q. (1954). *J. Opt. Soc. Am.*, **44**, 574.

Duntley, S. Q., Boileau, A. R. and Preisendorfer, R. W. (1957). *J. Opt. Soc. Am.*, **47**, 499–506.

Duntley, S. Q., Johnson, R. W. and Gordon, J. I. (1964). Tech. Document Rep. AL-TDR-64-245, U.S. Air Force Systems Command, Wright-Patterson Air Force Base, Ohio, U.S.A.

Duntley, S. Q., Edgerton, C. F. and Petzold, T. J. (1970). SIO Ref. 70–27, University of California, San Diego, Scripps Institution of Oceanography, San Diego, California, U.S.A.

Edgerton, C. F. (1967). SIO Ref. 67–27, University of California, San Diego, Scripps Institution of Oceanography, San Diego, California, U.S.A.

Fean, C. (1961). SIO Ref. 61–27, University of California, San Diego, Scripps Institution of Oceanography, San Diego, California, U.S.A.

Gordon, J. I. (1969). SIO Ref. 69–20, University of California, San Diego, Scripps Institution of Oceanography, San Diego, California, U.S.A.

Gordon, J. I., Harris, J. L. and Duntley, S. Q. (1963). SIO Ref. 63–62, University of California, San Diego, Scripps Institution of Oceanography, San Diego, California, U.S.A.

Greaves, J. R., Spiegler, D. B. and Willand, J. H. (1971). NASA CR-61345, NASA Contract Rep., NASA-George C. Marshall Space Flight Center, Alabama, U.S.A.

Guttman, A. (1968). *Appl. Opt.*, **7**, 2377–2381.

Henderson, S. T. and Hodgkiss, D. (1963). *Brit. J. Appl. Phys.*, **14**, 125–131.

Hopkinson, R. G. (1954). *J. Opt. Soc. Am.*, **44**, 455–459.

Jerlov, N. G. (1968). "Optical Oceanography." Elsevier Publishing Company, Amsterdam.

McClatchey, R. A., Fenn, R. W., Selby, J. E. A., Volz, F. E. and Garing, J. S. (1971). AFCRL–71–0279, Environmental Res. Papers, No. 354, Air Force Cambridge Res. Lab., L. G. Hanscom Field, Bedford, Mass., U.S.A.

Miller, D. B. and Feddes, R. G. (1971). "Global Atlas of Relative Cloud Cover, 1967–70, Based on Photographic Signals from Meteorological Satellites," U.S. Dept. of Commerce (National Oceanic and Atmospheric Admin., National Environmental Satellite Serv.), and U.S. Air Force (Air Weather Serv. [MAC], USAF Environmental Tech. Application Center), Washington, D.C., U.S.A.

Moon, P. (1940). *J. Franklin Inst.*, **230**, 583–617.

Nickerson, D. (1960). *J. Opt. Soc. Am.*, **50**, 57–69.

Richardson, R. A. and Hulburt, E. O. (1949). *J. Geophys. Res.*, **54**, 215–227.

Rozenberg, G. V. (1966). "Twilight, A Study in Atmospheric Optics." Plenum Press, New York, U.S.A.

Rozenberg, G. V. and Yu.-A. R. Mullamaa (1965). *Izv., Acad. Sci., USSR, Atmos. Oceanic Phys. Series*, **1**, 282–290. (Translated by P. A. Keehn.)

Tyler, J. E. and Smith, R. C. (1970). "Measurements of Spectral Irradiance Underwater." Gordon and Breach, Science Publishers, New York, U.S.A.

Chapter 15

Light and Photosynthesis of Different Marine Algal Groups

PER HALLDAL

Botanical Laboratory, University of Oslo, Blindern, Oslo, Norway

I. Introduction

The main primary production in marine environment occurs through benthic and planktonic algae and some higher plants. It is confined to the euphotic zone to a depth where about 1% of daylight is left. The other primary production process, chemosynthesis through certain specialized microorganisms is of little quantitative importance in the sea. Phytoplankton is by far the most important group in this respect and is responsible for about 90% of the production. This topic is extensively discussed by Steemann Nielsen in Chapter 17.

In order to capture light and transform its energy into energy-rich organic chemical compounds, photosynthetic organisms contain certain pigments confined to refined subcellular structures called chloroplasts in photosynthetic plants including algae, and photosynthetic lamellae or "chromatophores" in photosynthetic bacteria. The photosynthetic process of both bacteria and plants are extensively studied today and several recent reviews which deal with the process have been published (Goodwin, 1965, 1966, 1967; Vernon and Seely, 1966; Shibata and co-workers, 1968; Halldal, 1970; Sybesma, 1970). To discuss the topic around light and marine algae we need only a small fraction of the vast information now available on the photosynthetic process. In order to put the treatment into the right perspective some knowledge of fundamental general photosynthesis is definitely needed. It is also necessary briefly to sum up the action of light in the process. Rather than dealing with these subjects separately it is more practical to incorporate them into the portion of the chapters where they logically belong.

Algal photosynthesis from a primary production point of view is dealt with by Steemann Nielsen in Chapter 17. I will deal with photosynthetic pigments in relation to light levels and spectral light quanta distribution and the flexibility of the photosynthetic apparatus to adjust for environmental submarine light conditions. In this discussion ecological aspects will naturally be included.

II. Light Energy, Light Quanta and Photobiology

The energy of light is distributed in units of energy packages called quanta. The energy content of these units differs with wavelength or frequency. The energy content ϵ of one quantum at frequency or wavelength λ is expressed as follows:

$$\epsilon = h\gamma = \frac{hc}{\lambda}$$

where h is Planck's universal quantum constant, γ the frequency, and c the velocity of light.

This means that shorter wavelengths quanta contain more energy than longer wavelengths quanta. As one quantum irrespective of energy content can excite not more than one molecule, this means that red light with a certain energy content has the potential to do more photochemistry than blue light with the same amount of energy provided all other conditions equal. To take a practical example and comparing equal amount of energy at 450 nm and 675 nm we calculate that the latter contains 67% more quanta. This red light thus has the potential to excite a considerably greater number of molecules than the blue.

This leads us to the conclusion that if we work as photobiologists and wish to study the pigment systems of organisms in respect to light responses, quanta have to be considered. If energy amount is used in the calculations the effect of red light will be overestimated.

An ecologist may wish to estimate energy balance and calculations on energy units may then be preferred. In most other aspects of light in ecology considerations around quanta are more natural and will be followed in this chapter.

III. Spectral Light-Quanta Distribution in Air

In Fig. 1 is shown the spectral light quanta and energy distribution at two different localities. (A) and (B) are from Gullmarsfjorden at the Marine Biological Station, Kristineberg on the west coast of Sweden;

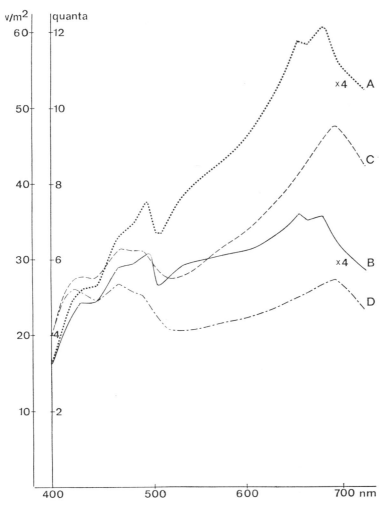

Fig. 1. The spectral distribution of light quanta and energy in air at two localities. (A) quanta and (B) energy distribution at Gullmarsfjorden, west coast of Sweden, (C) quanta and (D) energy distribution at Kings Bay, Spitsbergen, 79° N, 12° E. Scales W. m^{-2} 1 = 0·008 W. m^{-2} nm^{-1}. Quanta, 1 = 8·8 × 10^{12} quanta cm^{-2} s^{-1} nm^{-1}.

(C) and (D) are from Kings Bay, Spitsbergen at 79° N, 12° E. The measurements were performed with the prototype for the Incentive Research and Development Quantaspectrometer, QSM 2400. The daylight spectra at both localities were rather similar. At the particular times for the measurements the light at Kristineberg had a higher proportion of red light. This difference is, however, within daily observed variations at both localities. For example see Fig. 2. The spectrum from Kristineberg is also more rich in details. These differences have not been further analysed. Note that the spectral energy distribution was fairly uniform between 420 and 730 nm at both localities, while the numbers of red quanta were much higher than those of blue.

The daylight spectrum shows great variations with time of the day, weather and localities (see Bainbridge et al., 1966; Henderson, 1970; Linder and Halldal, 1972). One spectral quanta (and energy) distribution of particular interest is usually observed on clear days around dusk and dawn. This so called twilight effect, first reported by Johnson et al. (1967) results in a spectrum rich in blue and red and relatively poor in green and yellow (see Fig. 9). A possible ecological significance of this effect will be discussed in connection with the photosynthesis of green algae treated on page. 358.

Variations in the daylight spectrum caused by weather, sun altitude and other factors modifying it, only influence submarine natural light spectra to a small degree. The absorption characteristic of water itself and light attenuation caused by dissolved matter and particles in the water are, to a very high degree, dominating factors. Details in daylight spectra observed particularly on sunny days shown in Figs. 1, 2 and 3 are effectively modified and smoothed out and later eliminated some few metres below the surface. Because of this it is possible to demonstrate the presence of phytoplankton pigments in natural waters through their absorption bands which cause distinct changes in underwater spectral curves at particular wavelengths as have been demonstrated by Tyler (1964, 1965), and by Halldal and Halldal (1973).

IV. Light Attenuation in Different Water Masses

It is felt that it is of importance for the particular topic presented in this chapter briefly to sum up a few facts, and to include in the discussion a few new measurements which I have carried out in connection with marine algal photosynthesis. In particular some recent analyses from

Kristineberg (west coast of Sweden) and Kings Bay (Spitsbergen), all performed with the Quantaspectrometer prototype, will be presented.

At Kristineberg the 570 nm penetrated deepest, while light, particularly at shorter wavelengths, was rapidly reduced (Fig. 2). At 10 m, no light at 400 nm was left, and no light below 450 nm could be detected with the instrument below 30 m. Submarine light at Kristineberg has the typical quanta distribution of European coastal waters. The water is to a great degree removed of light below 450 nm, and even the amount of light below 500 nm is low. At longer wavelengths practically no light above 650 nm is present above 20 m. The coastal water in the Kristineberg area acts more or less like a spectral filter which peaks at about

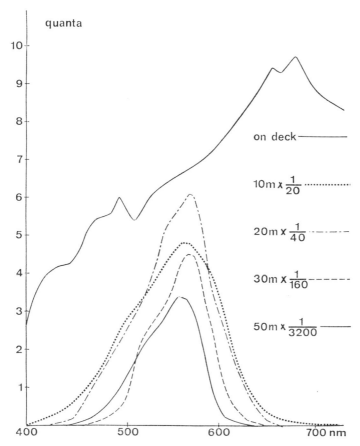

Fig. 2. The spectral distribution of light quanta in air, and at four different depths from Gullmarsfjorden at the west coast of Sweden, July 1971. Scale for quanta, $1 = 44 \times 10^{12}$ quanta cm^{-2} s^{-1} nm^{-1}.

570 nm with a half band width of about 80 nm. The water at Kristineberg belongs to an optical type in the neighbourhood of 9 of Jerlov (1968).

The total light attenuation is given in Table 1.

TABLE 1. Total light quanta attenuation at Kristineberg and Kings Bay. Based upon calculations from Figs. 2 and 3. The figures are the percentage remaining.

Depth in m	Kristineberg	Kings Bay
5	10	16·4
10	1·27	6·18
25	0·24	0·58
50	0·0037	0·053

It is seen that light was rapidly reduced. At 5 m only about 10% was left (interpolated value), at 10 m 1·27, at 25 m ca. 0·24 (interpolated value), and at 50 m as little as 0·0037%. Note that these values necessarily must be considerably different from those obtained by measuring "per cent light left" with a selenium photovoltic cell (luxmeter). A Se-cell has a spectral sensitivity which peaks around 560 nm and a distinct sensitivity curve somewhat resembling that of the human eye. Such instruments are therefore not very well suited for measurements of daylight intensity and submarine light attenuation, as considerable variations in solar spectra and optical characteristics of water masses occur.

Kings Bay has a different coastal water type from that at Kristineberg (Fig. 3). It contained more blue radiation and even at 50 m 410 nm light could be measured with the instrument. Precise calculations with the aim to characterize the water type have not been performed, but the type seems to be close to No. 7 or 6 of Jerlov. One factor seriously distorting such calculations in the Kings Bay area are the great number of particles from the glacier rivers where local areas may be heavily influenced. At Kings Bay light at 570 nm penetrated most effectively to 25 m. At this depth an indication of a spectral shift was observed around 500 nm, and at 50 m the maximum penetration occurred at this wavelength indicating a mixture of two different water masses, one coastal in the upper metres, and one atlantic somewhat deeper.

For Kings Bay the total light attenuations are also presented in Table 1. Compared with the Kristineberg water the water at Kings Bay was considerably clearer.

Most analyses of algal photosynthesis have been performed on benthic algae. In this chapter I will, therefore, concentrate on the description on submarine light conditions to coastal water and some recent measurements to illustrate this have been chosen. As oceanic

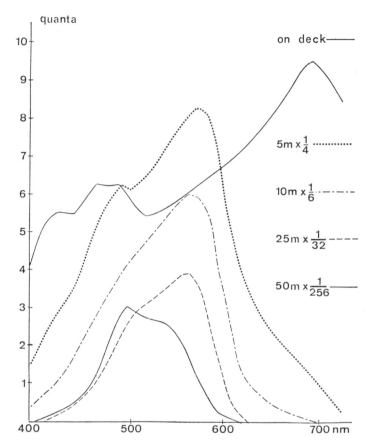

Fig. 3. The spectral distribution of light quanta in air and from four different depths at Kings Bay, Spitsbergen, July 1971. Scale for Quanta, $1 = 8 \cdot 8 \times 10^{12}$ quanta $cm^{-2}\, s^{-1}\, nm^{-1}$.

waters do not contain significant amounts of yellow substance and usually contain much less particles than coastal waters the submarine light conditions will be very different. On this subject references should be made to the book "Measurements of Spectral Irradiance Underwater" (Tyler and Smith, 1970).

V. Absorption and Action Spectra of Different Algal Groups

An *in vivo* absorption spectrum of an alga gives information on the combined absorption of the total pigment content in the organism. It also tells us how effectively the different spectral regions of light are absorbed. It does not, however, give information of the function of these pigments, nor can we directly tell which pigments are responsible for the absorption characteristics. To find this out, extraction, separation and identification of the different components are necessary. It must be kept in mind that extracted pigments in many cases have absorption characteristics which greatly differ from the pigments of live cells. In algae particularly chlorophylls and carotenoids undergo such changes. Chlorophyll *a in vivo* has a red absorption peak around 670 to 680 nm while the red absorption peak for extracted chlorophyll *a* in ethyl ether lies at 662 nm (French, 1960). Chlorophyll *a* of living cells also occurs in several different forms (see French *et al.*, 1970) which participate in different partial reactions of photosynthesis (see Halldal, 1970). It is assumed that the particular absorption of *in vivo* pigments and of the different forms of chlorophylls is caused by pigment-macromolecular associations.

Since the works of Engelmann (1881, 1882) it has been known that chlorophyll and carotenoids participate in photosynthesis. In order to analyse for active and inactive pigments in the photosynthesis of algae, high precision action spectra or quantum yields measurements over the spectrum must be performed. Examples of such analyses are the pioneering works of Haxo and Blinks (1950). Later these measurements were made automatic by French *et al.* (1960), Halldal (1968, 1969), Taube and Halldal (1972). In the combined action spectrum instrument and spectrophotometer of Halldal (1969), action spectrum and absorption spectrum may be analysed from the same piece of alga under identical optical conditions. Such comparative analyses allow us to determine photosynthetically active and inactive pigments. Action spectra and *in vivo* absorption spectra for three different common algal groups are shown in Figs. 4, 5 and 6.

Figure 4 shows absorption and action spectrum of the green alga *Ulva taeniata* redrawn from Haxo and Blinks (1950). This alga inhabits surface waters and is exposed to a light very little affected by water absorption. It has a pigment composition, absorption and action spectrum very similar to slices from higher plants on land and in the water. A very good agreement exists between thallus absorption and photosynthesis measured as oxygen production. We can safely conclude that

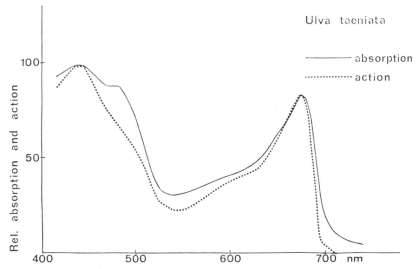

Fig. 4. *In vivo* absorption spectrum and photosynthetic action spectrum (oxygen evolution) from the green alga *Ulva taeniata* (after Haxo and Blinks, 1950).

chlorophyll is involved, and because of the relatively high effect around 500 nm where chlorophyll absorption is low, we conclude that carotenoid(s) must be involved. We can also conclude that certain yellow pigments, carotenoids and others, are inactive in photosynthesis by screening of the light as the curve for action always lies behind that of absorption when the curves are adjusted for equal chlorophyll content by making them fit in the red at 675 nm. The rapid drop of photosynthetic activity above 680 nm, discovered by Emerson and Lewis in 1943, was explained by Emerson *et al.* (1957) and by Blinks (1957) and is caused by the "two pigment system of photosynthesis" (see Halldal, 1970).

The brown alga *Laminaria saccharina* was collected from about 10 m and measured by automatic recording at the Marine Biological Station at Kristineberg. It was exposed to light conditions related to that of 10 m in Fig. 2. Action and absorption spectra are presented in Fig. 5. Note that *Laminaria* utilizes the green and yellow light much more efficiently than *Ulva*. The shoulder around 500 to 540 nm reflects the photosynthetically active carotenoid fucoxanthin. More details were resolved in the absorption spectrum which was measured in a thinner piece than that of the action spectrum.

Both green and brown alga show good agreement between action and absorption spectra. In this respect it should also be pointed out that the important phytoplankton groups diatoms and dinoflagellates

15

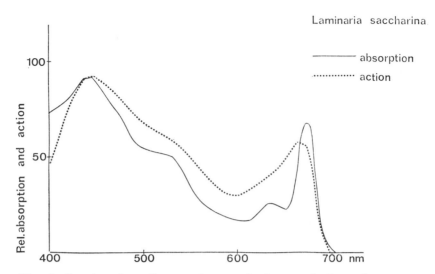

Fig. 5. *In vivo* absorption spectrum and photosynthetic action spectrum (oxygen evolution, automatic recording) from the brown alga *Laminaria saccharina* (Halldal, 1969).

show spectral photosynthetic response very similar to those of brown algae (Haxo, 1960; Halldal and Haxo, 1965, unpublished results). The diatoms contain the same accessory carotenoid, fucoxanthin, as the brown algae, and the dinoflagellates have the accessory carotenoid peridinin with related absorption characteristics as fucoxanthin.

The red algae differ markedly from green and brown both with respect to pigment composition and photosynthetic spectral response. This algal group contains with certainty only chlorophyll *a*, in some cases also chlorophyll *d* may be detected (see Halldal, 1970). Green algae, like higher plants, have chlorophyll *b* as accessory pigment, and the brown algae, the diatoms and the dinoflagellates different chlorophyll *c* (see Halldal, 1970).

In *in vivo* absorption spectra of red algae, absorption bands of chlorophyll *a*, carotenoids and phycoerythrin are clearly visible. Usually the other phycobilin, phycocyanin, occurs in less amounts but it is readily visible in the absorption spectrum of *Porphyra umbillicalis* where it may be isolated in significant amount (Eriksson and Halldal, 1965). In contrast to the other algal groups treated above, the red algae absorb light more effectively over a wider spectral region. Light in the green to yellow is especially much more efficiently captured. As was mentioned in the introduction, *in vivo* absorption curves do not tell us which of the pigments takes part in the photosynthetic process, neither the relative effect of those which are active, which clearly is demonstrated through the action spectrum of Fig. 6. The blue chlorophyll *a* peak at 435 nm is completely absent from the action spectrum and the red one is visible as a faint shoulder. Even the carotenoids seem to be completely ineffective. Whether carotenoids are effective accessory pigments in red algae or not is for the moment an unanswered question. Duysens (1952) and Goedheer (1969) believed they were, while Halldal (1970) came to the conclusion that they were completely inactive. Öquist (1972) supports the conclusion of Duysens and Goedheer. At any rate, red algae, especially those which grow on greater depths, show a spectral response for photosynthesis which to an amazing degree fits the light conditions in the environment (Fig. 7).

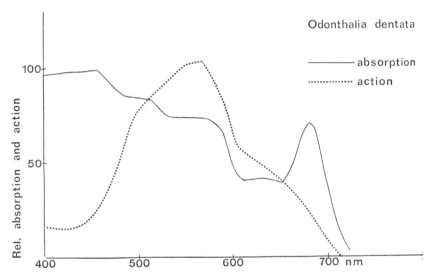

Fig. 6. *In vivo* absorption spectrum and photosynthetic action spectrum (oxygen evolution, automatic recording) from the red alga *Odonthalia dentata* (Measurements by Ö. Taube, Kings Bay, Spitsbergen, July 1971).

The first extensive analyses along these lines were performed by Levring (1947) and the problem was further analysed in the laboratory by Haxo and Blinks (1950). The results were later confirmed on several occasions (see Halldal, 1970). Red algae inhabiting the sublittoral, littoral or supralittoral zone, e.g. *Porphyra umbillicalis* and others, also have a very small effect in the blue and red region around the chlorophyll *a* peaks, though the relative effect is considerably lower than for the other marine algal groups. This shows that it is more to this phenomenon than pure environmental adaptation. Phylogenetic mechanisms, in

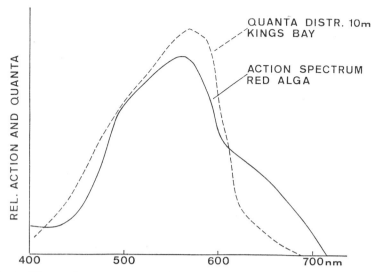

Fig. 7. The action spectrum of the red alga *Odonthalia dentata* from Fig. 6 compared with the quanta distribution of Fig. 3, 10 m.

addition to light induced pigment activations, are combined effects which determine the pigment composition of red algae, and in particular their adjustment to the light at greater depths where the pure red coloured red algae live and photosynthesize. These responses have been tested in the laboratory by Yocum and Blinks (1958), by Brody and Emerson (1959a,b) and by Brody and Brody (1962).

VI. Ecology

A. VERTICAL DISTRIBUTION OF BENTHIC ALGAE

Everyone studying algal distribution knows that it is impossible to separate definite vertical zones where certain types of algae distinctly occur. A mixture of green, brown and red algae occurs at practically all depths, though, particularly at greater depths, the pure red coloured red algae often completely dominate. Even here, however, green algae may be found. An example of this is *Chaetomorpha melagonium*. At the Swedish west coast this species is found down to 20 m (Kylin, 1949). Here only about 0·6% of the visible daylight was recorded to be left in July 1971 (see Fig. 2 and Table 1). Specimens found at such depths are very rich in chlorophylls and deep green in colour. The greatest mixture of algae is found in the littoral zone where several representatives for all the three major algal groups are found. In this respect the colour of the red algae is interesting. Practically all species of this algal group is brownish in colour due to a great amount of photosynthetically non-effective carotenoids. The function of these pigments undoubtedly is to balance the photochemical steps of the photosynthetic apparatus according to the succeeding dark reaction in photosynthesis. Several laboratory experiments have shown that in general, both for red and blue green algae, the amount of carotenoids increases in algae exposed to bright light (Halldal, 1958, Öquist, 1969). Öquist also showed that carotenoids in the unicellular green alga *Chlorella* are used to balance the rate of photosynthesis in the blue region of the spectrum.

Action spectra of photosynthesis from natural phytoplankton populations have to the author's knowledge not been reported. However, such analyses have been performed on diatoms and dinoflagellates in cultures (Haxo, 1960; Haxo and Halldal, 1965, unpublished) and on the symbiotic dinoflagellate *Symbiodinium mediterranea* (Halldal, 1968). We do not know the photosynthetic spectral response of diatoms, dinoflagellates and coccolithophorids from samples collected at say, 50 m. As, however, the pigment ratio and the photosynthetic spectral response of the chlorophyll *a*, chlorophyll *c*, fucoxanthin pigment system seem to be very stable, indicated by the constancy of the spectral response of brown algae collected from zero metres (Halldal, 1965, unpublished) from 10 m at the west coast of Sweden (Halldal, 1969) and from 80 m at Spitsbergen (Taube and Halldal, 1972), we believe that this also is the case for phytoplankton.

It is, therefore, assumed that the photosynthetic spectral response of phytoplankton collected from greater depths has about the same spectral response as the plankton cultured in the laboratory. Such a spectrum is presented in Fig. 8. If this is true it is interesting to note how inefficient phytoplankton organisms are using the photosynthetic pigment system at depths below 25 m. In Fig. 8 are drawn two typical curves of the quanta distribution at 25 m in coastal and oceanic water. Under these light conditions excitation of the red chlorophyll *a* absorption band is completely absent both in coastal and oceanic water. The red chlorophyll *c* peaks are excited in both areas, but in coastal water the chlorophyll *a* and *c* blue absorption bands are not excited while the light condition in oceanic water effectively excites chlorophylls in this spectral region. The conclusion is therefore that only a fraction of the pigment system of natural phytoplankton populations is excited in coastal water below 25 m, that the situation is greatly improved in oceanic water, but that in neither of these water types the red chlorophyll *a* absorption band is used for photosynthesis.

When the spectral distribution of light quanta and photosynthesis of alga in different parts of the spectrum are compared some interesting

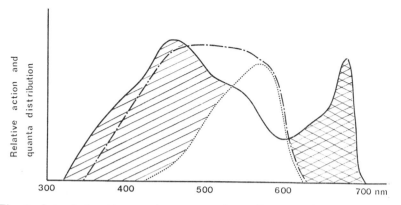

Fig. 8. A typical action spectrum curve for a diatom and a brown alga —.
Quanta distribution in coastal - - - and oceanic —·— water, depth 10 m. Photosynthetic non-effective part in both water types × × , and in coastal water ///.

features are observed. In the littoral zone and normal coastal waters
the spectral distribution of natural light in air undergo small changes.
Here we thus have a spectrum rather similar to that in air. In the upper
2 to 4 metres a rather uniform quanta distribution occurs (unpublished
results). The brown and the green algae inhabiting this region have two
photosynthetic high activity bands, one in the blue to green region,

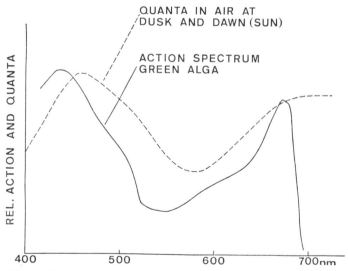

Fig. 9. The action spectrum of a green alga compared with the "twilight effect"
(Johnson *et al.*, 1967, Measurement P. Halldal).

another in the red; thus these algae are from an ecological point of view not particularly adjusted to the prevailing light. One interesting quanta distribution occurs, however, at some minutes around dusk and dawn where light, rich in blue and red light and with very little energy in the green to yellow region, illuminates the plants. It is interesting that this spectral distribution nicely fits the absorption spectrum of green and brown algae. This "twilight effect", first observed by Johnson *et al.* (1967) and many times later seen by myself and collaborators, fits the absorption characteristic and the photosynthetic action spectra of green and brown algae (Fig. 9). Whether this has any ecological significance has not been analysed. Personally I feel that this is an interesting coincidence. It certainly should be kept in mind during ecological considerations.

REFERENCES

Bainbridge, R., Evans, G. C. and Rackham, O. (1966). "Light as an Ecological Factor." *Brit. Ecol. Soc. Symp., No. 6*. Blackwell, Oxford.
Blinks, L. R. (1957). *In* "Research in Photosynthesis" (H. Gaffron, ed.), pp. 444–449. Interscience, New York.
Brody, M. and Brody, S. S. (1962). *Arch. Biochem. Biophys.*, **96**, 354–359.
Brody, M. and Emerson, R. (1959a). *Amer. J. Bot.*, **46**, 433–440.
Brody, M. and Emerson, R. (1959b). *J. Gen. Physiol.*, **43**, 251–264.
Duysens, L. N. M. (1952). Transfer of Excitation Energy in Photosynthesis. Doctoral thesis, University of Utrecht, The Netherlands.
Emerson, R. and Lewis, C. M. (1943). *Amer. J. Botany*, **30**, 165–178.
Emerson, R., Chalmers, R. and Cederstrand, C. (1957). *Proc. Nat. Acad. Sci. U.S.A.*, **43**, 133–143.
Engelmann, Th. W. (1881). *Botan. Z.*, **39**, 441–450.
Engelmann, Th. W. (1882). *Botan. Z.*, **40**, 419–428.
Eriksson, C. E. A. and Halldal, P. (1965). *Physiol. Plant.*, **18**, 146–152.
French, C. S. (1960). "Encyclopedia Plant Physiology" (W. Ruhland, ed.), Vol. V/1, pp. 252–297. Springer-Verlag, Berlin.
French, C. S., Myers, J. and McLeod, G. C. (1960). *In* "Comparative Biochemistry of Photoreactive Systems" (M. B. Allen, ed.), pp. 361–365. Academic Press, New York, and London.
French, C. S., Brown, J. S., Wiessner, W. and Lawrence, M. C. (1970). Carnegie Institution of Washington Year Book **69**, 1969–70, pp. 662–670.
Goedheer, J. C. (1969). *Biochim. Biophys. Acta*, **172**, 252–265.
Goodwin, T. W. (1965). "Chemistry and Biochemistry of Plant Pigments." Academic Press, London and New York.
Goodwin, T. W. (1966). "Biochemistry of Chloroplasts", Vol. I. Academic Press, London and New York.
Goodwin, T. W. (1967). "Biochemistry of Chloroplasts", Vol. II. Academic Press, London and New York.
Halldal, P. (1958). *Physiol. Plant.*, **11**, 401–420.
Halldal, P. (1968). *Biol. Bull.*, **134**(3), 411–424.

Halldal, P. (1969). *Photochem. Photobiol.*, **10**, 23–34.

Halldal, P. (1970). *In* "Photobiology of Microorganisms" (P. Halldal, ed.), pp. 17–55. Wiley-Interscience, London and New York.

Halldal, P. and Halldal, K. (1973). *Norw. J. Bot.*, **20**, 99.

Halldal, P. and Haxo, F. T. (1965). Unpublished results.

Haxo, F. T. (1960). *In* "Comparative Biochemistry of Photoreactive Pigments" (M. B. Allen, ed.), pp. 339–360. Academic Press, New York and London.

Haxo, F. T. and Blinks, L. R. (1950). *J. Gen. Physiol.*, **33**, 389–422.

Henderson, S. T. (1970). "Daylight and Its Spectrum". Adam Hilger, London.

Jerlov, N. G. (1968). "Optical Oceanography." Elsevier, Amsterdam, London and New York.

Johnson, T. B., Salisbury, F. B. and Connor, G. I. (1967). *Science, N.Y.*, **155**, 1663–1665.

Kylin, H. (1949). *Lunds Univ. Årssk. N.F. Avd.*, **2**, *Bd* 45(4).

Levring, T. (1947). *Göteborgs Kgl. Vetensk.-Vitterhets-Samhälles. Handl. Ser. B5.*

Linder, S. and Halldal, P. (1972). Manuscript.

Shibata, K., Takamiya, A., Jagendorf, A. T. and Fuller, R. C. (1968). "Comparative Biochemistry and Biophysics of Photosynthesis." University of Tokyo Press, Tokyo, and University Park Press, Pennsylvania.

Sybesma, Chr. (1970). *In* "Photobiology of Microorganisms" (P. Halldal, ed.), pp. 57–93. Wiley-Interscience, London and New York.

Taube, Ö. and Halldal, P. (1972). Manuscript.

Tyler, J. E. (1964). *Proc. Nat. Acad. Sci.*, **51**, 671–678.

Tyler, J. E. (1965). *J. Opt. Soc. Amer.*, No. **55**, 800.

Tyler, J. E. and Smith, R. C. (1970). "Measurements of Spectral Irradiance Underwater." Gordon and Breach, New York and London.

Vernon, L. P. and Seely, G. R. (1966). "The Chlorophylls." Academic Press, London and New York.

Yocum, C. S. and Blinks, L. R. (1958). *J. Gen. Physiol.*, **41**, 1113–1117.

Öquist, G. (1969). *Physiol. Plant.*, **22**, 516–528.

Öquist, G. (1972). Unpublished results.

Chapter 16

Light and Primary Production

E. STEEMANN NIELSEN

Freshwater Biological Laboratory, University of Copenhagen, Denmark

I. INTRODUCTION

Plankton algae are practically the only producers of organic matter in the open ocean and are the main topic of this chapter in which the primary production of organic matter due to the marine plants is explained.

Light provides the energy necessary for the transformation of inorganic matter into organic matter by the plankton algae, as of course also by all other photoautotrophic plants. It is only the light absorbed by the pigments active in photosynthesis that is used for this transformation, and this generally is only a minor part of the submarine light.

In the photic layers of most areas of the seas the replenishment of the nutrient salts containing N and P is the essential factor determining the magnitude of the annual primary production. It would be wrong, however, to assert that only the replenishment of the nutrients is important, because the Antarctic Ocean and North Pole Basin are areas in which the replenishment of nutrients seldom limits the size of the primary production.

For most areas, however, the illumination is adequate to provide the energy necessary for an organic production equivalent to the annually available amounts of P and N in the photic layer. But light is by no means a factor of only minor interest at medium and high latitudes. The annual variations in the rate of primary production here are largely due to the annual variations in the intensity of illumination. The illumination conditions determine together with the general hydrographic conditions the depth of the photic zone. It is of considerable importance for the grazing zooplankton whether a certain rate of primary production takes place in a deep or in a shallow photic zone.

Talling (1971) has presented an exhaustive discussion on the light climate as the controlling factor in the production ecology for freshwater phytoplankton. This valuable contribution is also of importance for oceanographers.

For some purposes it is advisable to present the illumination in a way that the wavelength composition is also taken into consideration. For other purposes, for example when measuring the illumination in connection with simulated *in situ* measurements (see page 385),—it is necessary to measure the rate in a simple way. Jerlov and Nygaard (1969) have described and produced a quanta and energy meter for photosynthesis studies. By means of moving between two filters it is possible for any light source to obtain, in the range 350–700 nm, either the number of quanta or the energy. Although measurements in quanta are theoretically the correct means of presenting the illumination rate in connection with photosynthesis, the use of energy units gives nearly the same results (Steemann Nielsen and Willemoes, 1971). In this chapter illumination will be presented as $quanta \times 10^{15} \times cm^{-2} \times s^{-1}$ (shortened to $quanta \times 10^{15}$). In some cases a presentation in energy units is added ($mW \times cm^{-2}$) or the illumination rate is only given in $mW \times cm^{-2}$.

II. The Incident Light Reaching the Surface of the Sea at Different Latitudes and Seasons

The intensity and quality of light reaching the chloroplasts of the plankton algae depend on the optics of the water and on the incident light reaching the surface of the sea, both of which must be considered in order to understand oceanic primary production.

Only a part—about 50%—of the solar radiation, corresponding roughly to the visible part, can be used for photosynthesis. The intensity of solar radiation received at the surface of the sea varies with the latitude, the season of the year, the time of the day and the cloudiness.

The annual totals at high and at low latitudes are not so different as is often thought. According to Kimbal (1935) at Fairbanks, Alaska (65° N) it is about 50% of that at Miami (26° N).

Whereas the seasonal variation in the tropics and subtropics is rather insignificant, it is very large at high latitudes. According to Kimbal (1935) the average daily insolation during the week that includes winter solstice at Miami is 55% of that during the week that includes summer solstice, whereas at Fairbanks it is only 0·8.

With an overcast or partly overcast sky the radiation received at the sea surface is reduced compared with that on days with a clear sky. By means of continuous recordings made in Copenhagen for two years and published by Romose (1940), Steemann Nielsen and Hansen (1961) presented (a) the integrated daily illumination rates for every month, (b) the corresponding rates for the three brightest days and (c) the corresponding rates for the three darkest days (Fig. 1). In Fig. 2 the illumination rates on the darkest days as a percentage of the rates on the brightest days are presented for the various months. For the period April to September the percentage is about 36 with the exception of July and August, where, most likely because of many thunderstorms, it is only 30. During midwinter the percentage is 19. Primarily, this is

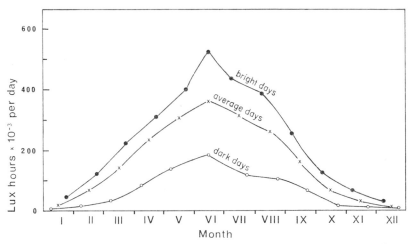

Fig. 1. The variation of the integrated daily illumination throughout the year in Copenhagen for bright, average and dark days.

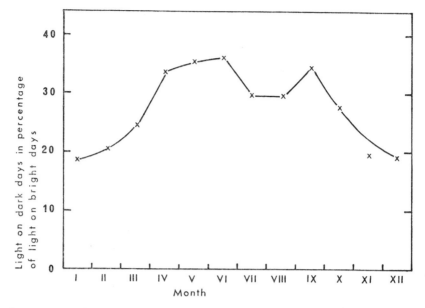

Fig. 2. The illumination on dark days as a percentage of the illumination on bright days throughout the year in Copenhagen.

likely to result from the low altitude of the sun increasing the average path of the light rays through a cover of clouds. The influence of the variation in cloudiness on the rate of primary production in the sea will be discussed in Section VII.

III. Photosynthesis as a Function of Illumination Intensity. Adaptation to Different Illumination Intensities and Temperatures

Two kinds of processes take part in photosynthesis—photochemical and enzymatical. The rates of the photochemical processes depend on the light quanta absorbed by the photosynthetic pigments, and therefore both on the concentrations of these pigments and on the illumination intensity. On the other hand, the rates of the enzymatic processes depend both on the concentration of the enzymes active in photosynthesis and on the temperature. The rate of the over-all process may be limited either by the rate of the photochemical or of the enzymatic process.

The most important consequence of adaptation (or physiological adjustment to the surrounding conditions) is the possibility for the algae to match to some extent the two kinds of processes under the prevailing ecological conditions.

As will be discussed in the next section, too high photochemical rates compared with the enzymatic rates give rise to light inhibition, whereas too high concentration of the enzymes, which constitute a major part of the organic matter in unicellular algae, results in more organic matter being produced than is really necessary.

The shape of the curve representing the rate of photosynthesis as a function of illumination (Fig. 3) gives important information about the adaptation. The slope of the initial part of the curve is a function of the photochemical part of photosynthesis, which again is the function of the concentration of the photosynthetic pigments. On the other

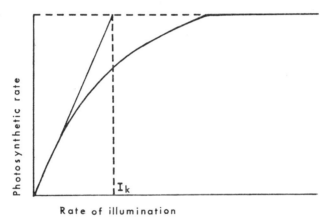

Fig. 3. The rate of photosynthesis as a function of the rate of illumination showing the position of I_k (redrawn from Talling, 1957).

hand, the horizontal part represents the maximum rate of the enzymatic processes at the present temperature and is thus for this specific temperature a function of the concentration of enzymes. Therefore the illumination intensity at the point where the initial slope and the horizontal part of the curve intersect—designated I_k by Talling (1957)— describes to a certain degree the ratio between the two kinds of processes and is thus an important means of describing the physiological adjustment to an algal population. The algae ordinarily are so adapted that they are able both to tolerate the short term variations in their habitat and to have the best economy at the environmental conditions ordinarily found here.

The adaptation of the algae is brought about by varying either singly or in combination (1) the concentrations of pigments per cell, and (2) the concentration of enzymes (in principle the enzymes active in photosynthesis, but in fact, because the rate of all processes have to match each other, all enzymes in the cell). For the stabilization of a new state of adaptation it is necessary that at least one new cell generation is produced (Steemann Nielsen *et al.*, 1962; Jørgensen, 1964). Various algal species differ greatly in their ability to adapt.

In Fig. 4, according to Steemann Nielsen *et al.* (1962), curves showing the rate of photosynthesis as a function of illumination intensity are presented for *Chlorella vulgaris* growing at continuous illumination of either $7\cdot5\times10^{15}$ quanta $(= 2\cdot7 \text{ mW}\times\text{cm}^{-2})$ or $75\cdot0\times10^{15}$ quanta $(= 27 \text{ mW}\times\text{cm}^{-2})$. The rate is given (a) per number of cells, (b) per unit weight of chlorophyll $a+b$, and (c) per unit weight of dry matter. Alternative (c) provides the best means of showing production, but for natural plankton it is normally not possible to use this form of presentation, as dead organic matter is often the major part of the organic matter. Comparing all the curves in Fig. 4 it is obvious that *Chlorella vulgaris* during light adaptation varies the chlorophyll concentration per cell and not the concentration of enzymes. However, as the alga decreases its size with increasing illumination rate, per unit of weight the concentration of enzymes increases. Thus, if we consider the rate of photosynthesis per dry weight of the alga—the most adequate form of presentation—it is obvious that the concentrations of both pigments and enzymes are altered when the illumination during growth is altered.

If the adaptation takes place due to a shift in temperature, usually only the concentrations of enzymes per cell is altered. Figure 5(a) shows for the diatom *Skeletonema costatum* the rate of photosynthesis per cell number as a function of the rate of illumination at 20°C, the temperature at which the diatom had been cultured. It also shows the response of these algae immediately after their transfer to 7° or 2°C (Steemann Nielsen and Jørgensen, 1968). As expected on theoretical grounds the optimum rate at 7°C is only about one-third of that at 20°C. The curves in Fig. 5(b) are for algae that had been grown at 20°, 8° and 2°C for three days and thus had adapted to this particular temperature. These results demonstrate that the amount of the enzymes active in photosynthesis increases per cell if the temperature is lowered. The amount of protein per cell was twice as high in the diatoms grown at 8°C as in those grown at 20°C. This agrees with the known fact that most of the proteins in cells constitutes enzymes. Furthermore, the concentration of all the other enzymes seemed to have increased at 8°C in the same way. Hence, the rate of respiration per cell was the same at 20° and 8°C for algae grown at these specific temperatures.

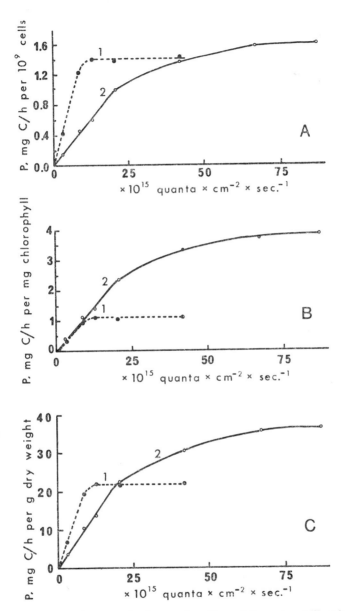

Fig. 4. The rate of photosynthesis as a function of the rate of illumination in *Chlorella vulgaris* grown at $7\cdot5 \times 10^{15}$ quanta (1) or 75.0×10^{15} quanta (2). A, calculated per unit number of cells; B, per mg chlorophyll, $a+b$, and C, per g dry weight. 21°C.

The adaptation occurring in Nature seems a little bit more complicated although in principle we find exactly the same as in laboratory experiments. In order not to get complications from varying growth states of the algae, the cells in the experiments presented above had been grown in continuous light, producing thereby a mixture, although not a perfect one, of all growth stages.

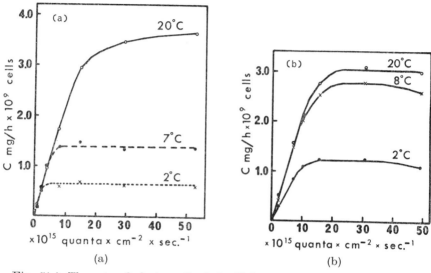

(a)   (b)

Fig. 5(a). The rate of photosynthesis in *Skeletonema costatum* per number of cells as a function of the rate of illumination. The algae were grown at 7.5×10^{15} (continuous illumination) at 20°C, but transferred for 30 min to 20°, 7° and 2°C. Fig. 5(b). The rate of photosynthesis in *Skeletonema costatum* per number of cells as a function of the rate of illumination. The algae were grown at 7.5 quanta $\times 10^{15}$ (continuous illumination) at either 20°, 8° and 2°C. The temperature during the experiment was the same as during growth.

In *Skeletonema costatum* grown at 5.1×10^{15} quanta of continuous light and 20°C, the I_k was about 15×10^{15} quanta (Steemann Nielsen and Jørgensen, 1968a). If the same alga was grown for 12 h at 5.1×10^{15} quanta and for 12 h in the dark, the I_k varied between 12·3 and 24·5 quanta $\times 10^{15}$. If *Skeletonema* was grown at 17.0×10^{15} quanta in the 12 h light period instead, the I_k varied between 25·0 and 35·4. In both cases some synchronization of the cells resulted, although by no means was it complete. If we only consider I_k in the light period, no overlapping between the two series of I_k is found. *Skeletonema costatum* occurs abundantly in temperate waters both during summer and winter. Figure 6, according to Jørgensen and Steemann Nielsen (1965), shows curves for rate of photosynthesis as a function of the rate of illumination for algae adapted to 2°C but with varying proportions between the hours of light and dark. It is obvious that a long night is necessary at low temperatures for giving a high rate of photosynthesis, which is why the species does not occur in the Arctic.

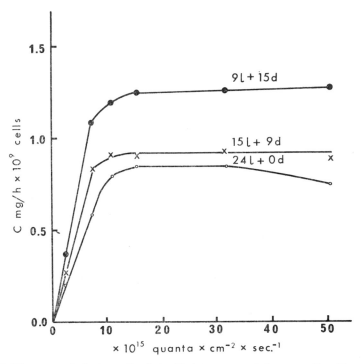

Fig. 6. The rate of photosynthesis in *Skeletonema costatum* as a function of the rate of illumination. The algae were grown at 7.5×10^{15} quanta given either continuously or in periods of 15 h light and 9 h dark, or 9 h light and 15 h dark, 20°C.

Illumination and temperature do not vary independently in Nature. Illumination varies according to latitude, season and depth. If the watermasses within the photic layer are homogeneous and more or less constantly mixed vertically, no differentiation, or only very little, in adaptation according to depth occurs. This is the case, for example, in several coastal regions with strong tidal currents and in many freshwater lakes where the depth of the euphotic layer and the depth of the epilimnion are identical. When vertical stability of the watermasses of the photic zone is practically lacking, the algae from all depths are nearly equally sun-adapted (Fig. 7).

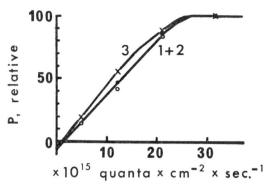

Fig. 7. Rate of relative photosynthesis as a function of the rate of illumination at a station west of the Faroe Islands during summer: (1) plankton from the surface; (2) plankton from 18 m; (3) plankton from 36 m, the lower limit of the photic zone. Temperature at 0 and 18 m 9·2°C, at 36 m 9·0°C. After Steemann Nielsen and Hansen (1959).

In most oceanic regions, e.g. in the North Atlantic during summer, the thermocline is located above the lower boundary of the photic zone. At Dana St., 10986 longitude west of Ireland, in August the temperature of the water was a constant 16·0°C down to 30 m from where it decreased to 12·7°C at a depth of 60 m, which was the lower boundary of the photic zone (Steemann Nielsen and Hansen, 1959). Figure 8 presents the rate of photosynthesis as a function of illumination rate in water from the surface (○), from 28 m, (●) where 10% of the surface light was found, and (×) from 60 m where 1% was found. The

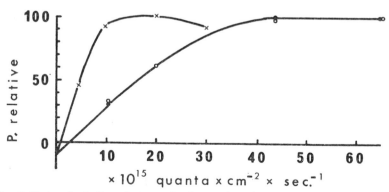

Fig. 8. Rate of relative photosynthesis as a function of the rate of illumination at a station in the Atlantic west of Ireland during summer. (○) plankton from the surface; (●) plankton from 28 m; (×) plankton from 60 m, the lower limit of the photic zone. Temperature at 0 and 28 m 16·0°C, at 60 m 12·7°C.

temperature of ($\circ$) and ($\bullet$) during the experiment was 16°C, whereas it was 13°C for ($\times$). It is obvious that the adaptation of the plankton from the surface and from 28 m was identical, I_k in both cases being 30×10^{15} quanta. The plankton from 60 m, however, was shade-adapted, with an I_k of 10×10^{15} quanta. The temperature, a mere 3°C less than at the surface, can have had only a minor effect.

However, in some cases it would seem that the difference in adaptation between the surface algae and the algae from the lower part of the photic zone is only apparent, the difference in shape of the curves resulting from the different temperatures in the upper and the lower part of the photic zone. Saijo and Ichimura (1962) have presented a striking example of this from the Oyashio area near Japan. The temperature at the surface was 20°C and at a depth of 20 m 8°C. Illumination rate-photosynthesis curves made at the correct temperature gave an I_k of 12 klux (about 20×10^{15} quanta) at the surface and of 6 klux at 20 m. However, if the surface plankton was also measured at 8°C, the I_k was 7 klux, thus practically the same as for the plankton occurring at 20 m.

If the stabilization of stratification is primarily due to differences in salinity, as, for example, in the Sound off Helsingør in late autumn (Steemann Nielsen and Hansen, 1961), the temperature in the lower part of the photic zone may be the same as or even higher than at the surface. Figure 9 shows an example from October, where the temperature was exactly the same (12°C). Whereas I_k of the surface water was $13 \cdot 6 \times 10^{15}$ quanta, it was only 5·1 at 18 m where 2% of the surface light was found. Figure 10 shows curves at 4°C from December from the surface and from a depth of 13 m. The same I_k, 5×10^{15} quanta was found. Light inhibition (see p. 374) was found in the plankton from 13 m but not in that from the surface.

Microbenthic algae occurring at very shallow depths during summer have a considerably higher I_k than surface plankton. Thus, for a Danish marine locality Gargas (1971) found an I_k for the microbenthic algae of 36×10^{15} quanta (15 mW) in contrast to 20×10^{15} quanta (8·6 mW) for the surface plankton. In principle the same was found by Burkholder et al. (1965).

From terrestrial ecology we know that "shade" leaves have the best economy at low intensities of illumination and "sun" leaves at high intensities (Boysen Jensen, 1932). The advantage of the sun leaves in full light lies in the fact that they can utilize high illumination rates due to the high level of their light saturation. At low intensities the rate of real photosynthesis per leaf area is the same in sun and shade leaves. In ordinary leaves the rate of photosynthesis per leaf area is independent of chlorophyll concentration, as the pigment in both cases

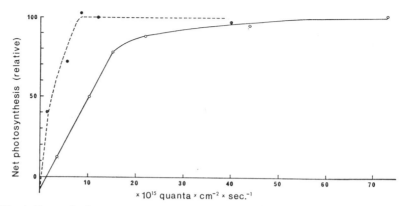

Fig. 9. Rate of relative photosynthesis (net) as a function of the rate of illumination in the Sound off Elsinore in October. —: plankton from the surface; - - -: plankton from a depth of 18 m, where 2% of the surface light was found. Temperature at both depths 12·0°C.

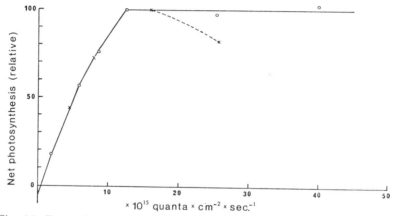

Fig. 10. Rate of relative photosynthesis (net) as a function of the rate of illumination in the Sound off Elsinore in December. —: plankton from the surface; - - -: plankton from a depth of 13 m, where 2% of the surface light was found. Temperature at both depths 4°C.

is in excess. When shade leaves have a better economy than sun leaves at habitats with low illumination, it is due to the lower rate of respiration. Both for higher terrestrial plants and for unicellular algae the rate of respiration is proportional to the rate of light-saturated photosynthesis (Steemann Nielsen and Jørgensen, 1968a). All the various processes in plants are adjusted in such a way that they give rise to growth in the most economical way at the special condition prevailing at the habitat.

In contrast to the leaves of terrestrial plants, unicellular algae have one more means of increasing their economy at low light intensity. They are able to raise the rate of photosynthesis per cell by increasing the quantity of photosynthetic pigments. At the most, growing algae assimilate about 0.40–0.60 mg C h^{-1} mg^{-1} chlorophyll at 1 klux more or less independently of the concentration of chlorophyll in the cells (Steemann Nielsen and Jørgensen, 1968). When Philips W/33 fluorescent light was used, about 0.2–0.3 mg C h^{-1} mg^{-1} chlorophyll is assimilated at 10^{15} quanta. The fact that unicellular algae growing at high illumination rates ordinarily have only small concentrations of the pigments must be considered first of all as a measure against light damage (page 374).

Shade-adapted algae found in the lowest part of the photic zone have at a low illumination rate a higher rate of photosynthesis per cell than the corresponding sun-adapted algae near the surface. At the same time the rate of respiration is lower per cell. As a result the compensation point occurs at a lower rate of illumination. The compensation point of *Chlorella pyrenoidosa* grown at a low rate of illumination is about 0.5×10^{15} quanta but is more than 1.7 when grown at a high illumination rate, as for example, 34×10^{15} quanta. Due to shade adaptation, the algae in the lower part of the photic zone are able to grow at illumination rates only one-third to one-fourth of those that would be necessary if the algae were more or less sun-adapted (Steemann Nielsen and Hansen, 1959). In an ocean where 1% of the surface light is found at a depth of 100 m, the depth of the photic zone is thus about 100 m. However, if the algae were not shade-adapted the photic zone would extend down only to the depth where 3–4% of the surface light penetrates. The photic zone would then be only about 75 m deep instead of the ordinary 100 m.

If the number of algae is more or less the same throughout the whole photic zone, the importance of shade adaptation of the algae in the lower part of the photic zone is relatively slight for the production of organic matter calculated per surface unit of the sea or a lake. By far the major part of the organic production results from photosynthesis by the sun-adapted algae in the upper half of the photic zone (Ichimura *et al.*, 1962; Talling, 1966). The situation is quite different, however, if the bulk of the algae is found in the lower part of the photic zone, a situation quite frequently encountered in the sea, as in Danish waters as shown by Steemann Nielsen (1964a).

For the algae themselves dark adaptation in the lower part of the photic zone must be of definite importance. As the concentration of all enzymes is relatively low, the amount of protein, the major part of the organic matter, is low (see page 366). Thus less photosynthesis is necessary to produce a new generation. Algal species not able to become dark-adapted can hardly compete with those able to do so.

IV. THE INFLUENCE OF HIGH ILLUMINATION INTENSITIES, INCLUDING ULTRAVIOLET RADIATION

It is well-known that many photo-autrophic plant species are seriously affected by illumination intensities higher than normally found at their habitats. Stålfelt (1960) has presented a review of the literature.

Species of higher plants seem to be influenced only by intensities considerably higher than those occurring in their habitat. Plankton algae behave differently. In most cases plankton photosynthesis on bright days is depressed near the surface. If bottles containing surface water are suspended at various depths within the photic zone during a bright day, the maximum rate of photosynthesis is found at the depth where about 50% of the surface light occurs. This is observed both in the sea and in freshwater, both in the tropics and at higher latitudes during summer (Fig. 11). The depression of photosynthesis by high

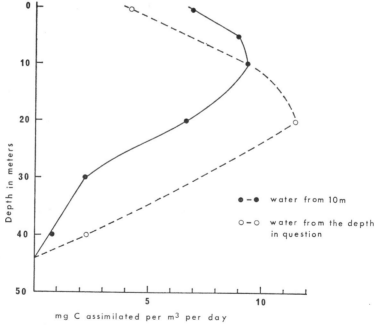

Fig. 11. The rate of production as a function of the depth in the Benguela Current off Angola: - - -. In addition a curve: — is presented showing the production if water from a depth of 10 m is used at all depths.

illumination rates is one of the causes—but not the only one—of the so-called afternoon depression first called attention to by Doty and Oguri (1957).

Because the depression of the photosynthetic rate in bright days near the surface is an ordinary phenomenon, it can hardly be considered harmful to the algae. In algae a special mechanism has been found (Steemann Nielsen, 1949, 1962), which protects the cell constituents against light energy absorbed by the photosynthetic pigments but not used in photosynthesis. This "surplus light energy" would otherwise be used for photo-oxidation, a dangerous situation for the cells. The

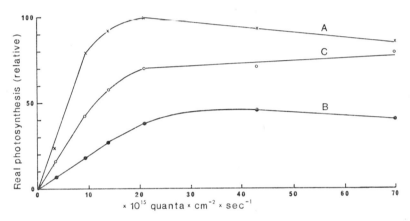

Fig. 12. The rate of photosynthesis as a function of the rate of illumination for *Chlorella vulgaris* grown at 7.5×10^{15} quanta, 20°C. A: directly from 7.5×10^{15} quanta; B: after three hours at 75.0×10^{15} quanta; C: after one subsequent hour in the dark.

principle of this mechanism is inactivation of a part of the photochemical reaction. This inactivation takes some time. Initially some photo-oxidation takes place. A part of the enzymes participating in photosynthesis, therefore, are destroyed, and the rate of light-saturated photosynthesis decreases. At a low illumination rate or in the dark, both a reactivation of the photochemical process and a production of the enzymes to the original level takes place. This takes some hours.

Figure 12 (from Steemann Nielsen, 1962) illustrates a series of experiments where *Chlorella vulgaris* grown at 7.5×10^{15} quanta was transferred to an illumination intensity 10 times as high for 3 h and subsequently transferred to the dark for 1 h. Curve A presents the rate of photosynthesis as a function of illumination intensity before the transference to the high illumination rate, curve B after 3 h at this rate and curve C after one subsequent hour in the dark. Curve A is a typical curve for algae grown at a rather low illumination intensity. The photosynthetic rate at 75×10^{15} quanta shows typical light inhibition, by about 15%. As all these photosynthesis experiments lasted only 15 min, the inhibition at this high illumination must occur relatively rapidly. Curve B which presents experiments similar to those represented in curve A, but after 3 h at high illumination, is very unlike curve A. The initial slope is only 20% of the original one, showing that about 80% of the chlorophyll, the concentration of which was unaltered, was inactive. The rate of photosynthesis at 75×10^{15} quanta was only 40% of the original light saturation value shown in curve A. Curve C shows that the algae recovered but not completely after 1 h in the dark after the 3-h exposure to high illumination. The initial slope is about half of that of A. The light saturated rate is about 75% of that of A.

As mentioned above no destruction of chlorophyll took place at 75×10^{15} quanta, although at still higher illumination intensities chlorophyll may be destroyed. In *Chlorella vulgaris* grown at 7.5×10^{15} quanta it takes about 2 h or more for the destruction to begin at 250×10^{15} quanta (Steemann Nielsen and Jørgensen, 1962).

In nutrient deficient algae or in algae influenced by small concentrations of a heavy metal like copper, light inhibition is much more pronounced than in ordinary cultures (Fig. 5 in Steemann Nielsen, 1962, or Fig. 3 in Steemann Nielsen and Wium-Andersen, 1971).

If the illumination rate is sufficiently high, light at all wavelengths is able to depress photosynthesis. Ultraviolet light, however, has a special influence. As shown by Steemann Nielsen (1964) the rate of photosynthesis decreases in full sunlight both in surface plankton and in dark-adapted plankton placed without any cover of glass at the surface. If by means of a neutral filter made of black netting the illumination is reduced to 30%, the presence of the ultraviolet part of the light is of no importance for surface plankton enclosed in ordinary, thin-walled glass bottles, but in dark-adapted plankton the walls of the bottles are not sufficient to give complete protection against the ultraviolet rays. However, a plate of clear glass 3-mm thick is sufficient to protect also the dark-adapted plankton. This is of importance when the simulated *in situ* technique is used to measure the rate of primary production. In addition to neutral filters made of black netting the samples from the lower part of the photic zone must be covered by a relatively thick plate of clear glass able to absorb the ultraviolet radiation more or less completely (see page 386).

V. The Utilization by the Algae of the
Light Penetrating the Sea Surface

It is now well established that about 8 quanta are needed for the assimilation of one molecule of CO_2. This means that in the red part of the spectrum about 50% of the light energy absorbed by the algae can be transferred into chemical energy under optimum conditions. By optimum conditions we understand (1) a low illumination intensity and (2) an optimum physiological state of the algae.

Only at low illumination intensities where the rates of the photochemical reactions limit the overall rate of photosynthesis is the optimum utilization of absorbed light energy obtained. According to the curves shown in Fig. 8 this would mean intensities up to $10-20 \times 10^{15}$ quanta. With increasing intensities the utilization steadily decreases and, as a result of light inhibition, the utilization is reduced further. In Nature at noon the light utilization near the surface is often about one-tenth or less compared with the utilization at low illumination intensities. At the surface of the sea light is distributed throughout the whole spectral region of importance for photosynthesis—350–700 nm. Photosynthesis is a quantum process. Compared with red light, natural surface light, which is a mixture of quanta of various sizes, will be utilized only by a factor of about 0·75 if measured in energy.

In Nature a still lower utilization of the light is found for the whole photic layer. This is not primarily due to a poor utilization of the light absorbed by the pigments in the algae. The cause is first of all that generally only a minimal percentage of the light penetrating the surface of the sea is absorbed by the pigments of the algae. The rest is absorbed by the water, by particulate dead matter, by non-autotrophic organisms and by coloured dissolved organic matter, first of all "yellow matter". For all oceans in average during a year only about 0·2% of the energy of the incident light (300–700 nm) is transformed into chemical energy (Müller, 1960).

A high utilization of the incident light is possible only if the plankton algae are concentrated in a shallow photosynthetic layer. The light absorption by the water is thus minimized. The highest natural rates of photosynthesis both in the sea and in freshwater (up to about 6 g C m^{-2}×day) have been found where the depth of the photic zone is extremely small (Steemann Nielsen and Jensen, 1957; Mathiesen, 1971).

In a theoretical ocean of absolutely pure water 1% of the blue light (475 nm) would according to Jerlov (personal communication) be found at about 160–165 m, and 1% of green light (525 nm) at about 90–95 m.

If we presume that the lower limit of the photic zone would be at the depth where 1% of the blue+green light is found, this would be at a depth of about 140 m. In such a plankton-free and thus production-free

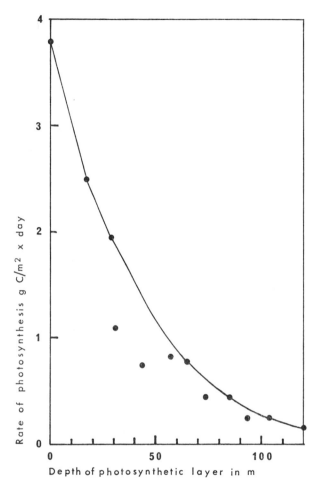

Fig. 13. The maximum rate of photosynthesis per m² surface as a function of the depth of the photosynthetic layer (see text).

ocean the photic zone is not much deeper than found in the most transparent parts of the ocean, such as the Sargasso Sea where the depth of the photic zone is about 120 m. Accordingly, only a very low standing stock of plankton can be found here, which explains the very low production rates.

The curve in Fig. 13 shows the maximum rate of photosynthesis as a function of the depth of the photic zone (according to measurements in the ocean during the Galathea Expedition). All stations were grouped into 10-m classes according to the depth of the photic layer. The highest rate of production was taken from each class, and hence the curve represents a ceiling or upper limit. Transparency is generally only a useful indicator of productivity in the ocean beyond the influence of land. Near the coast, where the algae themselves do not provide the chief mechanism for lowering the transparency—the water itself not being taken into consideration—the production rates are generally definitely lower than the values shown in Fig. 13. In freshwater lakes the production rates in many cases are much below these values, indicating a relatively high percentage of light absorption by mechanisms other than living algae and water. In fact it is often found that the relatively clear lakes produce as much organic matter per surface area as many of the less transparent lakes.

Generally photosynthesis and respiration in plankton algae per 24 h compensate each other at a depth where about 1% of the surface light occurs. This is true in the tropics, the subtropics and in temperate waters during summer. Not always the same part of the light spectrum has been considered in such statements. Steemann Nielsen and Jensen (1957), however, defined the compensation depth as the depth at which the sum of the blue and the green light is equal to 1% of the same sum measured at the surface of the sea. In coastal waters often only the green part of the light is considered. Red light, which in the sea is of practically no importance in the lower part of the photic layer, is ordinarily not considered by marine investigators. In freshwater lakes, where the red light often penetrates as deep as the blue and the green light, the problem is different. It must be stressed, however, that the actual location of the compensation depth can be determined only approximately.

At very low temperatures, such as found in high arctic and antarctic marine areas and in lakes in the high mountains, plankton algae must be able to grow at much lower illumination intensities than ordinarily (Pechlaner, 1971). At least down to intensities of about $2\cdot5\text{--}5$ fc $\simeq 0\cdot015\text{--}0\cdot03$ mW $\times$ cm^{-2}), photosynthesis per day may equal respiration per day (Bunt according to Allen, 1971). As the quantum efficiency cannot be increased, such low compensation levels— if correct—can be explained only by a very low respiration rate. This again means that the growth rate must be exceedingly low, and the grazing thus must also be low. Furthermore photosynthesis at low illumination is limited by the speed of the photochemical reactions, which is independent of temperature. Mixotrophy, such as suggested by Lund (1959), presents another possibility for explaining the presence of algae at very low illumination intensities. As growth of photoautrophic algae has never been found at such low illumination intensities at higher temperatures, we must conclude that very low temperatures are absolutely necessary.

By suspending dense cultures of various plankton algae in cuvettes 1 cm deep, Steemann Nielsen 1962 showed that in green alga up to about 900 mg and in diatoms up to 400 mg chlorophyll must be present per m^2 if light (400–700 nm) falling vertically on the surface is to be reduced to 1%. Such large amounts of chlorophyll cannot of course be found in the photic zone in nature, where the attenuation of the light is brought about not solely by the plankton algae. Ichimura (1956) has shown that the quantity of chlorophyll m^2 never exceeds 200 mg m^{-2} in the photic layer of Japanese lakes. Considerably higher amounts of chlorophyll have been measured, but under these circumstances most of the chlorophyll has been present below the compensation depth. In some marine coastal areas with a thin photic zone, concentrations of chlorophyll as high as in freshwater lakes may be found, even up to 277 mg m^{-2} according to Lorenzen (1972).

VI. THE INFLUENCE OF VERTICAL WATER MOVEMENTS ON THE LIGHT UTILIZATION OF THE PHYTOPLANKTON

A plankton alga has to follow the watermasses carrying it. If the watermasses make extensive vertical movements the algae are bound to do the same, and the light utilization by the algae is thereby influenced. Ordinarily, but by no means always, the photic zone in the open ocean has a greater vertical extension than the layer in which the watermasses daily are vertically mixed. In freshwater lakes, on the other hand, the photic zone often coincides more or less with the epilimnion, if the lake is not so shallow that the water is mixed down to the bottom.

In the oceans the size of the algal population may differ greatly between the daily vertically mixed surface layer and the rest of the photic zone. However, the physiological state of the algae will also differ (see p. 370). At higher latitudes the thermocline breaks down during winter both in the sea and in lakes. However, if a strong halocline is found, such as in the waters between the saline North Sea and the brackish Baltic Sea, the watermasses are vertically strongly stabilized the whole year round. If the sea does not freeze, it is possible for the plankton to grow during the whole winter (Steemann Nielsen, 1964b).

On the bright days in December more than $1.7 \, mW \times cm^{-2}$ can occur in 3 h plus about $0.9 \, mW$ in additional 2 h. As the I_k of the surface plankton in the Sound was $1.4 \, mW \times cm^{-2}$ (Steemann Nielsen and Hansen, 1961), on bright days a considerable rate of photosynthesis may take place near the surface. The compensation depth for 24 h was found at about 5 m.

It is usual in the sea to find that vertical stabilization is caused solely by a thermocline. At higher latitudes during winter the stabilization, therefore, will break down, and no growing plankton is found. A growing phytoplankton requires a minimum rate of respiration. Under these conditions the rate of respiration in the whole mixed layer would exceed that of photosynthesis. Plankton algae may still be found in small quantities, but they are very likely to be in a more or less dormant stage until the ecological conditions during spring make it possible for them to be transferred into an active stage again. The vertically mixed layer in the North Atlantic at latitudes of 55–60° is up to nearly 600 m deep in winter.

Braarud and Klem (1931) were the first to point out that there must be a critical depth. A growing phytoplankton population can occur only if the depth of the mixed layer is less than the critical depth. Sverdrup (1953) tackled the problem theoretically. With certain assumptions it was possible for him to compute the critical depth during the spring at the weather-ship "M" in the Norwegian Sea and to follow the outburst of phytoplankton, which in its final state took place during the second week of May.

A lack of vertical stabilization of the watermasses may be counteracted by a shallow depth such as seen west of the Faroe Islands over the isolated Faroebank, where the depth is only about 100 m. In the beginning of May 1934 the concentration of diatoms, peridinians and coccolithophorids was about $100 \, ml^{-1}$, whereas the concentration was very low in the ocean surrounding the bank (Steemann Nielsen, 1935).

Patten (1968) rejected Sverdrup's "critical depth" theory. He claimed that any natural system can adjust in such a way that net community production is positive in a completely mixed water column, no matter how deep, by changing its system parameters. Vollenweider (1970) has discussed Patten's vs. Sverdrup's theory and is of the opinion that the truth lies in between the two positions. As Sverdrup's theory concerns growing plankton I can hardly agree with Patten's ideas. An I_k being lower than about 3.5×10^{15} quanta was not observed during winter in the Danish waters (cf. Steemann Nielsen and Hansen, 1961).

VII. EXPERIMENTAL AND THEORETICAL DEPTH PROFILES OF
PRIMARY PRODUCTION

If the plankton, as a result of vertical mixing, is evenly distributed throughout the photic zone, it is sufficient to take water from a single depth and suspend it in a series of bottles at various depths of the photic zone as from noon to sunset. Such an experimental depth profile only approximates the real one. Due to being enclosed in the bottles, the algae have to stay at a definite depth, and this very likely has some bearing on the extent of the light inhibition near the surface. Rodhe (1958) has shown that for the top layer of the photic zone during a bright day the sum of five short-time exposures is higher than from a bottle exposed during the whole day.

If, as is most often found in the ocean, the plankton is not evenly mixed vertically, then the depth profile will of course depend on whether water samples from the various depths are resuspended at the same depths or whether water samples from one single depth are used for all bottles. The curves presented in Fig. 11 clearly show the difference from these two procedures.

In order to derive a widely applicable expression for the total or integral photosynthesis of a natural population beneath a unit area of surface, Talling (1957) used depth profiles obtained by suspending bottles containing the freshwater diatom *Asterionella formosa* either artificially cultured or caught fresh in the lake. The integral photosynthesis obtained by this method was related, under a wide range of conditions, to the logarithm of the surface light intensity. This relationship was used to present a relative logarithmic scale of light intensity applied to the calculation of the integral photosynthesis.

Figure 14 is based on Talling's photosynthesis-depth profiles in lake Windermere calculated for various values of the ratio I_o/I_k. I_o is the incident light at the surface, whereas I_k represents a value for the onset of light saturation. The depth is given in "optical depth" units. Each unit corresponds to the depth causing a halving of the light of the wavelength absorbed least strongly by the water. The influence of light inhibition is not considered. Besides this limitation, it must be noted that the theory breaks down for low subsurface light intensities. Furthermore, Talling's expression is used for calculating the instantaneous rate integrals of photosynthesis as functions of I_o/I_k. Under *in situ* conditions instantaneous rates continually change with changing environmental conditions. In Talling's theoretical approach the assumption of a physiological and quantitatively homogeneous population is an important part. In most areas in the oceans, where this is not the case, a theoretical approach will turn out to be more difficult.

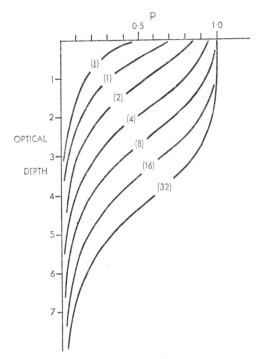

Fig. 14. Photosynthesis-depth profiles in Windermere calculated for various values (shown in brackets) of the ratio I_0/I_k. Depth is given in "optical depth" units.

Other scientists, first of all Vollenweider (1965, 1970), have tried to tackle theoretically both light depression and the calculation of daily rates of photosynthesis. Being no trained mathematician the author will abstain from discussing the contributions already mentioned and also the important contributions by, for example, Rodhe (1965); Rodhe et al. (1958); Steele (1965); Patten (1968); Talling (1971).

Steemann Nielsen and Jensen (1957) were able to present a simple empirical formula valid for the tropical, the subtropical and, during summer, even for the temperate parts of the ocean. This formula was based on simultaneous in situ experiments and experiments at light saturation in the laboratory on board a ship. Four oceanic stations varying considerably both in transparency and rate of production agreed very well inter se. Rodhe et al. (1958) theoretically arrived at a formula that was practically the same as the one found empirically by Steemann Nielsen and Jensen.

The variations from day to day in the light penetrating the surface is especially important when the maximum energies are small, such as during winter in temperate waters. During summer at higher latitudes and at low latitudes during the whole year the importance of variations of the light due to varying weather conditions is much less. During winter even in bright weather light saturation occurs only during a short time of the day. During the summer season the situation is quite different.

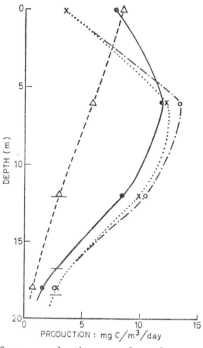

Fig. 15. The rate of gross production per m³ per day as a function of the depth in the Great Belt at the end of April. A, · · · ·: according to *in situ* experiments; B, —: calculated for a day with average illumination; C, -·-·-: calculated for bright days; D, - - -: calculated for dark days. The position of the compensation depth is shown.

Figure 15 presents according to Steeman Nielsen (1964a) a typical example from the Great Belt at the end of April. During a day with bright weather, *in situ* measurements (the dotted curve) were made from noon to sunset. At the same time photosynthesis was also measured in the laboratory using artificial light. By means of curves presenting the rate of photosynthesis as a function of illumination intensity and by means of the measurements of the vertical penetration of the light and the light measurements published by Romose (1940) for April (see also

Steemann Nielsen and Hansen, 1961), the rates of photosynthesis for every hour of the day for the four depths in question were calculated (a) for an average day, (b) for a bright day (average for the three brightest days during two years in April) and (c) for a dark day (average of the three darkest days in April). By summing up the rates of photosynthesis for all the hours of the day curves for a "bright", an "average" and a "dark" day were drawn showing photosynthesis as a function of the depth for the "bright", the "average" and the "dark" day. During the "bright" day the rate of production was only 12% higher than during the "average" day, whereas it was about the half during the "dark" day. The actual *in situ* measurements gave a result in between the "bright" and the "average" day.

If the bulk of phytoplankton is found in the lowest part of the photic zone—by no means a rare situation in the sea—day to day variations in the light situation are very important, the rate of photosynthesis at these depths always being more or less proportional to the intensity of the illumination.

VIII. *"In situ"* AND *"Simulated in situ"* MEASUREMENTS

In situ measurements are usually made during half a day and mostly from noon to sunset. For many reasons we cannot expect to obtain exactly the same result if the period from sunrise to noon is used instead. The algae during the day are growing and going through various stages while differently exposed to the environmental factors. Grazing must also be considered. It is not constant.

Vollenweider (1965) has considered if another time schedule for making experiments on production might be better. According to him the best period in a day divided into five equal periods should, for most areas, be the second one. The production rate of that period can be expected to be about 30% of the total day rate. Ordinarily it is not advisable to make experiments from sunrise to sunset or for 24 h. The enclosement of the water in a bottle is an artificial interference. The duration therefore should not be too prolonged. This is especially the case in very oligotrophic water, where the production of bacteria is strongly affected by the walls of the bottles (Steemann Nielsen, 1958a).

Most often *in situ* measurements will turn out to be too expensive in the oceans. Here a simulated *in situ* method may be used instead. Water samples from the different depths are suspended on the deck of the ship in a tube through which surface water is flowing. The light conditions at the different depths are simulated by neutral filters, e.g. black nylon netting or coloured glass above the bottles.

16

If the penetration of green light has been measured in coastal (green) waters it has been possible to use neutral filters (Fig. 16). This is not possible in blue ocean water (Jitts, 1963). By means of blue glass he was able to simulate the light conditions. The introduction of Jerlov's quanta meter especially made for simulated *in situ* measurements has facilitated the performance of such measurements both in

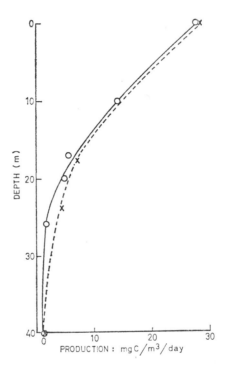

Fig. 16. The rate of gross production per m³ per day as a function of the depth. Off the mouth of Godthaabsfjord, Greenland. —: according to *in situ* experiments; - - -: according to simulated *in situ* experiments.

oceanic and coastal waters (Steemann Nielsen and Willemoës, 1971). The advantage is that it is not necessary to use filters the spectral properties of which correspond exactly to those in the sea. Cheap, commercial filters of plastic can thus be used. As dark-adapted algae seem to be very sensitive to the longer wave part of the u.v. radiation, it is necessary to protect the samples from the lower part of the photic zone with a sheet of ordinary glass of sufficient thickness.

REFERENCES

Allen, M. B. (1971). *Annu. Rev. Ecolog. Syst.*, **2**, 261–276.

Boysen Jensen, P. (1932). "Die Stoffproduktion der Pflanzen." Jena.

Braarud, T. and Klem, A. (1931). *Hvalrådets Skrift.*, **1**, 1–88.

Burkholder, P. R., Repark, A. and Sibert, J. (1965). *Bull. Torrey Bot. Club*, **92**, 378–402.

Doty, M. S. and Oguri, M. (1957). *Limnol. Oceanogr.*, **2**, 37–40.

Gargas, E. (1971). *Ophelia*, **9**, 107–112.

Ichimura, S. (1956). *Bot. Mag. Tokyo*, **69**, 7–16.

Ichimura, S., Saijo, Y. and Aruga, Y. (1962). *Bot. Mag. Tokyo*, **75**, 212–220.

Jerlov, N. G. and Nygård, K. (1969). Københavns Universitet, *Rep. Inst. Fys. Oceanogr.*, **10**, 1–19.

Jitts, H. R. (1963). *Austr. J. Mar. and Freshwater Res.*, **14**, 139–147.

Jørgensen, E. G. (1964). *Physiol. Plant.*, **17**, 136–145.

Jørgensen, E. G. and Steemann Nielsen, E. (1965). *Mem. Inst. Ital. Idrobiol.*, **18** (Suppl.), 39–46.

Kimbal, H. H. (1935). *Mon. Weather. Rev.*, **63**, 1.

Lorenzen, C. J. (1972). *J. Cons., Cons. Perma. Int. Explor. Mer*, **34**, 262–267.

Lund, S. (1959). *Medd. Groenland*, **156**, 1–72.

Mathiesen, H. (1971). *Mitt. Int. Verein. Limnol.*, **19**, 161–181.

Müller, D. (1960). "Encyclopedia of Plant Physiology" (W. Ruhland ed.), Vol. 2, 255–268.

Patten, B. C. (1968). *Int. Rev. Gesamten. Hydrobiol.*, **53**, 357–408.

Pechlaner, R. (1971). *Mitt. Int. Verein. Limnol.*, **19**, 125–145.

Rodhe, W. (1958). *Rapp. Proces-Verb. Reunions, Cons. Perma. Int. Explor. Mer*, **144**, 122–128.

Rodhe, W. (1965). *Mem. Inst. Ital. Idrobiol.*, **18**, (Suppl.), 367–381.

Rodhe, W., Vollenweider, R. A. and Nauwerck, A. (1958). In "Perspectives in Marine Biology", (Buzzati-Traverso, ed.), 299–325. University of California.

Romose, V. (1940). *Dan. Bot. Ark.*, **10**, H.4, 1–134.

Saijo, Y. and Ichimura, S. (1962). *J. Oceanogr. Soc. Jap.*, **20th Anniv. Vol.**, 687–693.

Steele, J. H. (1965). *Mem. Inst. Ital. Idrobiol.*, **18**, (Suppl.), 383–398.

Steemann Nielsen, E. (1935). *Medd. Komm. Dank. Fisker. Havundersφg. Ser. Plankton III*, **1**, 1–93.

Steemann Nielsen, E. (1949). *Physiol. Plant.*, **2**, 247–265.

Steemann Nielsen, E. (1952). *J. Cons., Cons. Perma. Int. Explor. Mer*, **18**, 117–139.

Steemann Nielsen, E. (1958). *Rapp. Proces-Verb. Reunions, Cons. Perma. Int. Explor. Mer*, **144**, 141–148.

Steemann Nielsen, E. (1958a). *Ibid.*, **144**, 38–46.

Steemann Nielsen, E. (1962). *Physiol. Plant.*, **15**, 161–171.

Steemann Nielsen, E. (1964). *J. Cons., Cons. Perma. Int. Explor. Mer*, **29**, 130–135.

Steemann Nielsen, E. (1964a). *J. Ecol.*, **52**, (Suppl.), 119–130.

Steemann Nielsen, E. (1964b). *Medd. Dan. Fisker. Havundersφg. N.S.*, **4** 31–71.

Steemann Nielsen, E. and Hansen, V. K. (1959). *Physiol. Plant.*, **12**, 353–370.

Steemann Nielsen, E. and Hansen, V. K. (1961). *Physiol. Plant.*, **14**, 595–613.

Steemann Nielsen, E., Hansen, V. K. and Jørgensen, E. G. (1962). *Physiol. Plant.*, **15**, 505–517.

Steemann Nielsen, E. and Jensen, E. Aa. (1957). *Galathea Rep.*, **1**, 49–136.
Steemann Nielsen, E. and Jørgensen, E. G. (1962). *Arch. Hydrobiol.*, **58**, 349–357.
Steemann Nielsen, E. and Jørgensen, E. G. (1968a). *Physiol. Plant.*, **21**, 647–654.
Steemann Nielsen, E. and Jørgensen, E. G. (1968b). *Physiol. Plant.*, **21**, 401–413.
Steemann Nielsen, E. and Willemoës, M. (1971). *Int. Rev. Gestamer Hydrobiol.*, **56**, 541–556.
Steemann Nielsen, E. and Wium-Andersen, S. (1971). *Physiol. Plant.*, **24**, 480–484.
Stålfelt, M. G. (1960). "Encyclopedia of Plant Physiology" (W. Ruhland, ed.), Vol. 2, 186–212.
Sverdrup, H. W. (1953). *J. Cons., Cons. Perma. Int. Explor. Mer*, **18**, 287–295.
Talling, J. F. (1957). *New Phytol.*, **56**, 133–149.
Talling, J. F. (1966). *J. Ecol.*, **54**, 99–127.
Talling, J. F. (1971). *Mitt. Int. Verein. Limnol.*, **19**, 214–243.
Vollenweider, R. A. (1965). *Mem. Inst. Ital. Idrobiol.*, **18**, Suppl., 425–457.
Vollenweider, R. A. (1970). Proceedings of the IBB/PP Technical Meeting, Třeboň, 1969, 455–472.

Chapter 17

Remote Spectroscopy of the Sea for Biological Production Studies[*]

GEORGE L. CLARKE AND GIFFORD C. EWING

Woods Hole Oceanographic Institution, Woods Hole, Massachusetts, U.S.A.

I. INTRODUCTION

The fact that daylight backscattered from beneath the surface of the sea has undergone spectral changes caused by specific materials in the water provides us with a useful tool for the remote sensing of these materials and the conditions in the ocean associated with them. We wish to measure certain dissolved and particulate substances that have been formed by living organisms, or have been stirred into the water from the shore or bottom, or have been added to the sea by man's activities.

One of the materials which has a characteristic signature, or effect on the spectrum of backscattered light, and hence affects the color of the sea in a specific way, is chlorophyll. The measurement of chlorophyll in the upper layers of the sea is of particular significance since the concentration of this substance is an index of the amount of phytoplankton present. Regions with abundant phytoplankton can support large populations of herbivores which serve as food for higher links in the animal food chain. Many of these are of economic importance to man. Thus, abundant chlorophyll indicates the presence of a potentially

[*] Contribution No. 2901 of the Woods Hole Oceanographic Institution Supported under contract N62306–71–C–0195 for the Space Oceanography Project, U.S. Naval Oceanographic Office, and grant NGR 22–014–016 from Advanced Applications Flight Experiments Program, NASA/Langley Research Center.

productive area. Because measurements from aircraft or spacecraft can be made over a greater range and much more rapidly than from ships, they make possible the procurement of synoptic surveys of varying distributions of chlorophyll, and of other oceanic properties, over extensive areas of the sea, and the repetition of such surveys as frequently as desired.

It is the purpose of this chapter to review the research that has been conducted by the authors on the measurement of the color of the sea, with particular reference to the use of remote spectroscopy for studies on oceanic productivity. The studies conducted in 1968 and 1969 were carried out in coastal water near Woods Hole and in offshore waters extending from the Sargasso Sea across Georges Bank to the Gulf of Maine. In 1970 measurements were undertaken in the Gulf of Mexico, the Caribbean and along both coasts of Central America.

II. PROCEDURE

The spectrometer used for these remote measurements of ocean color from aircraft was designed by Peter White of TRW Systems, Inc. and described by L. A. Gore (1968). R. C. Ramsey of TRW operated the instrument and took part in the reduction of the data and in the interpretation of the results. The TRW spectrometer is an electro-optical sensor of the off-plane Ebert type with an RCA 7265 (S-20 response) photomultiplier. The spectral range is 400 to 700 nm with a spectral resolution of 5 to 7·5 nm, a scan time of 1·2 s and a field view of 3° by 0·5°. Thus, with the instrument looking straight down from an altitude of 305 m, light was received from an area of sea surface measuring 16 m × 3 m. A continuous curve of the spectrum is provided by a Sanborn recorder for each scan. The spectrum of the incident light from the sun and sky was determined on the ground before or after a series of measurements by recording the light reflected from a horizontally placed Eastman Kodak "gray card" with a non-selective reflectivity of 18%. In 1970 the observed readings were used to calibrate the spectral sensitivity of the instrument in the field.

The aircraft used in 1968 and 1969 was the C-54-Q research plane operated by the W.H.O.I. In 1970 the DC-3 research plane of Scripps Institution of Oceanography was employed under charter. In both cases our spectrometer was mounted in the belly of the plane with its receiving window directed downward at the nadir or at selected angles from the vertical. When the angle was other than nadir, the instrument was oriented away from the sun so as to avoid receiving the glitter due to the reflection of direct sunlight from the sea surface. A polarizing filter, oriented at right angles to the major axis of polarization, was placed over the receiving aperture of the spectrometer in certain instances. The airplane was operated at altitudes ranging from 152 to 4,580 m. Surface water temperatures were obtained from the aircraft by P. M. Saunders using a Barnes infrared radiometer.

Ground truth measurements of chlorophyll concentration and temperature were obtained from aboard the R.V. *Crawford* in 1968 and the R.V. *Gosnold* in 1969 by C. J. Lorenzen, using a continuous flow Turner fluorometer and a thermister. During the latter cruise optical measurements above and below the sea surface from shipboard were available through the collaboration of J. E. Tyler, R. C. Smith and R. W. Austin of the Visibility Laboratory of the Scripps Institution of Oceanography.

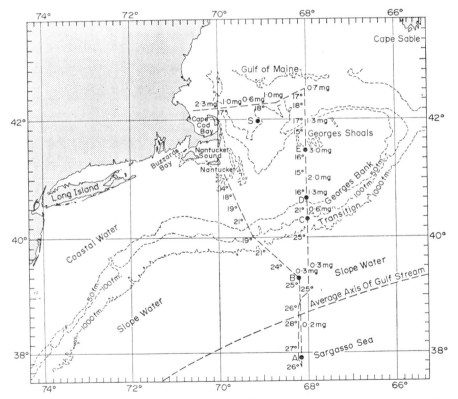

Fig. 1. Course of flight of 27 August 1968 and location of Stations A to E. Surface water temperatures are shown to the left or below the flight path; chlorophyll concentrations are shown to the right or above flight path (from Clarke *et al.*, 1970a).

A representative demonstration of the changes in the spectrum of backscattered light from waters of different chlorophyll content is available from our survey flight made at an altitude of 305 m on 27 August 1968. This covered areas north and south of Cape Cod and included a 520-km transect from the Sargasso Sea, across the Gulf

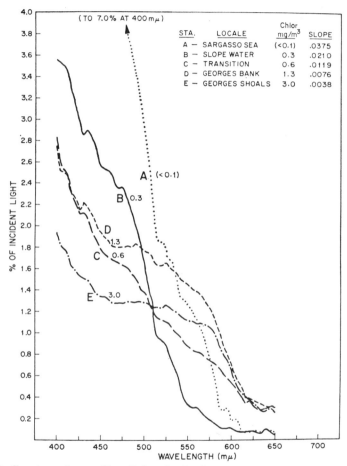

Fig. 2. Spectra of upwelling light obtained at 305 m on 27 August 1968 at Stations A to E shown in Fig. 1 (from Clarke *et al.*, 1970a).

Stream, the slope water, a transition zone, Georges Bank, Georges Shoals, to the Gulf of Maine (Fig. 1). Ground truth was obtained by our ship during a period that included the day of the flight. Surface temperatures ranged from 14° to 28°C, and chlorophyll concentrations ranged from < 0·1 to 3·0 mg m⁻³. Curves showing the backscattered light as a percentage of the incident irradiance (Fig. 2) reveal a progressive change from the Sargasso Sea, where values drop regularly from highs in the blue to lows in the red, to Georges Shoals where the curve tends to level off in the green before dropping again in the red.

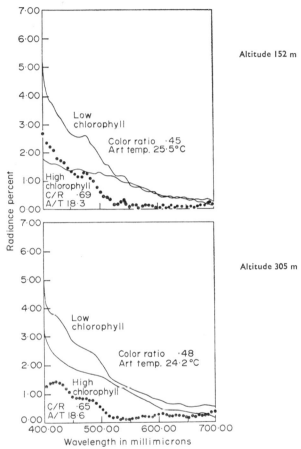

Fig. 3. Spectra of upwelling light obtained at 152 m and 305 m across transition from low to high concentrations of chlorophyll south of Georges Bank on 21 August 1969 (modified from Clarke et al., 1970b). Color ratio, C/R, is the ratio of reflectance at 540 nm to that at 460 nm. The dotted curve shows the difference between the curves for high and low chlorophyll.

On a previous flight over Buzzards Bay an actual increase in the percentage of backscattered light in the green region had been observed associated with the higher chlorophyll concentration of 4 mg m^{-3} found there (Clarke et al., 1970a).

During a more extensive study conducted in August 1969, in the same general region, it was found that the transition between the sterile

Gulf Stream water to the south and the fertile water associated with Georges Bank was particularly sharp, allowing repeated measurements of the contrasting areas to be made from altitudes ranging from 152 to 3050 m. The surface values for temperature and chlorophyll changed over a distance of 14 km from 25°C and 0·07 mg m⁻³ south of the

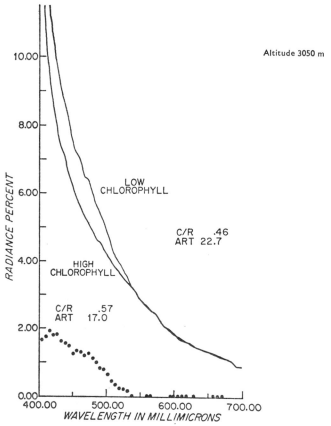

Fig. 4. Spectra of upwelling light obtained at 3050 m from same locations as Fig. 3.

transition to 18°C and 0·22 mg m⁻³ north of it. Curves contrasting the percentage backscattered light for the low and high chlorophyll waters recorded at 152 m, 305 m, and 3050 m are reproduced in Figs. 3 and 4. A comparison of the contrast ratios due to the differences in the spectra across the transition at all altitudes is shown in Fig. 5. The effect of increasing airlight on the upwelling light received is seen with increasing altitude. But since the *difference* in the spectra remains much the same, the interference of the airlight has not prevented the detection of the differences in the signals from beneath the water surface characteristic of the areas that were poor and rich in chlorophyll, as discussed by Clarke *et al.* (1970b). Curran (1972) has presented calculations showing

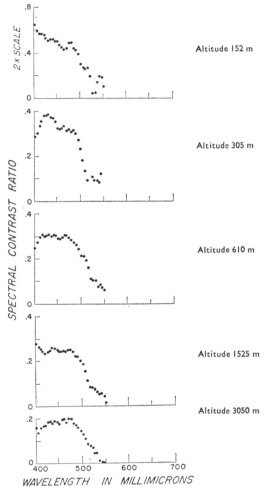

Fig. 5. The ratio of each wavelength of radiance difference to their mean.

that corrections for the scattering and absorption properties of the atmosphere may be made which will allow chlorophyll concentration to be determined to within one standard deviation of from 0·5 to 2·5 mg m^{-3} and, by sensing the aerosol optical depth to a greater accuracy, to detect chlorophyll concentrations to an uncertainty approaching 0·1 mg m^{-3}. Thus, wide-scale plotting of chlorophyll abundance in the surface waters of the sea should be feasible from satellites as well as from aircraft.

III. Spectral Measurements Above and Below the Surface

During 1969 and 1970 special effort was made to procure spectral measurements of downwelling and upwelling light both above and below the surface of the sea in the same area during the same time period. The observations carried out by John Tyler and co-workers enabled simultaneous data to be obtained from ship and aircraft at some stations. On other occasions when this was not possible, measurements from closely similar situations were used for comparisons.

A review of the subdivision of the incident light into various components by the action of the atmosphere, the water and the surface between them is shown schematically in Fig. 6. Incident irradiance consists of direct sunlight and skylight. When the solar altitude is greater than 30°, as in all observations considered in this report, most of the energy reaching the sea surface comes directly from the sun. Therefore, in Fig. 6, the incident light is represented by arrows indicating parallel solar radiation although it is understood that, in addition, skylight is reaching the sea surface from all parts of the upper hemisphere. Some of the sunlight and skylight is backscattered by the atmosphere and by moisture and other materials in the atmosphere. Airlight does interfere to some extent with the signal emanating from beneath the surface, but as we have seen in the previous section, under favorable conditions, the characteristic differences in the water, which we wish to measure, can be detected.

As shown in Fig. 6, part of the incident irradiance is reflected from the sea surface itself, but most of it penetrates into the water. In each successive water layer, part of the penetrating light is absorbed, part is scattered, and part is transmitted to the layer below. At a specific depth the ratio of the upwelling (or backscattered) irradiance to the downwelling irradiance at the same depth (or the reflectance) has been found to range from about 0·3 to 3% over the visible spectrum for water of moderate clarity, but rising as high as 10% in the blue region for extremely clear water. For the purposes of our diagram we shall take a representative average value of 1% for the underwater reflectance of the visible spectrum as a whole. Within the water the upwelling irradiance approaches the sea surface from all angles within the lower hemisphere. Since the radiation is attempting to pass from a denser to a less dense medium, it is not surprising that about half of it has been found to be internally reflected (Payne, 1972). This leaves about 0·5% of the original incident light to escape upwards from the sea surface as emergent light, as observed by Payne (1972).

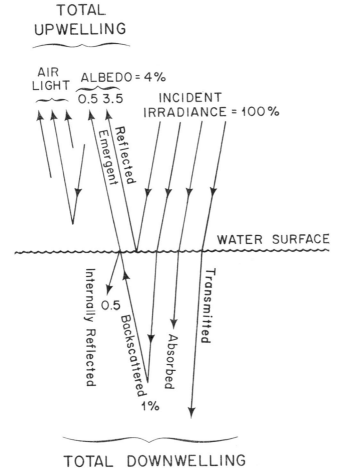

Fig. 6. Schematic representation of the fate of the incident radiation in the air above the surface of the sea, at the surface itself and in the water beneath the surface. The percentages are based on the findings of Payne (1972) for conditions at the entrance of Buzzards Bay, Mass.

The emergent light is added to light reflected from the surface itself to form the albedo of the sea surface. For waters at the entrance of Buzzards Bay, Payne (1972) measuring radiation from 300 to 700 nm, reported an albedo of 4% for clear weather and a solar altitude of 70°.

This means that the albedo is made up of 3·5% of the incident light due to surface reflection and 0·5% due to backscattered emergent light. An albedo of 6% was found for cloudy days with an atmospheric transmittance of 0·1. Taking into account average cloudiness and wind, Payne (1972) calculated that the average albedo of the Atlantic Ocean would be 6% on the equator and at 40° N latitude it would range from 6% in June to 11% in December. Just above the sea surface the total upwelling light consists chiefly of the emergent plus the reflected light. At higher altitudes more and more airlight is added until eventually it forms the largest component of the upwelling irradiance. Wherever possible, our measurements of nadir radiance were made in such a way as to avoid the reflection of the direct rays of the sun.

Using our actual spectral measurements we may now consider how the foregoing relationships are affected in various water bodies by the different concentrations of chlorophyll that characterize them. Simultaneous measurements of the spectrum of upwelling irradiance were made at three altitudes in the air and at a depth of 0·20 m in the water at a station in Buzzards Bay on 13 August 1969. Downwelling irradiance at the sea surface was measured on 14 August 1969 near the same position but with somewhat different conditions of haze in the atmosphere (Fig. 7). The NASA Standard Curve for downwelling irradiance outside the atmosphere (Thekaekara and Drummond, 1971) is also shown. A comparison of these curves reveals that about half the radiant energy was lost in passing through the atmosphere on this occasion. Another curve of the spectrum of the irradiance incident upon the sea surface off our Atlantic Coast was obtained by Tyler and Smith (1970) for the Gulf Stream near Florida. This curve is similar in shape to the NASA Standard Curve and indicates that only about 35% of the energy was lost as the light travelled through the clearer atmosphere off Florida.

The curves for the downwelling irradiance both reach a maximum about 480 nm, and the curve for the upwelling irradiance measured at a depth of 0·20 m beneath the surface peak at about 550 m (Fig. 7). This is consistent with the expected effect of the concentration of 2·3 mg m^{-3} of chlorophyll observed at that station in Buzzards Bay. A similar maximum for the curve of upwelling light was reported by Tyler and Smith (1967) for San Vicente Reservoir with about the same chlorophyll concentration. The curves for the upwelling irradiance measured from our plane at altitudes of 152 m, 305 m and 610 m are plotted on a relative scale because the different type of instrument used in the plane makes comparison in absolute energy units unreliable. The curves show an increasing skew toward the blue due to the pro-

gressive addition of airlight. Most of the effect occurred in the lower
300 m presumably due to the low haze which was present on that day.
By subtracting the effect of the atmosphere between 152 and 305 m
from the curve at 152 m, we obtain an extrapolated curve representing
the spectrum of the upwelling irradiance just above the surface, provided that the spectral effect of the airlight in the lower 152 m stratum

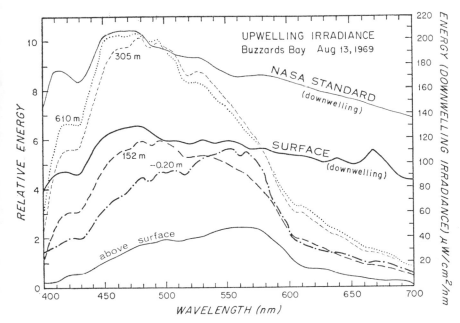

Fig. 7. Spectra of the energy of upwelling light on 13 August 1969 in Buzzards
Bay at the altitudes shown (measured by TRW spectrometer) and at 0·20 m
below the sea surface (measured by Scripps spectrometer) as compared with
downwelling irradiance of the NASA Standard spectrum outside the atmosphere,
and the downwelling irradiance measured on 14 August 1969 at the sea surface.
For the "above surface" curve see text. The energy scale on the right is for downwelling light; that on the left is in relative units, except that the values on the left
also represent $\mu w \, cm^{-2} \, nm^{-1}$ for the upwelling light at 0·20 m.

was the same as that for the stratum from 152 to 305 m. The extrapolated curve does in fact present a similar shape to that obtained at
−0·20 m and reaches a maximum between 550 and 570 nm. Although
the ordinate for this curve is only relative, the curve for upwelling light
above the surface should lie below the underwater curve because
about half of the upward light underwater is expected to be internally
reflected at the underside of the water surface.

The data from the study of Buzzards Bay on 13 August 1969 have been plotted on a percentage basis in Fig. 8 to permit them to be compared with observations from other oceanic areas. The curves for the spectrum of the backscattered light recorded by the aircraft at 610 m, 305 m and 152 m are shown as percentages of the incident irradiance,

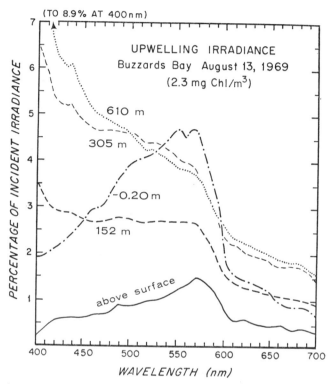

Fig. 8. Spectra of the upwelling irradiance in Buzzards Bay on 13 August 1969 plotted as a percentage of the downwelling irradiance at the surface. The values for altitudes of 152 to 610 m were related to the "gray card" measurements of incident irradiance made before the flight, and those for 0·20 m below the surface were related to incident irradiance measured on shipboard the next day.

which had been determined by directing the spectrometer at a horizontally placed "gray card" of known reflectivity exposed to sun and sky light before the start of the flight. The highest percentage for reflectance occurs in the blue region at 610 m, with progressively lower values at 305 m and 152 m, since less airlight is included at the lower altitudes. The effect of the stratum of haze below 305 m is clearly observed. The tendency for the curves to level off in the green region is seen, especially at 152 m, and this is characteristic of water rich in chlorophyll. An extrapolated curve for the upwelling light just above the surface was obtained by finding the difference between the 305 m and 152 m curves at each wavelength and subtracting that value from the 152 m curve. This curve has a maximum at about 570 nm, a fact consistent with the high concentration of chlorophyll existing in the surface waters of Buzzards Bay. Since the curve averages something

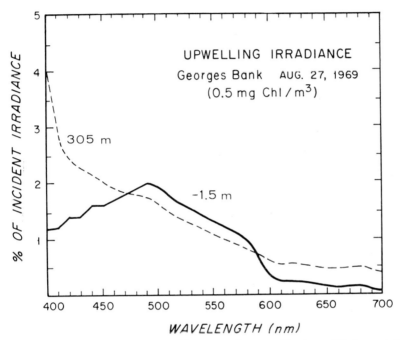

Fig. 9. Spectra of upwelling irradiance on Georges Bank on 27 August 1969 plotted as a percentage of the incident irradiance measured before the flight for the curve at 305 m, or measured on shipboard on 3 July 1967 (see text).

more than 0·5%, it is in reasonable agreement with the findings of Payne (1972) in view of the differences in the methods used. Finally, when the data for the measurements at 0·20 m beneath the sea surface are plotted on a percentage basis, a maximum is again found between 550 nm and 570 nm, with values throughout the spectrum which are higher than those calculated for above the surface. This is in agreement with theory, and the reflectance values obtained underwater are consistent with those reported for San Vincente with similar chlorophyll content.

17

The foregoing curves may now be compared with those based on data obtained on Georges Bank with chlorophyll concentration of 0·5 mg m^{-3} (Fig. 9) and in the Sargasso Sea and Gulf Stream where chlorophyll is characteristically less than 0·1 mg m^{-3} (Fig. 10). In the former case the upwelling light at a depth of 1·5 m was found to show

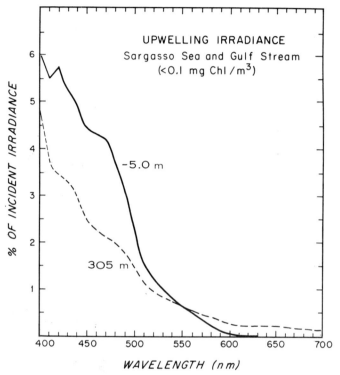

Fig. 10. Spectra of upwelling irradiance in the Sargasso Sea (39° 25′ N × 67° 52′ W) measured on 22 August 1969 at 305 m plotted as a percentage of the incident irradiance measured before the flight. The same for a depth of 5·0 m in the Gulf Stream (25° 45′ N × 79° 30′ W) measured 3 July 1967 and related to the incident irradiance measured the same day (data from Tyler and Smith, 1970).

a maximum of 2% of the irradiance incident on the surface at a wavelength of 490 nm. In the latter case the blue component of the upwelling light at a depth of 5 m reached a value of 6% at 400 nm. In both types of water the underwater curves lie above the curves for reflectance at an altitude of 305 m for part of their length, indicating the interplay of the addition of airlight and the loss of radiation by internal reflection. The differences in the shape of the curves again are related to the differences in the concentrations of chlorophyll present.

IV. Use of the Spectrometer on an Extended Survey

The work carried out in May of 1970 was directed toward examining the potential of remote spectroscopy for reconnaissance of the sea surface over a large area rather than, as in the previous work, the comparison of spectra of adjacent water masses of differing composition.

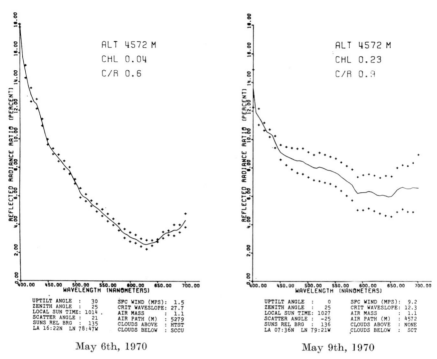

ALT 4572 M
CHL 0.04
C/R 0.6

ALT 4572 M
CHL 0.23
C/R 0.9

UPTILT ANGLE	:	30	SPC WIND (MPS):	1.5
ZENITH ANGLE	:	25	CRIT WAVESLOPE:	27.7
LOCAL SUN TIME:		1014 .	AIR MASS	: 1.1
SCATTER ANGLE	:	21	AIR PATH (M)	: 5279
SUNS REL BRG	:	135	CLOUDS ABOVE	: HTST
LA 16:22N LN 78:47W			CLOUDS BELOW	: SCCU

UPTILT ANGLE	:	0	SPC WIND (MPS):	9.2
ZENITH ANGLE	:	25	CRIT WAVESLOPE:	12.3
LOCAL SUN TIME:		1027	AIR MASS	: 1.1
SCATTER ANGLE	:	-25	AIR PATH (M)	: 4572
SUNS REL BRG	:	136	CLOUDS ABOVE	: NONE
LA 07:36N LN 79:21W			CLOUDS BELOW	: SCT

May 6th, 1970

May 9th, 1970

Fig. 11. Samples of data from high altitude with low and high chlorophyll concentration (0·04 and 0·23 mg m^{-3} respectively). The solid curve represents the average reflectance of five consecutive spectra with the 95% confidence boundaries shown by broken curves above and below. The color index (C/R) is the reflectance ratio at 540 nm to that at 460 nm. It is approximately the inverse of the ratio given by Jerlov. Note the flatter spectrum accompanying the higher ratio and higher chlorophyll concentration.

The experiment was conducted in a manner quite similar to that described above, save that the flight track was extended to some 10 000 km of over-ocean flying, traversing the northeast Gulf of Mexico, the Yucatan Straits, the central and western Caribbean to Jamaica and Panama, thence northward along the west coast of Nicaragua and the east coast of Honduras and Yucatan, and finally along the west coast of Mexico from Tehuantepec to the head of the Gulf of California.

Throughout the flight, spectra were taken at approximately every 60 km, mostly from altitudes of 1300 to 1500 m. As nearly as possible all spectra were taken under blue sky conditions with no visible haze above or below the aircraft. At each station, one or more groups of spectra were recorded, looking away from the sun at an angle of incidence between 0° and 40° as required to avoid the sun's glitter on the sea surface. Each group consisted of five unfiltered spectra. Examples are shown in Figs. 11 and 12. During the recording of a single group, the

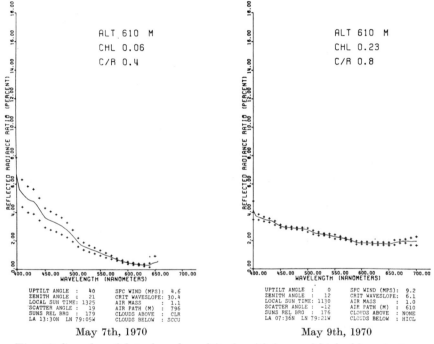

UPTILT ANGLE :	40	SFC WIND (MPS):	4.6
ZENITH ANGLE :	21	CRIT WAVESLOPE:	30.4
LOCAL SUN TIME:	1325	AIR MASS :	1.1
SCATTER ANGLE :	19	AIR PATH (M) :	796
SUNS REL BRG :	179	CLOUDS ABOVE :	CLR
LA 13:30N LN 79:05W		CLOUDS BELOW :	SCCU

May 7th, 1970

UPTILT ANGLE :	0	SFC WIND (MPS):	9.2
ZENITH ANGLE :	12	CRIT WAVESLOPE:	6.1
LOCAL SUN TIME:	1130	AIR MASS :	1.0
SCATTER ANGLE :	-12	AIR PATH (M) :	610
SUNS REL BRG :	176	CLOUDS ABOVE :	NONE
LA 07:36N LN 79:21W		CLOUDS BELOW :	HICL

May 9th, 1970

Fig. 12. Samples of data from low altitude with low and high chlorophyll concentration. Because of the shorter air path, the interference from air light in this situation is much less than in the previous figure.

aircraft advanced approximately 360 m, hence the individual spectra are not strictly coincident.

For an airspeed of 60 m s^{-1} (117 knots) and an altitude of 1300 m looking straight down, the area covered by each group of five spectra was 372 m along track by 68 m cross track. At frequent intervals the optical density of the water was estimated by means of a photometer dropped into the sea and monitored in the airplane by FM radio telemetry as it sank to a depth of 300 m. The photometer measured the downwelling irradiance under water over a band of wavelengths centered at 550 nm. These so-called bathy-photometers were designed and built by Mr. Foster L. Striffler and were operated by Mr. Edward A. Denton who also maintained a continuous record of the infrared radiation temperature of the ocean as measured at flight altitude. Whenever the opportunity presented itself, the aircraft overflew the United States Coast and Geodetic Survey R.V. *Discoverer* which, under the scientific leadership of Mr. John E. Tyler, of the University of California's Scripps Institution of Oceanography, was engaged in evaluating the daily photosynthetic production by the carbon-14 method and the concentration of chlorophyll *a* by fluorometric and spectroscopic measurements. During five over-flights of the *Discoverer*, spectra were made at altitudes from 150 to 4572 m. The data furnished by the *Discoverer* enabled us to relate the shape of the observed spectra to the attenuance by chlorophyll.

Before each flight, as in previous work, spectra were made of a horizontally placed Kodak gray card as a measure of the response of the photometer to reflected vertical radiance. The observed energy distributions were compared to values for direct sunlight calculated by the method of Moon (1940) with the addition of skylight calculated by the method of Fritz (1951). The appropriate meteorological moisture parameter was obtained from radiosonde soundings made at airports in New Orleans, Kingston, B.W.I., and Balboa, Canal Zone. The calculated transmission through one airmass and the equivalent upward irradiance from a horizontal diffuse reflector exposed to full sunlight and a clear sky are shown in Table 1 for the conditions encountered at Kingston, Jamaica. The comparison of observed to expected values serve as a field calibration of the equipment. The calculated irradiance values are in reasonably close agreement with those measured at sea by Dr. Raymond C. Smith on the deck of the *Discoverer* on a different occasion.

The total data bank acquired by these means consists of 160,000 discrete measurements representing approximately 2454 spectra in 312 groups. Such a volume of data is of little value until digested, and this is still being carried on with the assistance of Mrs. Evelyn L. Ziegler. Until further analysis, it is impossible to draw final conclusions. We are therefore limited herein to showing one or two samples of typical results. By way of illustration, Figs. 11 and 12 each show curves of two groups of five spectra averaged. The two groups are from water of low and high chlorophyll content. The color ratios shown (C/R) are an index used to describe the shape of the spectra as related to water

TABLE 1. Calculated values of irradiance on a flat surface at sea level, Kingston, B.W.I., May 7th, 1970 at 1447 GMT and measured by Tyler *et al.* on *Discoverer* at 02:58 S, 84:28 W, May 12th, 1970 at 2243 GMT.

Wave-length m^{-9}	Extra-atmospheric irradiance w^{-6}, cm^{-2}, m^{-9}	Calculated transmission coefficient[a]	Calculated irradiance sun+clear sky w^{-6}, cm^{-2}, m^{-9}	Measured irradiance w^{-9}, cm^{-2}, m^{-9}
400	142·9	0·468	105	138
405	164·4	0·480	122	—
410	175·1	0·492	131	—
415	177·4	0·504	133	—
420	174·7	0·515	132	145
425	169·3	0·526	129	—
430	163·9	0·537	126	—
435	166·3	0·547	129	—
440	181·0	0·557	141	155
445	192·2	0·567	151	—
450	200·6	0·576	158	—
455	205·7	0·586	163	—
460	206·6	0·594	165	174
465	204·8	0·603	164	—
470	203·3	0·612	164	—
475	204·4	0·620	166	—
480	207·4	0·628	169	172
485	197·6	0·635	162	—
490	195·0	0·643	160	—
495	196·0	0·650	162	—
500	194·2	0·657	161	155
505	192·0	0·664	160	—
510	188·2	0·671	157	—
515	183·3	0·677	154	—
520	183·3	0·683	154	144
525	185·2	0·689	156	—
530	184·2	0·695	156	—
535	181·8	0·701	155	—
540	178·3	0·707	152	143
545	175·4	0·712	150	—
550	172·5	0·717	148	—
555	172·0	0·722	148	—
560	169·5	0·727	146	144
565	170·5	0·732	148	—
570	171·2	0·737	149	—
575	171·2	0·742	150	—
580	171·5	0·746	150	138
585	171·2	0·750	150	—

TABLE. 1.—*continued*

Wave-length m^{-9}	Extra-atmospheric irradiance w^{-6}, cm^{-2}, m^{-9}	Calculated transmission coefficient[a]	Calculated irradiance sun+clear sky w^{-6}, cm^{-2}, m^{-9}	Measured irradiance w^{-9}, cm^{-2}, m^{-9}
590	170·0	0·754	149	—
595	168·2	0·759	148	—
600	166·6	0·763	147	124
605	164·7	0·767	145	—
610	163·5	0·770	145	—
615	161·8	0·774	144	—
620	160·2	0·778	142	124
625	158·6	0·781	141	—
630	157·0	0·785	140	—
635	155·7	0·788	139	—
640	154·4	0·791	138	120
645	152·7	0·794	137	—
650	151·1	0·797	136	—
655	149·8	0·800	135	—
660	148·6	0·803	134	108
665	147·1	0·806	133	—
670	145·6	0·809	132	—
675	143·8	0·812	130	—
680	142·7	0·814	129	114
685	141·4	0·817	128	—
690	140·2	0·820	128	—
695	138·5	0·822	126	—
700	136·9	0·825	125	89

[a] Airmass 1·0; water 6·6 cm; dust 300 P m^{-3}

discoloration, giving a convenient way to map results numerically. They represent the ratio of the reflectance at 540 nm to that at 460 nm. This "color ratio" was used by us in 1969 and has also been adopted by Curran (1972) and by Arvesen *et al.* (1971). The scheme was originated by Strickland (1962) in a more sophisticated form which normalized the blue and green to the energy level in the red. We have found that the simple ratio is sufficient for our purposes.* Beneath each graph are 12 items of environmental information. The second column shows the local weather as it affects the problem of glitter entering the instrument. In addition, we have calculated the slope of a critical wave facet sufficiently steep in the plane of reflection to reflect the sun's image into the instrument. By comparison, for the low wind speeds which prevailed

* The color index used here closely approximates the inverse of the index described by Jerlov.

throughout this trip, it is doubtful whether many actual facets steeper than 12° were present. Also shown are descriptions of those geometric parameters of the observation which relate to scattering of air light, namely the zenith distance of the sun, the time angle between the sun

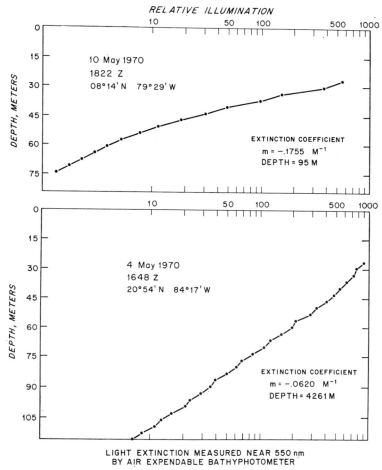

Fig. 13. Sample of optical turbidity of water measured *in situ* by air expendable bathyphotometer.

and the local meridian, and the angle between the direction of the sun's rays and the direction vector of the principal axis of the instrument. (The negative sign implies that the instrument was looking at the water more steeply than the sun.) The sun's relative bearing is shown positive to the right and negative to the left. Also shown are the (geometric) airmass, the length of the airpath traversed by emergent light between the water and the instrument, and the presence or absence of clouds.

The solid curves show the reflected light averaged in groups of five, expressed as a percentage of the upward radiance from a perfectly reflecting horizontal surface exposed to sun and sky under the conditions encountered at Kingston, B.W.I. at the time of calibration, calculated in the way described above.

Directly above and below the average, centered at each 10 nm wavelength, are shown the 95% confidence boundaries on the sample average. Figure 11 shows curves characteristic of spectra from water with very low and with moderately high chlorophyll concentration. With each are shown the chlorophyll concentration (CHL) and the color ratio (C/R). In the case of water with low chlorophyll concentration, the general slope of the curve is such that the ordinate decreases sharply from the blue to the green, giving 0·6 as a color ratio. The spectrum taken over water with moderately high chlorophyll is much flatter, and consequently its color ratio, 0·9, is larger. Attenuance by chlorophyll in the blue combined with the ability of plankton organisms to reflect in the green would account for this tendency for the color ratio to approach unity as more and more phytoplankton organisms are present. Figure 12 represents the same comparison at 610 m.

The high and low attenuance measured *in situ* by bathyphotometer is shown in Fig. 13, corresponding to the color ratio index for high and low chlorophyll concentration.

In Fig. 14 is shown the distribution of stations at which spectra were taken in flight. Adjacent to each station location is the number corresponding to the color ratio. Also shown are the locations of the *Discoverer* and the chlorophyll concentrations measured on board. Although we obtained data from many different altitudes, the figure has been drawn for those data obtained at 610 and 1524 m. The figure suggests that over most of the Caribbean and Gulf of Mexico the color ratios are quite uniform and of the type associated with low chlorophyll concentrations. The exceptions, where the shape of the spectra is similar to those associated with high chlorophyll, are shaded in the figure. These areas are:

1. Gulf of Panama
2. Shoals northwest of Cabo Gracias a Dios
3. Waters northwest of Havana
4. Bay of Campeche
5. Mexican coastal waters west of Acapulco
6. Gulf of California.

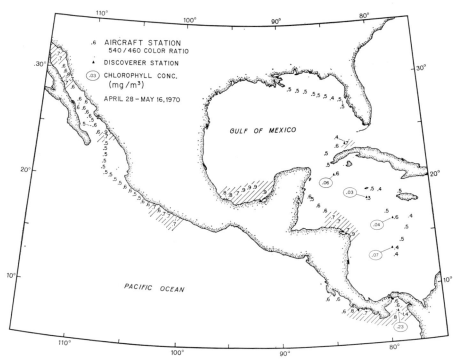

Fig. 14. Preliminary estimate of the distribution of discolored water along flight track. Shaded areas correspond to 540/460 nm reflectance ratios equal to 0·7 or more. Station locations for R.V. *Discoverer* are shown together with notation of the measured chlorophyll concentration.

The spectra obtained over the Campeche Bank, shown in Fig. 15, indicate high turbidity and high chlorophyll concentration. The high turbidity was confirmed in flight by bathyphotometers. The observed high color index is to be expected in view of the well-known productivity of this area which supports the major shrimp industry of the Caribbean. By way of further confirmation, a photograph of Campeche Bank, taken by astronauts Cooper and Conrad from GEMINI V on 2 August 1965, is shown as Fig. 16. This photograph from satellite altitude clearly shows currents emerging from Laguna de Terminos heavily laden with sediments. Further offshore in Campeche Bay, clouds of what appear to be drifting chlorophyll are to be seen, although the color contrast in the original is marginal. One cannot unequivocally determine whether the discolorations are caused by chlorophyll or by mud and silt. However, it is relevant that high productivity on the shrimp beds and high runoff from the coastal lagoons go hand-in-hand, the terrigenous mineral nutrients generating high rates of photosynthesis and hence of chlorophyll. Similarly, the Gulf of California is well known for its high fertility and abundant spring plankton blooms. Since Spanish times the resulting discoloration has given this area the fanciful name of the Vermilion Sea. Areas of turbidity can be seen in photographs (not shown here) of the northern end of the Gulf off the mouth of the Colorado

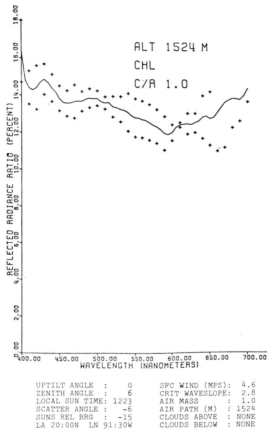

ALT 1524 M

CHL

C/R 1.0

UPTILT ANGLE :	0	SFC WIND (MPS):	4.6
ZENITH ANGLE :	6	CRIT WAVESLOPE:	2.8
LOCAL SUN TIME:	1223	AIR MASS :	1.0
SCATTER ANGLE :	−6	AIR PATH (M) :	1524
SUNS REL BRG :	−15	CLOUDS ABOVE :	NONE
LA 20:00N LN 91:30W		CLOUDS BELOW :	NONE

Fig. 15. Graph of data obtained over the shrimp beds of Campeche Bay. May 14th, 1970.

River (S–65–34673) and in the central section (S–65–45702) (NASA. 1967). Here again is found a highly developed sports fishery for marlin as well as an extensive shrimp industry. Other areas of high color ratio are also to be seen. In Panama Bay, the region around the Pearl Islands is famous for its sports fishing, and Chamay Bay is a steady source of bait for the Pacific tuna clippers. It therefore can be described as a

Fig. 16. Laguna de Terminos and Campeche Bay on the Yucatan Peninsula, Mexico, original in color taken by astronauts Cooper and Conrad from GEMINI V, 2 August 1965, at 2009 GMT.

very productive oceanic area. The discolored regions all were found in coastal areas but not necessarily in shallow water.

In summary, although one cannot assert that water color is an infallible indicator of high productivity related to fisheries, it is nonetheless encouraging that in this study as well as in earlier work on Georges Bank off the coast of Massachusetts, the spectrophotometer has apparently discriminated local areas of high productivity in shallow discolored water in striking contrast to the relative sterility of the deep blue ocean.

REFERENCES

Arvesen, J. C., Millard, J. P. and Weaver, E. C. (1971). "Remote Sensing of Chlorophyll and Temperature in Marine and Fresh Waters." *Proceedings of the XXII International Astronautical Congress*, Brussels, Belgium, 1971.

Clarke, G. L., Ewing, G. C. and Lorenzen, C. J. (1970a). *Science, N.Y.*, **167, 20,** 1119–1121.

Clarke, G. L., Ewing, G. C. and Lorenzen, C. J. (1970b). *Proc. 6th Int. Symp. Remote Sensing Enviro.*, 1969, **2**, 991–1001. University of Michigan Press, Ann Arbor, U.S.A.

Curran, R. J. (1972). Ocean Color Determination Through a Scattering Atmosphere. Preprint X–651–72–58. Goddard Space Flight Center, Greenbelt, Md., U.S.A.

Fritz, S. (1951). *In* "Smithsonian Meteorological Tables", prepared by R. J. List, p. 420.

Gore, L. A. (1968). Ocean Color Spectrometer. Unpublished report, Electronic Systems Div., TRW Systems, Inc.

Moon, P. (1940). *J. Franklin Inst.*, **230**, 583.

Payne, R. E. (1972). *J. Atmos. Sci.*, **29**, No. 5, 959–970.

Strickland, J. D. H. (1962). The Estimation of Suspended Matter in Sea Water from the Air. Unpublished paper.

Thekaekara, M. P. and Drummond, A. J. (1971). *Nature Phys. Sci.*, **229**, 6–9.

Tyler, J. E. and Smith, R. C. (1967). *J. Opt. Soc. Amer.*, **57**, No. 5, 595–601.

Tyler, J. E. and Smith, R. C. (1970). "Measurements of Spectral Irradiance under Water." Gordon and Breach Science Publishers, New York.

(U.S.) National Aeronautics and Space Administration (1967). Earth Photographs from GEMINI III, IV, and V. U.S. Govt. Printing Office, Washington, D.C., pp. 59 and 127.

Chapter 18

Underwater Light and the Orientation of Animals

TALBOT H. WATERMAN

Department of Biology, Yale University, New Haven, Connecticut, U.S.A.

I. Introduction

The proper placement of a marine animal in its environment is a primary necessity for survival. Consequently location and orientation in the biosphere constitute a sort of ecological geometry dependent on the organism's adaptive needs.

In general such placement involves six degrees of freedom of movement, three translational and three rotational. Thus location is defined by geographical position on the earth's surface (x, y) and depth in the water column (z). Spatial orientation (neglecting flexibility and posture) is defined by the angular relations of the animal's structural and functional axes of symmetry (θ, ϕ, ω for a bilaterally symmetrical form) to an environmental coordinate system (Fig. 1). The latter is usually established by the plane of the horizon and by the gravitational vertical. Typically of course the anteroposterior and transverse body axes are maintained horizontal and the dorsoventral axis vertical while the animal's heading is constantly controlled in azimuth. Furthermore since organisms exist in time this adds a seventh variable to the six spatial parameters.

No doubt the monitoring and control of these seven variables are formidable tasks for the organism (as they are in a submarine or rocket!).

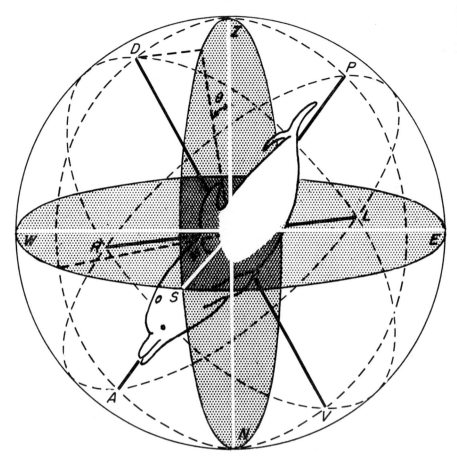

Fig. 1. Biological orientation in space depends on the relations between the organism's coordinate system and the environmental coordinate system. In the case of a diving porpoise its axes of symmetry are dorsoventral (DV) antero-posterior (AP) and right-left (RL). Their relations to a geographic horizontal plane (NESW) and a vertical plane (ZNNS) related to the zenith and nadir define the animal's spatial orientation (drawn by Virginia Simon).

Yet it is equally certain that the effective execution of these functions is biologically essential. As a result organisms must behave in a manner guaranteed to establish their proper location and orientation.

This behavior must take into account effects of time both on the environment and on the organism itself. Some environmental changes like turbulence and meteorological variability are random as far as the organism is concerned. Nevertheless appropriate compensatory responses must be made. Other changes, however, recur systematically on diurnal, tidal, lunar, seasonal, annual or other cycles. As a result the spatial location of the organism's optimal environment may shift accordingly. Hence shorter or longer migratory excursions must be undertaken to follow such periodic shifts.

In addition the animal's needs for food, mate, shelter, etc. usually vary with time and the stage of its development. As this happens different configurations of environmental variables will become most favorable and appropriate relocation of the organism will again be required to provide optimal conditions.

It is the objective of this chapter to examine both the static and dynamic aspects of such orienting behavior of animals in so far as they are related to light in the sea. Attention will be mainly directed to recent relevant work. Even so the topic is a very big one. However, two major subtopics may be largely omitted here since they are dealt with elsewhere in this volume, Bioluminescence in Chapter 19 (Boden and Kampa) and Underwater Visibility in Chapter 7 (Duntley).

For present purposes the marine organism can be considered as a black box which receives a number of sensory inputs from the environment and responds with a repertory of behavioral outputs relating to its position and its location. Five basic parameters of underwater light are known to provide input to such systems. They are spectral irradiance, irradiance, radiance distribution, degree of polarization and direction of the e-vector. These topics will form the major sections of this review.

Typically all five of these parameters are functions of time and space in the sea. Hence they generally change with the hour, season, depth, location on the earth, optical properties of the water column and meteorological conditions. As a result the fundamental parameters of submarine light vary with relation to these distributional variables in ways which are of great significance for animal orientation.

In turn the relative significance of the various optical parameters and their changes will differ for various organisms. Thus different animals will have particular input sensitivities relative to the fundamental quantities. As a consequence some cannot perceive e-vector direction, some are more sensitive to blue light than to yellow, in some light input may be mediated by extraoptic mechanisms rather than by the eyes, etc. Similarly the response repertories comprising the behavioral output will vary markedly depending on the particular ecological niche to which a given species or stage is adapted as well as on the mechanism of such adaptation.

Nevertheless these reactions will all be alike in the sense that they involve the control of the six degrees of freedom of movement mentioned

18

above. More specifically the responses related to spatial position (control of θ, ϕ and ω) are relatively simple "righting reactions" typically mediated in part visually and in part through gravity receptors. In contrast those responses relating to location (control of x, y and z) and its changes are much more complex. Basically they require the solution of the difficult problem of submarine direction finding (Waterman, 1972). They may also involve seemingly virtuosic migrations whose scientific explanation remains a stimulating challenge. Clearly such behavior presupposes the biological solution of a variety of additional orientational and navigational problems.

An animal which is going to orient using underwater illumination obviously must have appropriate light receptor mechanisms. In turn these require a visual pigment which absorbs photons and as a result undergoes some photo-chemical change. This is the primary event leading to visual orientation and has been the focus of much recent work and discussion. Consequently the striking relation between light in the sea and the visual pigments of marine animals provides a good starting point for this review.

II. SPECTRAL IRRADIANCE

Of course visual pigments are ubiquitous. In all cases studied (arthropods, mollusks, vertebrates) they conform to a single molecular type. This is a conjugated chromo-protein whose polypeptide moeity "opsin" bears as its prosthetic group "retinal" which is vitamin A aldehyde. The latter may be either retinal itself or 3 dehydroretinal. Despite their type-similarity the visual pigments differ widely in wavelength sensitivity within the visible and near ultraviolet. Their absorption peaks (λ_{max}) in various animals range from the near ultraviolet (345 nm) (Gogala et al., 1970) to reddish orange (620 nm) (Dartnall, 1969). These differences depend in part on the two alternative prosthetic groups but more widely on differences in the opsins.

Clearly the location of the λ_{max} will have important consequences for vision in the sea. Not only is the attenuance of light by seawater rapid relative to the depth of the ocean (Fig. 2) but the absorptance is strongly wavelength selective (Murray and Hjort, 1912; Jerlov, 1968). As a consequence probably more than three quarters of the ocean's volume is devoid of sunlight significant for vision at any wavelength even at noon (assuming that the average depth of the world ocean is 4000 m and the "absolute" threshold occurs around 1000 m in very clear water). Furthermore in the upper 1000 m the attenuance is markedly dependent on wavelength as demonstrated in Fig. 3.

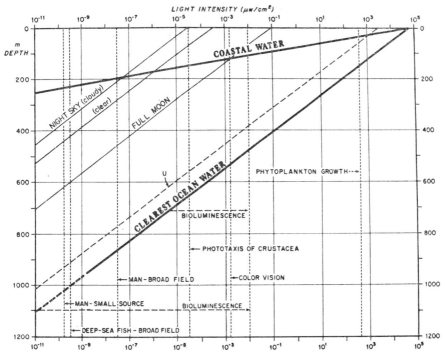

Fig. 2. Irradiance as a function of depth in the sea. The heavy lines diagrammatically indicate the penetration of sunlight into clear coastal water (k assumed to be 0·15 and constant throughout the water column) and into the clearest ocean water (k for the water column taken to be 0·033). In the latter case the sun's reflected upwelling irradiance is shown by the broken line (U). The penetration of moon and night sky light are also indicated. Estimates of relevant biological thresholds or boundaries are shown by the vertical broken lines (from Clarke, 1971).

The consequent enormous adaptive advantage of having a visual pigment's λ_{max} coincident with that for transmittance was pointed out long ago (Bayliss et al., 1936; Clarke, 1936; Waterman et al., 1939). However, only recently has sufficient relevant data been collected to show what a marked effect this aspect of underwater optics must have had on the evolution of visual pigments (reviews by Bridges, 1972; Lythgoe, 1972). Note that at this point we are concerned with the animal's detection of irradiance and brightness perception as they are affected by wavelength but not with its color discrimination if that is present. The latter requires two or more wavelength specific retinal cell types and is beyond the scope of this review.

Apparently Kampa (1955) was the first to discover a blue sensitive visual pigment (λ_{max} 462 nm) in a mesopelagic animal, the euphausiid

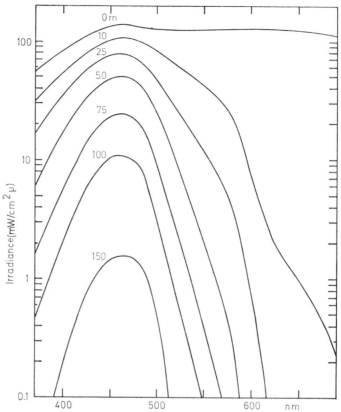

Fig. 3. Filtering effect of very clear ocean water on downwelling spectral irradiance. Sun's elevation at 56° (from Lundgren and Hojerslev, 1971).

Euphausia pacifica. However, she did not in the original paper mention its possible adaptive significance. Not long after Denton and Warren (1957) discovered "chrysopsins" blue sensitive rhodopsins in bathypelagic fish with their λ_{max} around 480 nm. This was recognized by the authors to be very close to the transmittance λ_{max} of the water column in which the fish were living. Subsequent work done by Munz, Wald, Dartnall, Lythgoe, Bridges among others has confirmed and greatly extended this work to many species of fish including comparative data for a variety of habitats (see reviews by Munz, 1971; Bridges, 1972; Crescitelli, 1972; Lythgoe, 1972).

In general freshwater teleosts have their λ_{max} at the longest wavelengths, (λ_{max} 520 nm), littoral and turbid water marine species are intermediate and generally resemble terrestrial vertebrates in this regard (λ_{max} *ca.* 500 nm) whereas clear water pelagic and especially meso-

and bathypelagic fish species have their λ_{max} near 470–480 nm. Actually no pigments in pelagic animals have been reported yet (Crescitelli, 1972) with their λ_{max} matching the wavelength of maximally transmitted or upwelling light around 450–460 nm in the clearest known natural waters (see Chapter 4 by Jerlov, and Tyler and Smith, 1970).

Adaptations of the λ_{max} of visual pigments may in fact be quite complex biologically since they can be intraspecific as well as interspecific. Thus ontogenetically, seasonally, migrationally and photically induced changes in λ_{max} are well documented in a number of cases. Perhaps the most spectacular examples are those correlated with long range migrations in salmonids and eels. For them different phases of the life cycle range from freshwater to marine littoral to open sea or mesopelagic habitats.

Thus the freshwater yellow eel shifts its λ_{max} from 523 nm to 487 nm in metamorphosing into the oceanic pelagic silver eel. The mechanism of such shifts is not yet certain. Earlier ideas that they were phylogenetic in origin or osmotically induced now seem inadequate. Direct effects of light (Bridges and Yoshikami, 1970) and possible hormonal control (Beatty, 1966, 1969) have been demonstrated in specific instances of pigment metamorphosis.

While the adaptive correlations of the visual pigments in teleost fishes are the most wide ranging among those studied parallel cases are known in other animal groups. Blue peaking rhodopsins have been found in three deepsea sharks (Denton and Shaw, 1963) and in two deepwater chimaerids (Holocephali) which have their λ_{max} at 477 nm and 484 nm (Denton and Nicol, 1964; Beatty, 1969; Crescitelli, 1969). Among mammals seven out of ten species of cetaceans tested have their visual pigment λ_{max} shifted toward the blue (481–487 nm) from the typical terrestrial vertebrate location near 498 nm (McFarland, 1971). A deepwater rhodopsin has also been reported from the pelagic deep feeding elephant seal *Mirounga* (Lythgoe and Dartnall, 1970). For the marine vertebrates then the adaptive challenge of the selective wavelength absorptance in clear deep water has evoked the evolution of special visual pigments at least in three major fish groups and in two groups of mammals.

Among the invertebrates the situation is not so clear no doubt because far fewer marine forms have been studied in this regard. It is true that among mesopelagic and bathypelagic crustaceans six euphausiids as well as two penaeid shrimps and one caridean have been found to have blue sensitive visual pigments (λ_{max} 460–480 nm) (references in Goldsmith, 1972). However, unlike the rule in vertebrates, one terrestrial isopod has been reported to have its λ_{max} at

480 nm and several littoral crabs have λ_{max}'s ranging from 476–484 nm. It is not yet known whether these represent some exceptional adaptations or depend on some secondary factors related, for instance, to retinal extraction procedures.

In the only other marine invertebrate group studied, the cephalopod mollusks, no evidence for specific visual pigment adaptations to the λ_{max} of ambient light has been made evident. There are of course no freshwater or terrestrial members of this group and all eight species so far studied have their λ_{max} between 470 and 500 nm. But there is no obvious habitat correlation among the littoral, benthic and pelagic species involved (for detailed references see reviews by Goldsmith, 1972; Hara and Hara, 1972).

Considering the marked adaptive advantage cited above, it would be surprising if any deepwater animal with important visual input related to penetrating daylight should fail to develop a deepsea rhodopsin. The deepsea cephalopods which are numerous and seem highly specialized should be subject to this generalization. Whether or not this is indeed the case can only be determined by the adequate study of a wider range of species.

An important adjunct to natural light in the sea is provided by bioluminescence which is reviewed in Chapter 19 of this volume. Suffice it to say here, however, that below some intermediate depth like 500 m, biologically produced light begins to exceed the level of penetrating daylight at noon even in the clearest water (Fig. 2). It continues to increase in intensity to 900 m but from there on down occurs with decreasing vigor to the greatest depth tested (3750 m). Bioluminescence is of relatively low intensity and usually has its λ_{max} near 470 nm which approximates the wavelength of greatest transmittance of clear oceanic water and the λ_{max} of deepsea rhodopsins.

The general coincidence of these two biological wavelength adaptations with the peak transmittance of the water column means that the causal relations involved are potentially complex. Thus the light production may act to reduce visibility of the organism from below against the downward penetrating daylight. Eyes may have blue-peaking rhodopsins to detect bioluminescence or make the best of the filtered sunlight present. Interpretations may therefore be ambiguous in the absence of detailed experimental data.

Preliminary study has been reported on a particularly interesting adaptive correlation which illustrates this point (Denton et al., 1970). In certain deepsea stomiatoid fish large subocular photophores have long been known to produce a deep red luminescence instead of the blue green or yellow green common to most light organs. Such red light would presumably be invisible or very dim to eyes having deepsea rhodopsins. However, one of the fish (*Pachystomias*) with a red sub-ocular photophore has a rhodopsin with its peak absorptance at a surprisingly long wavelength ($\lambda_{max} = 575$ nm). Hence they should see their own light.

The red photophores probably function in turn to unmask the protective coloration of many bathypelagic animals which are deep red or brown in color (Murray and Hjort, 1912) and perhaps potential prey by *Pachystomias*. Black in the deep penetrating 475 nm daylight and in the light of typical greenish blue bioluminescence these forms would become visually conspicuous in the red "searchlights" of such long wavelength sensitive predators. Of course red photophores could also alternatively be used as a means of intraspecific communication imperceptible to other species having ordinary deepsea rhodopsins.

It is notable that red (among other colors) photophores have also been reported in deepsea cephalopods (*Lycoteuthis*) but the corresponding visual or behavioral correlations are not known. However, light organs on or near the eyes are known in teleosts (Brauer, 1908; Munk, 1966a) in squids (Chun, 1903 and Berry, 1926) and in crustaceans (Chun, 1896). Such close juxtaposition of light sensing and light producing structures surely suggests important correlations but few relevant data are available.

III. Irradiance

A variety of visual adaptations to the low level of submarine irradiance have been attributed to deepwater animals although little experimental work has been done to confirm the presumptive functional significance of the anatomical or other detail observed. In various crustaceans and fish the photoreceptor organelles (rhabdoms and rod outer segments) have been found in deepwater forms to be longer and hence presumably having greater absorptance of low intensity light fluxes than their shallow water counterparts. Double and up to six layers of rods in tandem occur in the retinas of a variety of deepwater fishes (Vilter, 1953; Munk, 1966a). These have again been interpreted as a means of increasing sensitivity by providing potentially longer total light absorbing paths without sacrificing visual acuity. However, even the morphological evidence is ambiguous in this regard (Munk, 1966a) and no experimental data are available.

Fish from deeper water typically have pure rod retinas classically more sensitive to dim light. Also the degree of convergence of primary receptor cells onto ganglion cells is often greater than usual. This implies lower thresholds since it makes available larger receptive fields for catching photons. Screening pigments generally and the vertebrate pigment epithelium in particular are usually substantially reduced in deep pelagic species.

It is interesting that in deepwater teleosts either a choroidal or retinal tapetum is rare (Munk, 1966a). Such light reflecting layers are common in elasmobranchs as well as in mesopelagic crustaceans. Their near absence in deepwater teleosts may relate to a negative adaptive value of "eye-shine" in the presence of luminescent neighbours. But this effect should presumably be equally relevant to the sharks and crustaceans, too. Tapeta are of course frequent in nocturnal insects and vertebrates where they apparently function to improve vision in dim light (Walls, 1963; Denton, 1970). However, they may not be important for lowering the absolute threshold since the visual pigment density in tapetal eyes may be only about half that of eyes lacking this reflecting layer (Denton, 1970). However, deepsea fish visual pigments have been reported with densities 5–6 times those in ordinary terrestrial eyes like frog's.

Many of these ideas on deepsea adaptations stem from major earlier studies like those of Chun on bathypelagic euphausiid (1896) and squid eyes (1903) or those of Brauer on deepsea fish eyes and light organs (1908). However, there has recently been something of a renaissance of interest in deepsea fish eyes largely sparked by the work of Ole Munk of Copenhagen. Yet most of the inferences from morphology remain to be tested.

Thus large eyes, large lenses and large pupils have all been reported in bathypelagic fishes. In so far as these increase the light collecting power of the system (i.e. reduce the f-number) and increase the diameters of receptive fields these should lower the eyes' threshold.

Presumably similar optical effects are developed in the remarkable dorsally directed frontal eyes of the bathypelagic euphasiids (Chun, 1896). Instead of projecting an ordinary image on the retina like a camera compound eyes of this kind have been presumed to form superposition images which according to Exner's theory (1891) intensify brightness by a special optical device dependent on the systematic variation of refractive index in the cylindrical lenses of each ommatidium. The large elongate parallel ommatidia of the deepwater crustaceans in question would appear particularly strongly adapted to perceiving dim light according to such a mechanism (Waterman, 1961, p. 23 ff.). However, the whole concept of superposition has been summarily challenged (e.g. Kuiper, 1962; Miller et al., 1968; Horridge, 1969) even though the issue presently seems far from being closed (Kunze, 1970, 1972; Kunze and Hausen, 1971).

In camera eyes the adaptive enlargement of the eye lens may reach its extreme in the tubular eyes present in several families of deepsea fish and some cephalopods (Fig. 4). These are typically directed dorsally but in several cases anteriorly. Such eyes have been interpreted as a means of substantially increasing the diameter (and hence the light collecting power) of the lens without requiring anatomically prohibitive large eyeballs of more conventional shape.

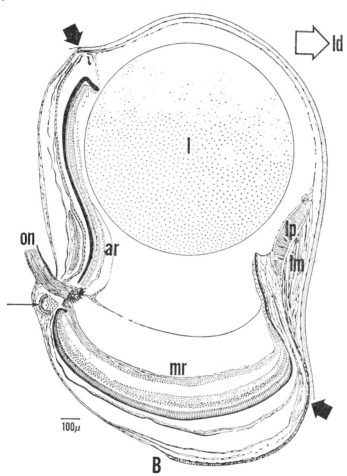

Fig. 4. Dorsally directed tubular eye of the midwater teleost *Scopelarchus guentheri*. The enormous size of the lens (*l*) relative to the eye diameter is evident. Abbreviations: *mr* main retina, *ar* accessory retina, *on* optic nerve, *lm* lens muscle. The small arrow (left) indicates an isolated part of the accessory retina, the heavy black arrows the margin of the cornea and the large open arrow (upper right) the lateral direction (*ld*) (from Munk, 1966a).

18*

Such an adaptation to liminal light intensities would of course require a corresponding sacrifice of field of view. Some support for this notion is lent by the extraordinary occurrence in tubular eyes of accessory retinas which apparently view out of focus images of moderately off-axis stimuli. Also they may have vesicular pockets of retinal tissue which would appear to monitor light far off the tubular axis through small windows in the eye's cylindrical wall (Munk, 1966a).

Since tubular eyes often have near parallel axes their visual field is strongly binocular. This could significantly (but not spectacularly) lower the absolute threshold (Weale, 1955). Also those directed upward seem well adapted to detect the "last" photons of the downwelling irradiance. Still more efficient threshold light detection might be achieved by reducing the ocular media. Aphakic apertures are in fact common in deepsea fish eyes (e.g. Munk, 1968). Indeed the most extreme

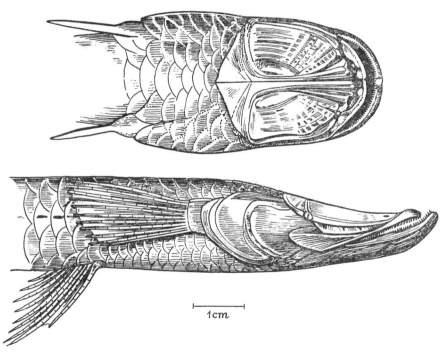

1cm

Fig. 5. The head of *Ipnops murrayi* in dorsal and lateral view. The remarkably modified eyes of this bathypelagic fish lie under a transparent bony plate covering the anterior dorsal half of the head. There are no lenses; other accessory structures are much reduced. The retinas with their typical vertebrate rods comprise large flattish plates looking dorsally. There are relatively few photoreceptor cells for a teleost and only about 500 fibers in each optic nerve (from Munk, 1959).

morphological adaptation, seen in *Ipnops* (Fig. 5), has essentially eliminated the lens, humors and iris (Munk, 1959, 1966a). The flattened retinas lie on the top of the head covered only by translucent dermal bone; they are obviously binocular in field and directed upward towards the last vestige of daylight.

Clearly, however, the visual adaptive strategies of various animals, even fairly closely related ones, are different. This often makes the interpretation of the relations of environmental light to adaptations quite difficult. Thus it is generally true that below the photic zone (*ca.* 1000 m) eyes begin to be relatively smaller (Murray and Hjort, 1912). Known cases of retinal degeneration occur at this and greater depths (Munk, 1969).

Yet *Omosudis* which is bathypelagic has large eyes. Most deepwater fish have mainly or only rods in their retinas (Brauer, 1908). *Omosudis* has a pure cone retina (Munk, 1965). Most large-eyed forms in deep water have photophores. *Omosudis* apparently has none! Another mesopelagic fish *Diretmus* has a mixed retina of rods *and* cones (Munk, 1966b). The latter are localized in a small area serving binocular vision in a rostral and slightly dorsal direction. This suggests prey pursuit (or mate recognition) visually guided by some bioluminescent mechanism involving either color discrimination or longer wavelengths than those present in the downwelling irradiance.

Indeed it is perhaps surprising that eyes begin to degenerate at depths greater than about 1000 m since photometric measurements have demonstrated nearly maximal bioluminescence at this depth and considerable animal light much deeper. However, at least in fishes photophore development and deployment seems less elaborate in deeper water (Nicol, 1967). Most likely there are important correlations between diet, activity pattern, diurnal vertical migrations, schooling, predators, camouflage and other factors which interact strongly with depth and ambient light factors. If so our inability to explain everything using only these last parameters is not surprising.

In the above discussion we have been concerned with the limiting influence of underwater light intensities on the visual orientation of animals. This was derived largely from eye morphology and comparative physiological evidence for species that have been studied experimentally. There are, however, at least two other aspects of the overall topic which arise in somewhat different contexts. These relate to the action of light first as a releaser of particular behavior patterns and second as a Zeitgeber in synchronizing biological rhythms. This is a large topic (Fraenkel and Gunn, 1940 (1961); Menaker, 1971) but may be exemplified by the influence of submarine illumination on diurnal vertical migration (DVM) where these two different effects may actually be difficult to isolate.

As a major diurnal variable the changes in underwater light have always been considered an important factor in the widespread DVM of pelagic animals in the photic zone. Indeed in all but the most superficial water layers it would seem to be the only parameter of the physicochemical environment whose variation could control such daily behavior cycles. Most other factors are too stable to change regularly on the required time scale.

Direct evidence correlating animals' depth in the water column with the light intensity at that same depth was first obtained by Clarke (1934) and by Russell (1934). Since then extensive correlations have been demonstrated between animals' depths in the water and the ambient light intensities (e.g. Kampa and Boden, 1954; Clarke and Backus, 1956, 1964; Boden and Kampa, 1967).

In some cases the coincidence of the maximum concentration of a particular organism (or more often in recent work of a sonic deep scattering layer (DSL) often of undetermined animal content (Farquhar, 1971)) and a particular isolume is striking. Evidence that this coincidence is at times the result of behavioral choice of a particular light intensity is available from the effects of cloud cover on organisms' depth distribution (Dietz, 1962; Blaxter and Currie, 1967) and from the displacing effect of an experimentally introduced artificial light (Blaxter and Currie, 1967). Furthermore in circumpolar latitudes where daily light changes are seasonally absent or much reduced the plankton instead of showing *daily* vertical movements may undertake comparable *seasonal* migration correlated with the levels of natural illumination (Bogorov, 1946; Hunkins, 1965).

To understand the possible physiological mechanism of such light intensity preferences one may postulate the following model. For a given responsive state of the animal a particular irradiance releases upwardly directed swimming. Its upward direction may be oriented either visually or by the direction of gravity. Suppose then that a stronger irradiance releases a downwardly directed swimming or inhibits swimming so that the animal sinks.

At some level between these two neither the upward nor the downward tendency predominates so that the organism is in effect trapped at that "steady state" intensity. When the depth of that isolume in the water column changes the above model would induce appropriate following responses. Such simple preferential behavioral patterns have been extensively studied in the laboratory on many kinds of animal (e.g. Fraenkel and Gunn, 1940 (1961); Harris, 1953; Harris and Wolfe, 1955).

In other cases, however, selecting a particular light level in the water column and following its changes in depth with time is an inadequate explanation of vertical migratory behavior. Thus organisms which have appeared to be following an isolume have been found clearly to diverge from it by moving up or down faster than that particular irradiance. This raises the question of whether they follow isolumes part of the time and other cues at other times. Obviously the extreme case occurs at night when the daylight intensity drops to a level where the apparently "preferred" isolume does not occur at any depth in the water column.

For example, there are many references in the literature to the increased depth dispersal of migratory forms as light fails in the evening coupled with their reaggregation at a specific depth when light intensities reach a certain level in the morning. However, in other cases there is also good evidence that aggregation at particular depths and continuing migration occur at night. Here isolume following obviously cannot be the explanation. Thus the evidence for DVM in mesopelagic organisms raises the question of whether isolume movements can directly control such behavior. For instance the calyptopis larvae of *Euphausia superba* (Fraser, 1936) and adults of the decapod prawn *Gennadas elegans* (Waterman *et al.*, 1939) show movements at least down to reasonable limits for the photic zone.

Furthermore in *Gennadas*, for example, their downward migration from 400 to 800 m occurs mainly in the very early morning hours before there is significant daylight even at the surface (Waterman *et al.*, 1939). Instances of upward movement continuing significantly after sunset and twilight are over could also be cited. In these examples and perhaps in the more general case one has to think of a different functional model.

To postulate one the organism's behavior may be assumed to be physiologically maintained on a diurnal or circadian rhythm by means of an internal biological timing cycle (Harris, 1963; Enright and Hammer, 1967; Ringelberg and Servas, 1971). These are almost universal (Bünning, 1964; Aschoff, 1965; Menaker, 1971). Typically, however, the endogenous rhythm is not precisely set on a 24 h period and needs some Zeitgeber to entrain its frequency and phase with the earth's rotation. Provided that for some even brief time of day a migrating organism were at a depth reached by perceptible downward irradiance its clock could be kept in synchrony.

However, since about three quarters of the seas' huge volume is below the photic zone, this poses an intriguing problem. Do vertical migrations cease as the bathypelagic zone is reached? If not how far down do they persist? Since light is the only likely physical or chemical

Zeitgeber in deep water how would synchronous DVM's be maintained below the photic zone?

A field study to answer the first two of these questions has been made under the author's leadership. A series of deepwater closing net hauls were made at four times of day and over four different depth ranges between 400 and 1850 m (only an abstract of the biological results has so far been published: Waterman and Berry, 1967). In the resultant 4×4 matrix there were 57 catches; from these 206 species were identified in seven major groups: teleosts 72 spp., amphipods 40 spp., decapods 35 spp., euphausiids 23 spp., medusae 17 spp., chaetognaths 14 spp. and mysids 5 spp. Of these the data for 25 individual species yield evidence of DVM.

Most interesting were certain decapod crustaceans and chaetognaths that moved from a night time maximum occurrence located around 1000 m to a day-time peak near 1400 m and other deeper living species of chaetognaths which peaked at 1725 m at midnight (Fig. 6). The latter seemed to go into still deeper layers below our fishing depths in the daytime. It would seem quite unlikely that any animals in the peak level for either group (1000 m or 1725 m at midnight, deeper at noon)

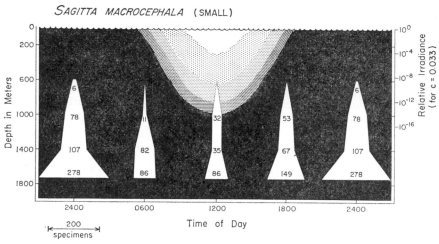

Fig. 6. Distribution of small individuals of the chaetognath *Sagitta macrocephala* collected with closing nets in the eastern North Atlantic. The histograms show the numbers captured at four times of day and at four depths fished. The penetration of sunlight is roughly diagrammed as a function of time and the attenuance of the irradiance from the surface is calibrated on the right hand ordinate. The data imply that individuals living even as deep as 1750 m at midnight are migrating downward during the day and returning to depths still well below the photic zone again at night (Waterman, in preparation).

could at any time of day receive perceptible irradiance from the sun. Yet they appear to be carrying out extensive synchronized DVM's.

It seems rather an enigma that they can do this. Documentation on the range and depth distribution of such DVM's are urgently needed and almost no critical data are available to answer the question of mechanism. Presumably, however, its answer will not be found in the ambient irradiation from the surface at the depths concerned where its visibility must be nil.

IV. RADIANCE DISTRIBUTION

Ideally the radiance distribution can provide animals with two important orientation cues.* One is the direction of the earth's vertical which is fundamental to spatial position. The other is azimuth information ultimately derived from celestial cues. The latter provide the only globally available visual data which an open water organism can use for maintaining compass directions. Hence it is basic not only for course steering by migrants but also for station keeping by nonmigrants which otherwise could get "lost".

Two phases of the radiance distribution need to be distinguished. The first depends primarily on refraction at the air-water interface and is predominant only in the few meters nearest the surface (see Chapter 5 by Smith). The second depends not only on refraction but also on the scatterance, reflectance and absorptance in the submarine light path. This phase includes nearly all of the photic zone.

Supposing only refraction to occur through a completely calm surface and no scatterance, absorptance or reflectance to be present all the incoming light underwater would form a 97° cone with the organism at its downward pointed apex. Under the ideal conditions specified this cone would be constant with depth. Its axis would lie in the vertical. If the sky were uniformly overcast the mean vector of the incoming light would also be vertical. If the sky were clear, the sun and sky including the latter's polarization would be visible within this cone (Waterman, 1954) or at night the moon and stars.

In this restricted phase the radiance distribution would indicate the vertical by the axis of its conical pattern. We already know that the dark light contrast at the edge of the Snell circle may be important in an animal's spatial orientation (Ringelberg, 1964, 1969). Azimuth indication would be provided by any of the four celestial parameters

* In this discussion orientation underwater by visual contact with landmarks will not be considered. Of course the strong absorptance and scatterance of light by water will very sharply reduce the range over which this can be effective. But otherwise such submarine pilotage no doubt is similar to that in air. See Chapter 7 on visibility by Duntley.

mentioned above provided the animal could perceive them and convert their apparent positions to biologically significant "compass" headings.

The second nearly ubiquitous phase of the radiance distribution occurs in practice as soon as there is any significant thickness of water between the sea surface and the point of observation. Scatterance, absorptance and reflectance by the water then become significant. Depending on the corresponding coefficients for the water mass as well as the state of the sea surface already at quite shallow depths the edge of the Snell circle becomes indistinct, sky details and the stars imperceptible. As depth increases further the inverted cone pattern changes to an ellipsoid with its major axis parallel to the refracted direction of the sun's rays underwater and the sun's disc is no longer visible as such (Jerlov, 1968).

The asymmetry of the ellipsoid decreases with depth down to the critical depth (at 300–500 m in clear water), where the radiance distribution is symmetrical about the vertical. The latter geometry holds even in shallow water when there is a uniformly overcast sky. Although ellipticity of the pattern decreases with depth even at the deepest levels measured the downwelling irradiance is greater than the upwelling irradiance (Jerlov, 1968). Thus below the equilibrium depth the vector sum of the radiance distribution would coincide with the upward vertical. At shallower depths it would generally include an azimuth component dependent on the sun's (or when applicable the moon's) bearing.

Biologically we know that many animals such as fishes and crustaceans regulate the spatial orientation of their dorsoventral axis by the so-called dorsal light reflex. Usually this behavior pattern maintains the middorsal part of the body in such a position that it is directed towards the mean vector of the incident light distribution. In general this would tend to align the dorsoventral axis with the vertical or point it toward the "center of gravity" of the radiance distribution (von Holst, 1950; Gordon and Cohen, 1971).

With regard to the dorsal light reflex this has been studied almost entirely in the laboratory. There the illumination ordinarily involves the use of a high intensity collimated light beam set in a dark featureless surround. Such a radiance distribution is certainly very rare in nature and some have considered that the behavior evoked thereby is highly abnormal (Verheijen, 1958). In fact we know that experimental modifications of the radiance distribution, even of a relatively simple sort, may critically alter the oriented response patterns of animals (Jander and Waterman, 1960; Waterman, 1960; Umminger, 1968, 1969). However, critical tests with reasonable facsimiles of natural radiance distributions remain to be made.

In terms of orientation the azimuth of the sun is perhaps the most important information a pelagic animal could derive from the radiance pattern. Provided the organism has some learned or instinctive knowledge of the diurnal changes in the sun's apparent position within the critical angle and at the same time has a biological clock, compass directions could be recognized by this mechanism. As the author has recently reviewed the evidence for submarine sun compass orientation particularly in fishes (Waterman, 1972) only brief reference is made to it here.

Although further discussion of underwater *visibility* will not be included here it seems appropriate to mention some data on *invisibility* which are directly relevant to the radiance distribution. These data relate to animal camouflage. For most pelagic animals it seems adaptively important to be nearly invisible. This could be advantageous to both prey and predator.

To be as inconspicuous as possible a free swimming animal should appear from all angles the same color and brightness as the surrounding water when that is seen from the same directions. This might be accomplished by the organism being as transparent as possible so that in fact the radiance distribution will undergo only negligible modification by the animal's body. Some larvae, medusae, heteropods and the like approach this closely. But with a large complex organism eyes, blood, gut contents, etc. cannot be that transparent. Hence another means must be used.

Generally in epipelagic forms this involves bluish dorsal pigmentation, (Herring, 1967) lighter or silvery lateral areas and pale or white ventral surfaces. The related optics of silvery reflecting surfaces in fish and their remarkably adaptive relation to the radiance distribution has been elegantly studied by Denton and Nicol, 1966; see reviews by Nicol, 1967 and Denton, 1970 (Fig. 7). From the reports of divers fish are most conspicuous when viewed from below even in relatively shallow depths where their light underbellies will be relatively well illuminated. In deeper water of the mesopelagic zone one would expect that the shadow cast downward will be even more noticeable.

In 1927 Rauther speculating on the biological significance of the often numerous ventrolateral photophores of many teleosts proposed that as their light had essentially the same spectral distribution as the downwelling irradiance, it could act to eliminate the fish's shadow as viewed from below. He suggested accordingly that at certain times of day a steady glow from the light organs would significantly decrease the visibility of the fish. Recent measurements of Denton (1970) on the pattern of light production by the mesopelagic teleost *Argyropelecus*

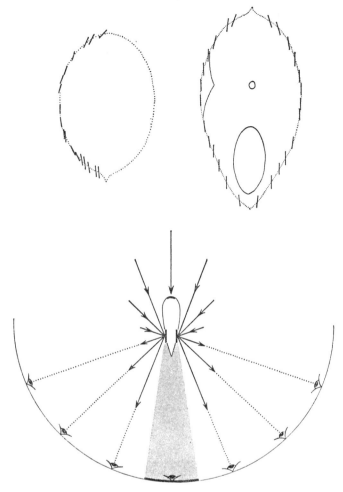

Fig. 7. Effect of silvery sides on fish visibility. The two upper drawings show the reflecting platelets in the skin of a salmon smolt (*left*) and a herring (*right*). The lower diagram indicates the way in which reflection of irradiance occurs relative to an observer in various positions indicated. The shaded sector is the fish's shadow (from Nicol, 1967).

provide considerable support for Rauther's suggestion which had since been taken up by a number of others.

The status of this hypothesis as it relates to fishes has been reviewed by McAllister (1967). He concludes that shadow reduction during the day may well be one function of ventral bioluminescence but that there may be at least two other adaptive explanations for such downward directioning of light production. It should be noted that in quite a few crustaceans and squid some, but not all, photophores are small and scattered mainly over the ventrolateral body surface (Harvey, 1952). Again this may be an adaptive convergence with the situation in teleosts which also may hold for some luminous sharks as well.

V. DEGREE OF POLARIZATION

Sunlight underwater is in part linearly polarized (Waterman, 1954). As in the sky the e-vector is everywhere perpendicular to the directional rays in the medium. Measurements *in situ* have shown that in clear water the degree of polarization (p) reaches a maximum of about 60%. Since the optics of submarine polarization are discussed by Ivanoff in Chapter 8 on Polarization Measurements and some of their biological implications have recently been reviewed (Waterman, 1972), these references should be consulted for further details and bibliography.

The biological interest in this parameter of underwater light arises from the sensory capacity of many animals to distinguish e-vector orientation in partially or fully polarized light (Waterman, 1966a). With particular regard to p there are two points to be made here. One relates to the threshold for distinguishing the plane of polarization. Relatively little is known about this but the few data available (none for marine animals!) suggest that in clear water the degree of polarization in the scattered sunlight is adequate throughout the photic zone for visual analysis by those animals possessing the basic analytical mechanism.

Thus in the laboratory *Daphnia* has been shown to orient polarotactically when $p = 20\%$ and sometimes when $p = 10\%$ (Waterman and Jander, unpublished; cited by Waterman, 1966b). Similarly the honeybee can orient effectively to the plane of polarization when $p \cong 10\%$ (the critical range is between $p = 7\%$ (disoriented) to $p = 15\%$ (well oriented), von Frisch, 1965, p. 411).

Another point of interest relating to p is that the presence of polarization has been shown in laboratory experiments to release endogenous diurnal rhythmic behavior in a copepod (Umminger, 1968). Polarizers with horizontal e-vector were placed around the transparent experimental vessel containing the animals. These filters were transilluminated with horizontal light sources and unpolarized vertical illumination was provided from above. When the intensities of the horizontal (polarized) and vertical (unpolarized) illumination were the same (approximating a deepwater radiance distribution) the copepods (*Cyclops vernalis*) swam mostly up and down in the water at the beginning and end of their 12 h light phase of a 12 : 12 LD cycle. Around "noon" they mainly swam horizontally. Vertical swimming predominated all day when the downward illumination had about 10 times the intensity of the lateral light. In the absence of the lateral polarized light diurnal rhythmic behavior was not detected.

Interestingly enough daily vertical movements in the experimental

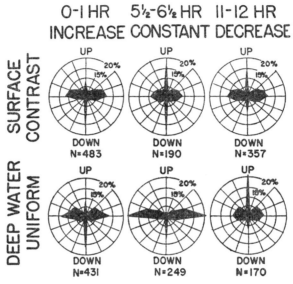

Fig. 8. Oriented behavior and vertical migration of the copepod *Cyclops vernalis* in the laboratory. The activity pattern depends on (1) the lateral presence of horizontal linear polarization, on (2) the time during the 12 h light period of a 12 : 12 LD regime (columns 1–3), on (3) a gradual (rather than abrupt) increase or decrease of light intensity at the beginning and end of the light period (columns 1 and 3), and on (4) the radiance distribution (ratio of vertical to horizontal radiance, it was 10 : 1 in the upper row, 1 : 1 in the lower row). Under the right combination of conditions the animals swam mainly downwards during the first lighted hour (left column), horizontally at midday (lower center) and upward in the last lighted hour (lower right) (from Umminger, 1968).

vessel analogous to those widely observed in the field appeared only if one more feature of the illumination was adjusted appropriately. This additional stimulus was a gradual increase in light intensity at the beginning of the light phase of the artificial day instead of a sudden single step off-on. With such a simulated dawn downward movement predominated in the vertical movements typical of the early part of the light phase. Both downward and upward swimming had occurred about equally with the sudden single-step "daybreak".

Similarly with a gradual decrease of light intensity instead of a single step blackout at the end of the light period predominantly upward swimming occurred. As a result of these light conditioned behavioral responses the copepods moved downward toward the bottom of the vessel in the early morning, swam there mainly horizontally during midday and moved upward toward the surface in the evening (Fig. 8). Note that even in this laboratory induced behavior appropriate values of three parameters of underwater light were required to release it: changes in irradiance, a particular radiance distribution and a given pattern of polarization.

VI. Direction of e-Vector

Because the directionality of underwater daylight is primarily determined by the sun when it is shining, the polarization pattern in the sea in turn changes as the sun moves through the sky. Since many kinds of animals from worms to vertebrates use the sun as a compass, underwater polarization is potentially available for the same use down at least to 200 m (Waterman, 1955, 1958).

A wide range of invertebrate animals has been shown to be capable of perceiving the e-vector of linearly polarized light (Waterman, 1966a). Their capacity to do so depends at least in cephalopods and crustaceans on the dichroism of the visual pigment and the fine structure of the rhabdom, the photoreceptor organelle of their retinas (Stockhammer, 1959; Moody, 1964; Langer, 1966; Tasaki and Karita, 1966; Waterman et al., 1969). More recently evidence has been accumulating that at least some of the lower vertebrates can also perceive the plane of polarization even though they have no rhabdoms and the visual mechanism involved has not yet been demonstrated (Waterman and Forward, 1970, 1972; Waterman, 1972).

Thus we know that submarine light is polarized in a pattern which could be an extension of a direct sun compass. We also know that many marine animals can perceive the natural e-vector pattern underwater. However, relatively few data are available to establish the correlation of these two factors. There is, however, some good evidence for direct sun-compass utilization by fishes (see Waterman, 1972 for review) and we have been carrying out relevant field studies on teleosts for several years. The main findings of these experiments are reviewed below (Forward et al., 1972; Waterman and Forward, 1972).

A West Pacific halfbeak fish *Zenarchopterus* has been used in this work carried out with SCUBA in Palau in the Western Caroline Islands. Its spontaneous orientation has been recorded when it was exposed to

natural illumination underwater and at the water surface. Also the effect
of imposing linear polarized light on the existing irradiance has been
systematically studied in both situations.

The results obtained so far confirm the earlier evidence for oriented
responses (polarotaxis) to the imposed polarization (Waterman and
Forward, 1970). In the underwater experiments the spontaneous
azimuth orientation also was correlated with the sun's bearing

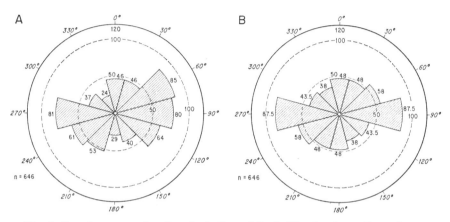

Fig. 9. Spontaneous azimuth orientation of the halfbeak teleost *Zenarchopterus*
exposed in shallow water to the sun, sky and natural submarine irradiance and
polarization as well as an imposed *e*-vector produced with a dichroic filter over the
experimental vessel. A. Responses plotted relative to the sun's bearing at 0° so
that the imposed *e*-vector is randomized in 30° steps around 180°. In this distribution the animals were significantly choosing directions between 60° and 90° as
well as 240° and 270° to the sun's direction. This indicates that a sort of sun
compass orientation is present. B. Same data replotted so that the *e*-vector is
coincident with the 0–180° axis. There the 90°–270° directions are highly significant. This shows that a polarotaxis with the preferred direction perpendicular to
the *e*-vector is also present (Waterman and Forward, 1972).

(essentially constant during the measurements) and hence the natural
polarization of the sky and water (Fig. 9). The water surface experiments
were carried out with two different (nearly 180° apart) bearings of the
sun. Nevertheless the spontaneous geographical headings of the fish
were essentially the same at both times. Hence the sun compass was
time-compensated. The preferred direction was parallel to the *e*-vector
orientation in the blue sky in the sun's vertical.

When a Polaroid filter with its *e*-vector aligned in this same direction
(90° to the solar bearing) was placed over the experimental vessel, the
fish oriented as it did with no filter present (Fig. 10). However, when
the imposed plane of polarization was oriented at other angles to the

sun's bearing the headings preferred were still parallel to the *e*-vector but had different relations to the sun's bearing and geographical directions. The evidence thus supports a time-compensated compass and possibly a sky polarization compass for *Zenarchopterus*. Its basic polarotactic behavior proves that a sensory basis for the latter is in fact present.

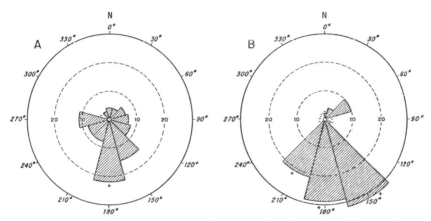

Fig. 10. Spontaneous orientation relative to North at 0° of *Zenarchopterus* swimming at the water surface exposed to clear sun and sky (plotted as percentages). A. Responses with no filter over the experimental vessel. On the two days of the experiments the sun's bearing was 90° and 270° but the preferred direction was South on both occasions. This direction was parallel to the sky polarization in the sun's vertical. B. Response under the same conditions but with a polarizer added with its *e*-vector at 90° to the sun's bearing (and therefor parallel to the natural polarization in the sun's vertical). Again the southerly quadrant was significantly preferred. The plus signs indicate those 30° sectors significantly ($p < 0.01$) preferred. The minus signs indicate those significantly avoided (Forward *et al.*, 1972).

Further work is planned to extend these observations and to correlate them effectively with the behavior and natural history of the fish concerned. In addition extensive laboratory experiments are being carried out in two areas. One uses spontaneous orientation and other behavioral responses to investigate the distribution of this sensory capability among fishes. The other uses electrophysiological techniques to analyze the perceptive mechanism concerned.

VII. Conclusions

Clearly our understanding of the visual orientation of animals in the sea remains in a rather elementary state despite the considerable interest of some problems involved and despite not inconsiderable relevant field and laboratory research. The phenomena are complex, however, and the real difficulty is to devise crucial experiments which will be fruitful in disclosing explanations rather than just adding some more data to the myriad already available.

Perhaps a weakness in our approach until now has been the failure to attack the biological problems involved on a broad enough front. From this review the relevance of most of the fundamental quantities of underwater optics should be quite clear. Yet rarely are either field or laboratory experiments carried out with an adequate control or even cognizance of these optical factors. Even relatively crude approximations like those made earlier by Harris and Wolf (1955) and more recently by Umminger (1968, 1969) have evoked in the laboratory a surprisingly rich spectrum of oriented behavior in planktonic animals.

Undoubtedly the collaboration of the optical oceanographers may be an important factor in allowing the biologists to make headway in this field. It would for example be most interesting and perhaps even important in terms of the ensuing results to simulate a natural radiance distribution in a laboratory set-up. Experience would indicate how carefully this would have to be done and whether or not it acts to release some heretofore unexpected behavior pattern.

Some approximations of the natural pattern of polarization in submarine light would also be most interesting but such simulation is a rather challenging optical problem. Conversely the experimental biologist can take his laboratory experiments out into the field as we have done on several occasions (Bainbridge and Waterman, 1957; Daumer et al., 1963 and the more recent work on Zenarchopterus cited above). Here the proper experimental design can take advantage of complex natural configurations of potential stimuli almost impossible to duplicate in the laboratory. However, only the crudest beginnings have been made in these directions.

Not only is it important for the biological oceanographer to collaborate with specialists in underwater optics but he also needs to have appropriate biophysical and biochemical knowledge to unravel the underlying sensory and behavioral mechanisms. The current richness of our insight into the adaptive relations of visual pigment λ_{max} provides a good case in point. Unfortunately, however, it is still a rather rare one since generally the physiologists, ethologists and comparative psychologists have not been drawn into these areas of oceanography as they should be.

One of the potential strengths of a symposium like this is its capacity of permitting or even provoking some of these necessary collaborations to take place for the mutual advantage of all concerned.

REFERENCES

Aschoff, J. (ed.) (1965). "Circadian Clocks", 479 pp. North-Holland, Amsterdam.

Bainbridge, R. and Waterman, T. H. (1957). *J. Exp. Biol.*, **34**, 342–364.

Bayliss, L. E., Lythgoe, R. J. and Tansley, K. (1936). *Proc. Roy. Soc., London*, **B120**, 95–113.

Beatty, D. D. (1966). *Can. J. Zool.*, **44**, 429–455.

Beatty, D. D. (1969). *Vision Res.*, **9**, 855–864.

Berry, S. S. (1926). *Biol. Bull.*, **51**, 257–268.

Blaxter, J. H. S. and Currie, R. I. (1967). *In* "Aspects of Marine Zoology" (N. B. Marshall, ed.), pp. 1–14. Academic Press, London.

Boden, B. P. and Kampa, E. M. (1967). *In* "Aspects of Marine Zoology" (N. B. Marshall, ed.), pp. 15–26. Academic Press, London.

Bogorov, B. G. (1946). *J. Mar. Res.*, **6**, 25–32.

Brauer, A. (1908). Die Tiefsee-Fische, 2. *Anat. Teil., Wiss. Ergeb., "Valdivia"*, **15**.

Bridges C. D. B. (1972). *In* "Photochemistry of Vision" (H. J. A. Dartnall, ed.), pp. 417–480. Springer-Verlag, Berlin.

Bridges, C. D. B. and Yoshikami, S. (1970). *Vision Res.*, **10**, 1315–1332.

Bünning, E. (1964). "The Physiological Clock", 145 pp. Academic Press, New York; Springer-Verlag, Berlin.

Chun, C. (1896). Atlantis. *Zoologica (Stuttgart)*, **19**, 1–260.

Chun, C. (1903). *Verh. Deut. Zool. Ges.*, **13**, 67–91.

Clarke, G. L. (1934). *Biol. Bull.*, **67**, 432–455.

Clarke, G. L. (1936). *Ecology*, **17**, 452–456.

Clarke, G. L. (1971). *In* "Proceedings of the International Symposium on Biological Sound Scattering in the Ocean, 1970" (G. B. Farquhar, ed.), pp. 41–50. Maury Center for Ocean Science, Department of the Navy, Washington, D.C.

Clarke, G. L. and Backus, R. H. (1956). *Deep-Sea Res.*, **4**, 1–14.

Clarke, G. L. and Backus, R. H. (1964). *Bull. Inst. Oceanogr.*, **1318**, 1–36.

Crescitelli, F. (1969). *Vision Res.*, **9**, 1407–1414.

Crescitelli, F. (1972). *In* "Photochemistry of Vision" (H. J. A. Dartnall, ed.), pp. 245–363. Springer-Verlag, Berlin.

Dartnall, H. J. (1969). University of California, *Los Angeles Forum Med. Sci.*, **8**, 235–256.

Daumer, K., Jander, R. and Waterman, T. H. (1963). *Z. Vergl. Physiol.*, **47**, 56–76.

Denton, E. J. (1970). *Phil. Trans. Roy. Soc. London*, **B258**, 285–313.

Denton, E. J., Gilpin-Brown, J. B. and Wright, P. G. (1970). *J. Physiol.*, **208**, 72–73P.

Denton, E. J. and Nicol, J. A. C. (1964). *J. Mar. Biol. Assoc. U.K.*, **44**, 219–258.

Denton, E. J. and Nicol, J. A. C. (1966). *J. Mar. Biol. Assoc. U.K.*, **46**, 685–722.

Denton, E. J. and Shaw, T. I. (1963). *J. Mar. Biol. Assoc. U.K.*, **43**, 65–70.

Denton, E. J. and Warren, F. J. (1957). *J. Mar. Biol. Assoc. U.K.*, **36**, 651–662.

Dietz, R. S. (1962). *Sci. Am.*, **207(2)**, 44–50.

Enright, J. T. and Hamner, W. M. (1967). *Science, N.Y.*, **157**, 937–941.

Exner, S. (1891). "Die Physiologie der Facettierten Augen von Krebsen und Insecten", 206 pp. Deuticke, Leipzig and Vienna.

Farquhar, G. B. (ed.) (1971). Proceedings of an International Symposium on Biological Sound Scattering in the Ocean. 629 pp. Maury Center for Ocean Science, Department of the Navy, Washington, D.C.

442   T. H. WATERMAN

Forward, R. B., Jr., Horch, K. W. and Waterman, T. H. (1972). *Biol. Bull.*, **143**, 112–126.
Fraenkel, G. S. and Gunn, D. L. (1940). "The Orientation of Animals", 352 pp. Clarendon Press, Oxford. (1961, new edition, paperback, 376 pp. Dover Publications, New York.)
Fraser, F. C. (1936). *Discov. Rep.*, **14**, 1–192.
Frisch, K. von (1965). "Tanzsprache und Orientierung der Bienen." 578 pp. Springer-Verlag, Berlin.
Gogala, M., Hamdorf, K. and Schwemer, J. (1970). *Z. Vergl. Physiol.*, **70**, 410–413.
Goldsmith, T. H. (1972). *In* "Photochemistry of Vision" (H. J. A. Dartnall, ed.), pp. 685–719. Springer-Verlag, Berlin.
Gordon, S. A. and Cohen, M. J. (eds.) (1971). "Gravity and the Organism", 473 pp. University of Chicago Press, Chicago.
Hara, T. and Hara, R. (1972). *In* "Photochemistry of Vision" (H. J. A. Dartnall, ed.), pp. 720–746. Springer-Verlag, Berlin.
Harris, J. E. (1953). *Quart. J. Microsc. Sci.*, **94**, 537–550.
Harris, J. E. (1963). *J. Mar. Biol. Assoc. U.K.*, **43**, 153–166.
Harris, J. E. and Wolfe, U. K. (1955). *Proc. Roy. Soc. London*, **B144**, 329–354.
Harvey, E. N. (1952). "Bioluminescence", 649 pp. Academic Press, New York.
Herring, P. J. (1967). *In* "Aspects of Marine Zoology" (N. B. Marshall, ed.), pp. 215–235. Academic Press, London.
Holst, E. von (1950). *Symp. Soc. Exp. Biol.*, **4**, 143–172.
Horridge, G. A. (1969). *Proc. Roy. Soc. London*, **B171**, 445–463.
Hunkins, K. (1965). *Deep-Sea Res.*, **12**, 879–881.
Jander, R. and Waterman, T. H. (1960). *J. Cell. Comp. Physiol.*, **56**, 137–160.
Jerlov, N. G. (1968). "Optical Oceanography", 194 pp. Elsevier, Amsterdam.
Kampa, E. M. (1955). *Nature (London)*, **175**, 996–997.
Kampa, E. M. and Boden, B. P. (1954). *Nature (London)*, **174**, 869.
Kuiper, J. W. (1962). *Symp. Soc. Exp. Biol.*, **16**, 58–71.
Kunze, P. (1970). *Verh. Deut. Zool. Ges.*, **64**, 234–238.
Kunze, P. (1972). *Z. Vergl. Physiol.*, **76**, 347–357.
Kunze, P. and Hausen, K. (1971). *Nature (London)*, **231**, 392–393.
Langer, H. (1966). *Verh. Deut. Zool. Ges. Göttingen, 1966 Suppl.*, **30**, 195–233.
Lundgren, B. and Hojerslev, N. (1971). *Kφbenhavns Universitet, Rep. Inst. Fys. Oceanogr.*, **14**, 1–33.
Lythgoe, J. N. (1972). *In* "Photochemistry of Vision" (H. J. A. Dartnall, ed.), pp. 566–603. Springer-Verlag, Berlin.
Lythgoe, J. N. and Dartnall, H. J. A. (1970). *Nature (London)*, **227**, 955–956.
McAllister, D. E. (1967). *J. Fish. Res. Bd. Canada*, **24**, 537–554.
McFarland, W. N. (1971). *Vision Res.*, **11**, 1065–1076.
Menaker, M. (ed.) (1971). "Biochronometry", 622 pp. National Academy of Science, Washington, D.C.
Miller, W. H., Bernard, G. D. and Allen, J. L. (1968). *Science, N.Y.*, **162**, 760–767.
Moody, M. F. (1964). *Biol. Bull.*, **39**, 43–86.
Munk, O. (1959). *Galathea Rep.*, **3**, 79–87.
Munk, O. (1965). *Vidensk. Medd. Dansk. Naturh. Foren.*, **128**, 341–355.
Munk, O. (1966a). *Dana Rep.*, **70**, 1–62.
Munk, O. (1966b). *Vidensk. Medd. Dansk. Naturh. Foren.*, **129**, 73–80.
Munk, O. (1968). *Galathea Rep.*, **9**, 211–218.
Munk, O. (1969). *Vidensk. Medd. Dansk. Naturh. Foren.*, **132**, 25–30.

Munz, F. W. (1971). *In* "Fish Physiology" (W. S. Hoar and D. J. Randall, eds.), vol. V., pp. 1–32. Academic Press, New York.

Murray, J. and Hjort, J. (1912). "The Depths of the Ocean", 821 pp. Macmillan, London.

Nicol, J. A. C. (1967). *In* "Aspects of Marine Zoology" (N. B. Marshall, ed.), pp. 27–55. Academic Press, London.

Rauther, M. (1927). *Bronn's Tierreich*, 6, Abt. 1, Book 2, 125–167. Leipzig.

Ringelberg, J. (1964). *Neth. J. Sea Res.*, 2, 319–406.

Ringelberg, J. (1969). *Verh. Int. Verein. Limnol.*, 17, 841–847.

Ringelberg, J. and Servas, H. (1971). *Oecologia (Berlin)*, 6. 289–292.

Russell, F. S. (1934). *J. Mar. Biol. Assoc. U.K.*, 19, 569–584.

Stockhammer, K. (1959). *Ergeb. Biol.*, 21, 23–56.

Tasaki, K. and Karita, K. (1966). *Jap. J. Physiol.*, 16, 205–216.

Tyler, J. E. and Smith, R. C. 1970. "Measurements of Spectral Irradiance Underwater" 103 pp. Gordon and Breach Science Publishers, New York, London, Paris.

Umminger, B. (1968). *Biol. Bull.*, 135, 239–251.

Umminger, B. (1969). *Crustaceana*, 16, 202–204.

Verheijen, F. J. (1958). *Arch. Neer. Zool.*, 13, 1–107.

Vilter, V. (1953). *C.R. Soc. Biol.*, 147, 1937–1939.

Walls, G. L. (1963). "The Vertebrate Eye and its Adaptive Radiation", 785 pp. Hafner Publishing Company, New York.

Waterman, T. H. (1954). *Science, N.Y.*, 120, 927–932.

Waterman, T. H. (1955). *Deep-Sea Res.*, 3 Suppl., 426–434.

Waterman, T. H. (1958). *In* "Perspectives in Marine Biology" (A. A. Buzzati-Traverso, ed.), pp. 429–450. University of California Press, Berkeley.

Waterman, T. H. (1960). *Z. Vergl. Physiol.*, 43, 149–172.

Waterman, T. H. (1961). *In* "The Physiology of Crustacea" (T. H. Waterman, ed.), vol. II, pp. 1–64. Academic Press, New York.

Waterman, T. H. (1966a). *In* "Environmental Biology" (P. L. Altman and D. S. Dittmer, eds.), pp. 155–175. Federation of American Societies for Experimental Biology, Bethesda, Maryland.

Waterman, T. H. (1966b). *In* "The Functional Organization of the Compoun Eye" (C. G. Bernhard, ed.), pp. 493–511. Pergamon Press, Oxford.

Waterman, T. H. (1972). *In* "Animal Orientation and Navigation" (S. R. Galler, K. Schmidt-Koenig, G. J. Jacobs and R. E. Belleville, eds.), pp. 437–456. National Aeronautics and Space Administration, Washington, D.C.

Waterman, T. H. and Berry, D. A. (1967). (Abstr.) *Am. Zoologist*, 7, 804.

Waterman, T. H., Fernández, H. R. and Goldsmith, T. H. (1969). *J. Gen. Physiol.*, 54, 415–432.

Waterman, T. H. and Forward, R. B., Jr. (1970). *Nature (London)*, 228, 85–87.

Waterman, T. H. and Forward, R. B., Jr. (1972). *J. Exp. Zool.*, 180, 33–54.

Waterman, T. H., Nunnemacher, R. F., Chace, F. A., Jr. and Clarke, G. L. (1939). *Biol. Bull.*, 76, 256–279.

Weale, R. A. (1955.) *Nature (London)*, 175, 996.

Chapter 19

Bioluminescence

BRIAN P. BODEN and ELIZABETH M. KAMPA

*Scripps Institution of Oceanography, University of California,
San Diego, La Jolla, California, U.S.A.*

I. Introduction

Practically all of the comments made in this chapter are repetitious of those of others or of ourselves, but this is inherent in the nature of a review article and no apology is offered. We have endeavoured to update the literature references in the field.

Little more can be said in an introduction of this sort except, perhaps, to quote Johnson (1966), who in his accolade to Harvey and his work proclaimed, in an unsurpassable burst of piety "May bioluminescence continue in progress".

II. The Nature of Marine Bioluminescence, Chemical Considerations

Approximately fifteen types of luminescence systems have been identified *in vitro* from different luminous organisms (Johnson and Haneda, 1966). Some of these are exhibited by terrestrial or fresh-water forms such as the firefly, earthworms, fungi or the fresh-water limpet *Latia*. These are considered to be outside the scope of this paper.

Cormier and Totter (1964) broke these down into four reaction-mechanism groups, and Hastings (1966, 1968) tentatively added a fifth on what appears to be good grounds (respecting the reservations he has made) to these reviewers. All are represented by marine forms. These reactions are as shown in Table I.

TABLE 1.* Types of biochemical reactions leading to luminescence

		Representatives	Peak emission nm
Type I	Direct oxidation: simple enzyme-substrate systems $LH_2 + O_2 \rightarrow$ light	*Pholas* *Cypridina* *Apogon* *Gonyaulax*	485 nm 460 nm 460 nm 470 nm
Type II	Substrate activation followed by oxidation: adenine nucleotide linked $\text{pre-}LH_2 \xrightarrow{\text{Activation}} LH_2 \xrightarrow{O_2} \text{light}$	*Renilla*	460 nm
Type III	Substrate reduction followed by oxidation: pyridine nucleotide linked $L \xrightarrow{\text{DPNH}} LH_2 \xrightarrow{O_2} \text{light}$	Bacteria	490 nm
Type IV	Peroxidation reactions $LH_2 + H_2O_2 \rightarrow$ light	*Balanoglossus* *Chaetopterus*	480 nm(?) 460 nm
Type V	Ion-activated: "precharged" systems $P \xrightarrow[\text{or } H^+]{Ca^{++}} \text{light}$	*Gonyaulax* *Aequorea*	470 nm 460 nm

* Modified from Hastings, 1966.

Dubois (1887) working with the boring clam *Pholas* demonstrated the first type of reaction and introduced the terms luciferin as a substrate for the enzyme luciferase. The mixing of these substances in the presence of oxygen resulted in the emission of light.

$$LH_2 + O_2 \xrightarrow{\text{luciferase}} \text{light}$$

For some time this was thought to be the basis of all luminescent systems, and the term "luciferin-luciferase" reaction crept insidiously into the literature, and the thinking of researchers, in the field. This is rather overwhelming phraseology in that it connotes a bioenergetic system common to all luminescent mechanisms.

Hastings (1966) emphasizes that the terms "luciferin" and "luciferase" are generic terms. Subsequently he notes that in Type I (enzyme-substrate) luminescence a reaction can be obtained between the luciferin of *Cypridina* (an ostracod crustacean) and the luciferin of *Apogon* and *Parapriacanthus* (teleosts). Since these are about as congeneric as Thallophytes and Mammals we hesitate to accept "generic" terminology and rather prefer his speculation that the chemical identities are almost always "specifically" different in each system. This may be considered as an exercise in semantics. However, any semasiological adventure in a rapidly burgeoning science should be rather rigidly scrutinized before a terminology, based on classical zoology, is accepted, and it seems possible that a completely new terminology may be necessary.

Thus it appears that the vocabulary adopted by workers in the field of bioluminescence is still somewhat confused (Johnson and Haneda, 1966). It will, undoubtedly, be clarified as progress is made in the biochemical purification and identification of the systems, which seems to be being accomplished largely by co-operative work between the Japanese and American workers.

The tentative establishment by Hastings of Type V luminescence (Ion-activated: "pre-charged" systems) may be compounding confusion. The protozoan *Gonyaulax* and the hydromedusan *Aequorea* are placed together in this group. A luciferin-luciferase system has been demonstrated in *Gonyaulax* cells (Hastings and Sweeney, 1957). These luminesce in the presence of water, NaCL, and O^2. In addition, separated, non-soluble, cell inclusions, called "scintillons", were found to luminesce when the pH was lowered from 8·00 to 5·70 in the presence of O^2. These particles, crystalline in appearance, do not contain luciferin or luciferase. Both systems require oxygen. In the hydromedusan *Aequorea*, Davenport and Nicol (1955) report that the luminescence is intracellular in the yellow, tightly packed polyhedral cells that bulge into the circular canal between the tentacular bulbs. No luciferase has been found and the substrate, named aequorin, is a non-enzymatic protein which luminesces in the presence of Ca^{++} or Sr^{++} (Shimomura *et al.*, 1962), but does not require oxygen for the reaction.

In view of this, however, we feel constrained to agree with Johnson (Johnson and Haneda, 1966) that aequorin should not be described as a type of luciferin, which requires oxygen, as it was by Nicol (1963) and Boden and Kampa (1964). At present it would indeed appear more appropriate to call it a type of "photoprotein" as suggested by Shimomura and Johnson (1966).

Thus while the coelenterate *Aequorea* belongs unequivocally in the Type V system in Hastings' classification, the protozoans *Gonyaulax* and *Noctiluca* (Eckert, 1966a,b) qualify for both Type I and Type V systems, as indicated by Hastings.

This rather brief summary of the chemical nature of bioluminescence was derived largely from the comprehensive reviews of the subject by Hastings (1966, 1968), Johnson (1955), Johnson and Haneda (1966) and Nicol (1962). They are unanimous in the opinion that progress in this aspect of the field is gratifyingly laudable. Nicol does not consider the breakdown of the various systems into five "types" and both Johnson and Hastings regard this classification as tentative.

Further enlightenment on the chemical nature of bacterial luciferase, its purification, general properties and functional differences within this group, inhibition and hybridization stems mainly from the following works: Yunsalus-Miguel *et al.* (1972); Nealson and Hastings (1972); Meighen *et al.* (1971). These extremely significant contributions to our understanding of the nature of luminescence and to cellular biology do not help us a great deal in our attempts to evaluate its role in the marine ecosystem.

III. THE NATURE OF MARINE BIOLUMINESCENCE, ANATOMICAL CONSIDERATIONS

Many marine animals possess highly developed, discrete light-emitting organs (photophores). However, some species utilize other mechanisms for luminescence such as symbiotic bacteria or the emission of luminescent clouds.

Both circadian and seasonal rhythms have been suggested particularly for the bacteria and dinoflagellates (Kelly and Katona, 1966; Yentsch and Laird, 1968; Seliger and McElroy, 1968; Hastings and Sweeney, 1958; Tett, 1969).

Even electron-microscopy has unveiled no structural features responsible for luminescence in bacteria (Johnson *et al.*, 1943). In this group luminescence is continuous but the *in vivo* intensity is apparently subject to seasonal rhythmic control (Turner, 1966). It is conjectured here that this may be a seasonal variation in population density. Apart from the bacteria adherent on detrital matter (Haneda, 1955) are those symbionts employed for luminescence by other marine groups, notably fishes and the sepiolids (squid) and the myopsid squid (*Loligo*). Herein lies a degree of nervous control in that the luminous bacteria are concentrated in photogenic cells that may be occluded at the will of the host by means of opaque membranes or by rotation of the light-emitting surface toward a dense, pigmented, cellular layer.

In this presentation we hoped to avoid any reflections on luminescence as related to phylogeny, since this has been so well covered by Harvey (1952), Nicol (1958b), and Boden and Kampa (1964) and others.

However, under this subhead a brief résumé seems to be necessary.

The continuous *in vivo* light emission of some marine bacteria is a respiratory phenomenon requiring oxygen. Luminescence has been demonstrated in cell-free extracts (Strehler, 1953) and the chemistry has been elucidated, *in vitro* (McElroy *et al.*, 1953; Cormier and Strehler, 1953; Strehler and Cormier, 1954; Strehler *et al.*, 1954; Johnson *et al.*, 1954; Strehler, 1955; Hastings and McElroy, 1955). As noted above there is a diurnal rhythm in the intensity of the emission.

In the dinoflagellates two systems appear to be utilized by the same plant, an intracellular, simple enzyme-substrate system that requires oxygen and a separate, particulate "scintillon" system for which no enzyme-substrate system has been demonstrated and which does not require oxygen (Hastings, 1966). It is at this stage in the phylogenetic scale that the first tangible, anatomical, luminescent particles appear. Diurnality in emission has also been noted (Haxo and Sweeney, 1955; Sweeney and Hastings, 1957; Hastings and Sweeney, 1957b, 1958).

Shimomura and Johnson (1966) report an interesting anatomical luminescent system in the tubiculous marine polychaete annelid *Chateopterus variopedatus*. The twelfth segment of this worm carries mucus-secreting, aliform notopodia and luminescent material generated by photocytes in the medial section are conveyed via mucus through the oral groove into the sea water. There are numerous posterior notopodia, which, while capable of brilliant luminescence in the intact animal, are incapable of secreting mucus.

Among the pelagic annelids the Tomopteridae and the part-time pelagic Syllidae are noted for their luminescent displays, apparently associated with sexual behaviour. The site of light production may originate in differentiated epidermal cells (Galloway and Welch, 1911), but no discrete photocytes have been detected.

In the planktonic crustacea, luminescence is a common characteristic. In the Ostracoda this phenomenon has been most vigorously investigated in the genus *Cypridina*, although three other genera (*Pyrocypris*, *Giganto* and *Conchoecia*) are also listed as luminescent (Harvey, 1952). Ostracod luminescence is extracellular, two types of granules being secreted simultaneously with a mucus emission and resulting in a luminous cloud.

19

There are numerous luminescent species in the Copepoda and these have been tabulated by Clarke *et al.* (1962). Representatives of five families of calanoid copepods are known to luminesce, and one family in the cyclopoids. There are dubious reports of luminescence in other families. David and Conover (1961) made histological studies of *Metridia lucens* and found cuticular, luminous glands on the anterior part of the head, the last thoracic segment and the posterior margin of each abdominal segment. Luminescence is extracellular. Two types of cells exist. These have a common duct and are presumed to be the individual sources of the luminous reactants which simultaneously discharge the substrate and enzyme into the surrounding medium.

Climbing the evolutionary ladder we find the euphausiid crustaceans, the first to have developed complicated photophores, apparently under neural control, and invested with a lenticular system. Only one of the eleven genera of the Euphausiacea does not luminesce, i.e. *Benth-euphausia*. In the others a photophore is situated in each eyestalk, one at the bases of each of the second and seventh thoracic appendages, and one medially and ventrally on each of the first four abdominal somites, except in the genus *Stylocherion* in which only the first abdominal somite is invested with a photophore (Boden and Kampa, 1964). Actual concentration or collimation of emitted light by the lens has not been satisfactorily demonstrated although Hardy (1962) has shown that the photophores can be rotated and he suggests that may be in order to direct beams of light in any desired direction.

Dennel (1940, 1954) has described the structure of the photophores of the penaeid and carid prawns in great detail. While the ectodermal distributional patterns of the photophores in these groups are similar, their histological structures are quite different. This is not surprising since the Penaeidea are only remotely related to the Caridea (Gurney, 1942). In the sergestids (Penaeidea) the ectodermal photophores are fibrous, lack a lens and have no demonstrable nerve supply. In the carideans (e.g. *Holophorus* and *Systellaspis*) the superficial photocytes are under both muscular and nervous control. Each group has developed endodermal photophores as modifications of the digestive gland. In the sergestids these are compact groups of luminous tubules that differ completely, structurally, from the modified, liver tubules of the carideans.

The nudibranch molluscs exhibit punctate luminescence (Harvey, 1952) which, according to Okada and Baba (1938) is intracellular mostly in the dermal papillae. There is, however, an additional extracellular mucous secretion. Some of the luminous pelagic dibranchiate cephalopods have developed several luminescent systems. In *Vampyro-*

teuthis three types of organs occur: those that can be occluded by a membrane, organs composed of clusters of photogenic cells but lacking any reflector surface and simple photocytes with reflectors of connective tissue. In some of the cephalopods luminescence is due to associated bacteria or by luminous secretion. Nicol (1962) reports that one pelagic squid (*Spirula*) glows continuously while, on the other hand, another (*Watasenia*) glows intermittently. This sporadicity is controlled by means of chromatophores that can expand to occlude the photophore. According to Bayer and Meyer-Arendt (1959) the pigments in these chromatophores absorb maximally in the spectral region of the emission of the photophores. In other forms the photophores themselves are innervated and are apparently under neural control.

The genus *Pyrosoma*, in the subphylum Tunicata has been exhaustively investigated (Nicol, 1958b). Luminescence is discontinuous and traverses through the zooids of the colony in waves. There is no apparent neural connection between the zooids. Symbiotic bacteria had been assumed to be the source of emission; however, Nicol presents good evidence of the presence of photogenic paraplasmic inclusions which may be similar to the scintillons reported in the dinoflagellates by De Sa *et al.* (1963).

Among the vertebrates the class Pisces presents the most formidable display of luminescence. Reviews of this subject have been presented by Haneda (1950, 1955), Harvey (1952, 1957b), Marshall (1954), Nicol (1962), Boden and Kampa (1964). Most luminescent teleosts are bathypelagic and widespread in distribution. Their photophores are diversiform. Both bacterial symbionts and intrinsic luminescent systems are utilized. The photocytes are, with very few exceptions situated dorsally, ventrally or internally (on the tongue, cheek or mouth-roof) or on anteriorly-directed, tasselated appendages (Bertelsen, 1943; Waterman, 1939). Those forms that employ bacterial symbionts, which glow continually, have developed numerous means to control light emission. The bacteria are contained in glandular structures that can be occluded by chromatophores with screening pigments, by opaque membranes, or can be rotated to face an optically dense surface. Harvey (1957b) lists eight families and one sub-order that employ bacterial symbionts. Probably the most elaborate endosymbiotic, bacterial system is that of the Leiognathid (pony) fish described by Hastings and Mitchell (1971), Hastings (1971). They have demonstrated that a glandular organ containing continuously luminescent bacteria surrounds the oesophagus and communicates with it via paired ducts. The gut tract of the fish loops into the swim bladder at the site of the photocyte. The organ, therefore, shines directly into the swim bladder, but it is vested with an occluding membrane. To complicate this optical system further the swim bladder is lined internally with highly reflecting guanine crystals. The ventral surface of the swim bladder is only partially reflective but has attached to it specialized, translucent, lenticular cells. Thus light from a relatively small source is magnified and diffused over most of the ventral surface of the body.

This confirms Haneda's (1950) observations on 15 species of Leiognathids in which he noted essentially similar mechanisms.

Many of the luminous teleosts do not depend on bacteria but sport self-luminous, innervated photophores. These are generally arranged on the head and along the body wall with the cephalic photophores being innervated by the cranial nerves and those along the body by the ventral root of the spinal nerve of the somite in which they are lodged. Extracellular luminescent secretions have been observed by Nicol (1958b) in the genus *Searsia*. He noted clouds of bright, blue-green, luminescent particles discharged from the post-cleithral process (post-clavicular, or shoulder, luminescent organ) of living animals. The luminous glands are dermal and this is the only instance of active secretion in teleosts. An extremely comprehensive bibliography on the bioluminescence of fishes has been published by Anctil (1971). This authoritative compendium is not repeated in this review, except as a reference, but it is undoubtedly the most up-to-date source of information on teleost luminescence.

Anctil (1972) also reports on the stimulation of bioluminescence in various lanternfishes (Myctophidae). In this work injections and applications of hydrogen peroxide, peroxidase and epinephrine were used as well as electrical stimulation of the spinal cord and of isolated photophores. There seems to be little correlation in the luminescent response in this group in the stages of their phylogeny. It does appear, however, that any one species may, possibly, employ more than one biochemical system since the cephalic organs, serial organs and caudal organs will respond differentially to similar stimuli.

IV. Visual Observations

The first records of bioluminescence were, naturally, of surface displays and stretch from Pliny to Boyle (1672) through Darwin (1833) to those of the more recent investigators such as Hardy (1956) and the monumental works of Harvey (1952, 1956). Johnson (1966) also notes observations by Bacon, Descartes, Hooke, Redi, Malpighi, Franklin, Spallanzani, Priestley, Davy, Faraday, Ehrenberg, Liebig, Pasteur, Lankester, Pfluger, Beijerinck, Dubois, Molisch and Kluyver, a fairly distinguished group. An interesting report on the classification of surface luminescent phenomena is that of Turner (1965). He also goes into the geographical distribution of surface luminescence which will be considered later in this review. However, at this point his classification of marine bioluminescent phenomena deserves some discussion.

TABLE 2. Suggested classification of marine bioluminescent phenomena

Site	Organisms	Stimulus	Appearance
1. Sea (1)	Bacteria(?)		"White water"
Sea (2)	Dinoflagellates	1. Mechanical	1. Apparently constant
	Copepods ⎫		illumination
	Ostracods etc. ⎭		(a) Extended bands
			(b) Blooms over
			large areas
			(c) Limited patches
			2. Flashing patches
			3. Fluctuating patches
			4. Disturbed water
			luminescence
		2. Seismic	1. Erupting lumines-
			cence
			2. Phosphorescent
			wheels
		3. Photic	Light-stimulated phos-
			phorescence
		4. Miscellaneous	Travelling luminescence
Sea (3)	Jellyfish ⎫		Luminescence of larger
	Ctenophores ⎬		planktonic organisms
	Pyrosoma etc. ⎭		
Sea (4)	Fish ⎫		Luminous nektonic
	Squid ⎭		animals
2. Air			Aerial luminescence

"White water" or "milky sea" is a uniform luminescence over large areas and its intensity is not exaggerated by perturbation. Peko (1954) attributes it to concentrations of coccolithophores. Minnaert (1954) and Dahlgren (1916) consider bacterial origin. The organisms responsible for this phenomenon have not been definitely identified. In his table Turner lists microplanktonic luminescent stimuli in four groups: mechanical, seismic, photic and unknown (Table 2). We think the first two stimuli are both perturbations, the third is a sympathetic reaction and agree that the fourth is unknown. We, too, are baffled about the origin of aerial marine luminescence. Rodewald (1954) and Tarasov

(1956) both speculate this may be spindrift of bacteria or dinoflagellates. While it is, apparently, true that such aerial luminescence occurs only above the luminescent wheels in the ocean, no samples have been taken, and the origin of these displays remains theoretical.

Submarine, visual observations of luminescence, because of their rarity, are really of greater interest than those of surface displays. Beebe (1934) was the first to describe luminescence at depth, observed from his suspended bathysphere. He was followed by Barham (personal communication) and Piccard and Dietz (1957) using the non-suspended submersible *Trieste*. Other submariners such as Monod (1954) and Bernard (1955) were primarily interested in the vertical distribution of plankton and made only brief notifications of luminescence from the bathyscaph F.N.R.S. III. Rechnitzer (1962) made continuous observations of bioluminescence during several ascending phases of dives by the *Trieste*, one of which was to the maximum known depth of the ocean, 35 800 ft in the Challenger Deep. He reports identifications of a number of luminescent organisms such as annelids, several types of crustaceans, molluscs and siphonophores. In his opinion the passage of the *Trieste* did not cause any agitation leading to increased emission. When passing through concentrations of luminescent flashing he turned on the searchlights but was usually able to see only masses of suspended matter and concluded that the flashing was due to the activity of protozoans and protophytes. He observed no spontaneous luminescence in the benthic fauna although many of these are known to be capable of it.

A. INSTRUMENTAL OBSERVATIONS

Instrumental scrutiny of luminescence is undoubtedly more objective than visual observations although it has the serious drawback that no, or only dubious, identification of the organisms under surveillance is possible. The objectivity achieved by instruments, despite their imperfections, is desirable since the more flamboyant displays of luminescence may bias the observations of even experienced oceanographers. However, as emphasized by Clarke and Denton (1962), in spite of the development of the multiplier phototube and its consequent refinements, no single instrument exists to satisfy the demands of biologists for understanding all the optical properties of the sea.

One specialized use of the phototube is demonstrated by Clarke and his colleagues (Clarke and Wertheim, 1956; Clarke and Backus, 1956; Clarke and Hubbard, 1959). This group developed an instrument of considerable sensitivity in order to detect luminescent strata at great depths. In the Atlantic (Clarke and Wertheim, 1956) this instrument initially recorded flashes of luminescence at about 200 m and between this depth and 300 m the flashes were 2 to 5 times greater than the downward irradiance. Below 360 m there was as much light at night as during the day and some of the flashes were 1000 times brighter than the background level. Clarke and Kelly (1965) using a photometer equipped with an agitator have demonstrated a considerable increase in luminescence at night, particularly in the upper layers, and contribute this to photosynthetic, epipelagic organisms reinforced by migratory, mesopelagic recruits. They conclude that the flashes they observed at depths to as much as 2000 m as due to much larger, nektonic animals. Their instruments require multiconductor cable between the undersea and deck units and this has the disadvantage of bulk, weight and possibility of leakage but the great advantage of continuous, simultaneous, monitoring of several parameters. They appear not to have suffered from any of the potential disadvantages.

The Sandia Laboratories of the Atomic Energy Commission have developed an extremely complicated radiometer (Fig. 1; Miyoshi, 1972). From a pre-print, kindly supplied by the author and his colleagues, we understand the instrument employs reflective optics over a large face as a collector and internal refractive optics, by means of aluminum mirrors, to concentrate the collected light on a very sensitive phototube. The threshold sensitivity of the system is rated at 10^{-17} W cm^{-2}, which is considerably greater than any hitherto reported for marine work. The output from the undersea unit is fed directly into computers and the results can be scanned momently. The system is extremely versatile in that, by command from the deck, the underwater unit can be changed, by occluding filters, from a radiance meter to an irradiance meter. By means of a filter wheel it is possible to obtain spectral information. At this stage of development no information is available of the type of filters in use.

Preliminary results obtained by this instrument in the San Diego Trough indicate that at levels of low bioluminescent activity the count-rate did not, at any time, approach the measured dark noise of the phototube. Measurements were also made of "moderate bioluminescent activity" defined as 75% of observation time and "high transient activity". With this system, by statistical analysis, it is possible to discriminate against bioluminescent activity except in the high transient activity category where quiet periods are virtually nonexistent and no reliable base-line can be established.

In spite of the scientific merits of this system it is doubtful that it will become a workhorse on any but large oceanographic vessels with elaborate deck-handling capabilities. The undersea system requires a

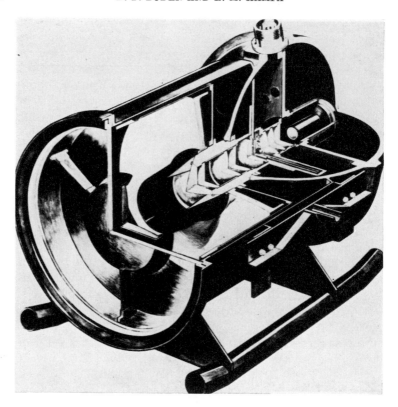

Fig. 1. Cross-section of the radiometer developed by the Sandia Laboratories illustrating the reflector collector surface and collimating refracting system.

30-ft aluminum vertical platform with 15-ft aligning fins and a 600-lb. anchor for added stability. The total water-weight of the system is 1200 lb. with an additional 1300 lb. if 2000 ft of cable are paid out. The electromechanical umbilicus between the underwater and deck units consists of a bundle of 12 no. 14 wires, 8 no. 22 wires, 4 no. 18 wires, and an RG-58 coax surrounded by 4 braids of stainless-steel wires. This has a tensile strength of about 20 000 lb. but, because of the sheer bulk of the cable, it necessitates a formidable winch.

In a search for a simplified sensor that would obviate the necessity for the use of large research vessels (Breslau, 1959) developed a portable bathyphotometer that requires no outside electric power source. The instrument has a threshold sensitivity of 10^{-7} μW cm^{-2}. It was used in Phosphorescent Bay, Puerto Rico and the Gulf of Naples (Clarke and Breslau, 1960) lashed to the gunnel of the ship, pointing downward, to measure the luminescence in the bow-wave. In the Bay in Puerto Rico the luminescence level proved to be 100 times greater than in adjacent waters, and attained a value of 3×10^{-2} μW cm^{-2}. This was attributed mainly to the dinoflagellate *Pyrodinium bahamense*. Clarke thinks that, after definitive background studies, it may be possible to determine the identity of the organisms responsible for the flashes by the recorded shape of the flashes. We concur with this hypothesis and regret that nobody seems to have had the opportunity to make the adequate background investigations.

The advent of sonar and the subsequent discovery of sonic-scattering layers and partial elucidation of their composition has added a new dimension to oceanographic observations of bioluminescence.

Clarke (1933) showed a correlation between the diurnal vertical migration of planktonic fauna and changes in submarine irradiation by means of a photovoltaic cell. Kampa and Boden (1954), using a multiplier phototube, found that one of the sonic-scattering layers in the San Diego Trough was closely related to an isolume during its twilight migrations. Clarke and Backus (1956) demonstrated that a sonic-scattering layer off New York split into two components during its twilight ascent toward the surface and that the recorded pattern of luminescent flashes differed in intensity and frequency with depth, time and geographical position (slope vs. coastal waters) and was apparently affected by the thermocline. Such a split had been demonstrated by Kampa and Boden (1954) but no bioluminescence observations were made. In the Mediterranean Boden and Kampa (1958) showed that a luminescent scattering layer was halted in its migration by a sharp thermocline. The first spectral evaluation of irradiance at the depth of scattering layers was made by Boden and Kampa (1957) in Hawaiian waters. Another boundary to the vertical migration of a sonic-scattering layer attributed to the crustacean, *E. pacifica*, has been attributed to an abrupt oxygen deficient layer that inhibits their downward twilight migration.

Mauchline (1960) found that the euphausiid, *M. norvegica*, luminesces brightly and spontaneously during the mating season when they congregate after a summer of indifferent dispersal. This was confirmed by David and Conover (1961) but not by Hardy and Kay (1964). The latter authors based their conclusion on laboratory work and maintain a reservation that under laboratory conditions spontaneity of luminescence may be depressed.

Discussion of these observations is undertaken in Boden and Kampa (1964) and is not reiterated here.

An effort to examine luminescence, *in situ*, that was not exaggerated by perturbation has been undertaken by means of an instrument designed by Snodgrass and described by Boden *et al.* (1965). The instrument consists of interconnected dual radiance meters mounted on a rack and surveying a common remote purview. The meters are battery operated and do not require any electrical connection with the deck. Each meter is capable of registering any flashes in its cone of vision. However, only flashes recorded coincidentally in their joint vision are considered to be spontaneous. Some reservations of this premise are retained. Several observations of bioluminescence in the Atlantic are reported by Boden (1969). In one of the experiments described three racks of instruments were lowered on hydrographic wire to positions below, in and above an observed scattering layer. The flashes recorded were of lesser magnitude than those recorded by a single irradiance meter, which must cause turbulence in its passage during raising or lowering or by surge while being held at a fixed wire-out position. There was, however, a great deal more luminescent activity in the layer than there was below or above it. Analysis of recordings of single flashes revealed a frequency within them and we hypothesize this may be a specific property and this supports Clarke's hypothesis.

B. GEOGRAPHICAL OBSERVATIONS

Several estimates of the relative geographical distribution of luminescent phenomena in different parts of the world have been made. These are somewhat subjective. Smith (1931) presents estimates based on data extracted from reports of merchant vessels comprising the Voluntary Observation Fleet. Turner (1966) points out that such estimates are skewed since they are made mostly by ships in densely travelled shipping lanes and gives evidence of this based on the distribution by Marsden squares of about 4500 reports from the Meteorological Office Phenomena Index for 1854–1956. He has formulated a coefficient to eliminate, as far as possible, such distortion. He concludes that luminescent activity is particularly high in the Arabian Sea and relatively less, but still reasonable in the tropical Atlantic. According to Smith (1931) this is probably seasonal. Certain areas of high activity such as the west coast of Africa and off Argentina are probably due to upwelling or current convergences. This is also true of the California Current where luminescence is often concentrated in local gyres.

Bityukov (1969) reports on traverses from the Balearic Sea to the Azores and from the North American coast to the Black Sea, during which a photometer was towed in the wake of the ship. He found the areas of least bioluminescent activity were in Sargasso Sea, the Canary currents along the continent and the Aegean Sea. The greatest activity was observed in the Straits of Gibraltar and east of the Azores in which area the recorded intensity was 600 times greater than that in the Gulf Stream and 5 times as high as in the Aegean Sea. He also reports apparently seasonal differences in activity between May and August.

V. BIOLUMINESCENCE AND VISION

A variety of hypotheses have been put forward to explain the advantages of bioluminescence to the oceanic animals exhibiting it. For pelagic animals these suggestions range widely—provision of greater illumination at depth, attraction or confusion of potential food organisms, camouflage, warning, intraspecific recognition to facilitate swarming and mating, and in photoregulated diurnal migrants a reference with which to assess changes in photoenvironment during migration. Each of these implies the existence of light-sensing systems in the animals themselves or in their neighbors capable of perceiving the emitted light.

For such an emitter-receptor combination to function, there must be a degree of compatibility (1) between the spectral characteristics of the two systems, and (2) between the intensity of the emission and the threshold level of the receptor.

Data to test the first of these criteria in marine animals are already available. The literature on the spectral distributions of bioluminescence in marine animals is summarized by Nicol (1958b, 1962, 1967) and by Boden and Kampa (1964). Summaries of the literature on spectral sensitivities of the eyes of marine animals, obtained either from visual pigment analyses or from electrophysiological studies, have been compiled by Crescitelli (1958), Dartnall (1964), Dartnall and Lythgoe (1965), and Goldsmith and Fernández (1966, 1968).

Comparison of the data for the two systems shows that there is a good correspondence between the bioluminescent emission spectra and the visual spectral sensitivities of animals living at similar depths in the water column. Epipelagic animals and those inhabiting shallower coastal waters exhibit blue to blue-green bioluminescence, and their eyes are most sensitive to blue-green light. In the meso- and bathypelagic zones, the bioluminescence is markedly blue, and the eyes of the animals have a pronounced sensitivity to blue light. Spectra for light emission and visual sensitivity are not necessarily identical, even within a single species, but the overlap is great enough that, if color were the only parameter to be considered, it could be said that any eyed animal in an oceanic community is capable of perceiving bioluminescence emitted by any other organism it may encounter.

TABLE 3. Measurements of intensity of luminescence of some pelagic animals

Group	Species	Stimulus	Radiant flux, μW cm^{-2} receptor surface at a distance of 1 m. Turner (1965) unpublished Ms.	Source
Dinoflagellata	*Noctiluca miliaris		0.016×10^{-9} μW	Nicol, 1960
Radiolaria	Cytocladus major and	Electrical†	0.6×10^{-9} μW	Nicol, 1958b
	Aulosphaera triodon	Electrical†	5.3×10^{-9} μW	Nicol, 1958b
Hydromedusae	Colobonema sericeum	Electrical†	7.2×10^{-9} μW	Nicol, 1958b
	Colobonema sericeum	Electrical†	9.5×10^{-9} μW	Nicol, 1958b
	Crossota alba	Electrical†	0.4×10^{-9} μW	Nicol, 1958b
	Aeginura grimaldii	Electrical†	9.3×10^{-9} μW	Nicol, 1958b
	Aeginura grimaldii	a.c.	112.5×10^{-9} μW	Clarke et al., 1962
	Aeginura grimaldii	a.c.	292.5×10^{-9} μW	Clarke et al., 1962
Siphonophora	Vogtia spinosa	Electrical†	13.7×10^{-9} μW	Nicol, 1958b
	Vogtia spinosa	Electrical†	320.9×10^{-9} μW	Nicol, 1958b
	V. glabra	Electrical†	120×10^{-9} μW	Nicol, 1958b
	Rosacea plicata	Electrical†	2.4×10^{-9} μW	Nicol, 1958b
	Rosacea plicata	Electrical†	13.7×10^{-9} μW	Nicol, 1958b
	Hippopodius hippopus	Electrical†	2.6×10^{-9} μW	Nicol, 1958b
	Hippopodius hippopus	Electrical†	4.2×10^{-9} μW	Nicol, 1958b
Scyphomedusae	Atolla wyvillei	Electrical†	0.3×10^{-9} μW	Nicol, 1958b
	Atolla wyvillei	Electrical†	199.9×10^{-9} μW	Nicol, 1958b
	Periphylla periphylla	a.c.	29.25×10^{-9} μW	Clarke et al., 1962
	Periphylla periphylla	a.c.	67.5×10^{-9} μW	Clarke et al., 1962

Group	Species	Stimulus	Light output	Reference
Ctenophora	*Beroë ovata	Electrical†	$16 \cdot 95 \times 10^{-9}$ µW	Nicol, 1958b
	*Beroë ovata	Electrical†	$8538 \cdot 5 \times 10^{-9}$ µW	Nicol, 1958b
	*Mnemiopsis leidyi	Transformer discharge	$12\,500 \times 10^{-9}$ µW	Clarke and Backus, 1956
	*Mnemiopsis leidyi	Transformer discharge	$18\,750 \times 10^{-9}$ µW	Clarke and Backus, 1956
Euphausiacea	Euphausia pacifica	NH₄OH (fatal)	160×10^{-9} µW	Kampa and Boden, 1957
	Euphausia pacifica	NH₄OH (fatal)	200×10^{-9} µW	Kampa and Boden, 1957
	Meganyctiphanes norvegicus	a.c.	9×10^{-9} µW	Clarke et al., 1962
	Meganyctiphanes norvegicus	a.c.	$29 \cdot 25 \times 10^{-9}$ µW	Clarke et al., 1962
	Meganyctiphanes norvegicus	Photic	$1 \cdot 3 \times 10^{-9}$ µW	Kay, 1965
	Meganyctiphanes norvegicus	Photic	200×10^{-9} µW	Kay, 1965
	Meganyctiphanes norvegicus	a.c.	3240×10^{-9} µW	David and Conover, 1961
Decapoda	Acanthephyra purpurea	Electrical†	$1 \cdot 9 \times 10^{-9}$ µW	Nicol, 1958b
	Acanthephyra purpurea	Electrical†	$8 \cdot 2 \times 10^{-9}$ µW	Nicol, 1958b
	A. pelagica	a.c.	2520×10^{-9} µW	Clarke et al., 1962
Copepoda	*Metridia lucens	a.c.	$38\,880 \times 10^{-9}$ µW	David and Conover, 1961
	*Metridia lucens	Mechanical	45×10^{-9} µW	Clarke et al., 1962
	*Metridia lucens	Mechanical	3240×10^{-9} µW	Clarke et al., 1962
	*Metridia lucens	a.c.	$4 \cdot 5 \times 10^{-9}$ µW	Clarke et al., 1962
	*Metridia lucens	a.c.	$173 \cdot 25 \times 10^{-9}$ µW	Clarke et al., 1962
	*Metridia lucens	Condenser shocks	$29 \cdot 25 \times 10^{-9}$ µW	Clarke et al., 1962
	*Metridia lucens	Condenser shocks	$580 \cdot 5 \times 10^{-9}$ µW	Clarke et al., 1962
	*Other copepods (8 spp.)	a.c. or condenser shocks	$2 \cdot 25 \times 10^{-9}$ µW	Clarke et al., 1962
	*Other copepods (8 spp.)	a.c. or condenser shocks	2115×10^{-9} µW	Clarke et al., 1962

TABLE 3 (continued)

Groups	Species	Stimulus	Radiant flux, μW cm^2 receptor surface at a distance of 1 m. Turner (1965) unpublished Ms.	Source
Tunicata	*Pyrosoma atlantica	Electrical†	$13\cdot8 \times 10^{-9}$ μW	Nicol, 1958b
	*Pyrosoma atlantica	Electrical†	$1361\cdot5 \times 10^{-9}$ μW	Nicol, 1958b
	*Pyrosoma atlantica	NH_4OH (fatal)	800×10^{-9} μW	Kampa and Boden, 1957
	*Pyrosoma atlantica	NH_4OH (fatal)	4000×10^{-9} μW	Kampa and Boden, 1957
Teleostei	Searsia schnakenbecki	Electrical†	150×10^{-9} μW	Nicol, 1958b
	Searsia schnakenbecki	Electrical†	430×10^{-9} μW	Nicol, 1958b
	S. koefoedi	Electrical†	2117×10^{-9} μW	Nicol, 1958b
	S. koefoedi	Electrical†	2808×10^{-9} μW	Nicol, 1958b
	Myctophum punctatum	Electrical†	$0\cdot925 \times 10^{-9}$ μW	Nicol, 1958b
	Myctophum punctatum	Electrical†	$52\cdot345 \times 10^{-9}$ μW	Nicol, 1958b
	Myctophum punctatum	a.c.	$19\cdot6 \times 10^{-9}$ μW	Clarke et al., 1962

* Known or potential contributors to visible surface phosphorescence.

† Nicol used either condenser shocks or square wave pulses as stimuli, but did not distinguish between the results. The former induce, at least in copepods, a simpler but stronger response (Clarke et al., 1962). The figures above were based mainly on data and calculations presented by Nicol (1958b).

The second criterion, compatibility between intensities of bio-luminescent emission and the intensities which the animal itself and its neighbors can see, is less readily satisfied from published laboratory findings. In Table 3 (Turner 1965) lists values for the observed intensities of luminescence in members of the eleven groups of marine organisms probably responsible for most of the biologically produced light in the sea. All of the results, however, were obtained from specimens, or parts thereof, removed from the normal habitats, subjected to confinement and then affronted by quite unnatural stimuli. This is no criticism of the researchers, for there is no other reliable recourse open to them at present. We (Kampa and Boden, 1957) were the worst offenders in taking our measurements from chemically induced death glows. It is reassuring, however, to note that our results with one species of euphausiid are within the same order of magnitude as those of Kay (1965) who used somewhat gentler photic stimuli to elicit luminescence in another euphausiids species.

Information on the absolute visual thresholds of oceanic animals is virtually nil. Physiologists, quite understandably, are inclined to select experimental animals that are readily available, thrive in captivity, and require at most simple additives such as bubbled O_2 (see Wald, 1968, regarding *Homarus americanus*) to survive an experiment. The visual electrophysiology that has been done with oceanic animals has been largely restricted to electroretinograms made either with wick electrodes on the surface of the eye or micro-electrodes inserted into the eye, and, as Wald (1968) so aptly puts it, "The erg is at best a coarse high-level response".

For this reason calculations of the possible distances at which luminescent flashes can be seen and arguments as to whether animals can perceive their own and others' luminescence have been based on numbers for the threshold limits of the human eye when exposed to various types of stimuli in various states of light or dark adaptation (Nicol, 1958b, 1962; Clarke and Denton, 1962).

Perhaps the best evidence for the ability of an oceanic animal to perceive bioluminescence is contained in the simple experiment described for the euphausiid *Meganyctiphanes norvegica* by Mauchline (1960). He found that by stimulating one specimen to luminesce, others in adjacent containers were stimulated to flash in response. Such a reaction is familiar to anyone who works in the dark with luminescent animals, particularly euphausiids. To have one animal glow in the middle of an experiment and set off all the others in the room is frustrating to say the least.

In addition to these laboratory observations, certain *in situ* studies of animals in their own habitats should be considered here. We are now convinced that there exists in all of the world's oceans a community of animals that can be detected vicariously by sonic techniques, and that this community seeks to remain within a particular photo-environment regardless of depth. We are also convinced that euphausiids are a part of this community—certainly in every location from which we have been able to obtain controlled biological collections.

The vertical distribution of this sonic-scattering community has been shown to be closely aligned with the depth at which the incoming irradiance from sun, moon, stars and sky is approximately $1 \times 10^{-4} \, \mu\text{W}$ $\text{cm}^{-2} \, \text{nm}^{-1}$ at 470 nm (Kampa and Boden, 1957; Boden and Kampa, 1967; Kampa, 1970). The light measurements were obtained just above the top of the community as shown by the echo sounder records taken simultaneously. The vertical thickness of the community at midday is usually more than 60 m, and from our midday attenuation curves, we know that the irradiance level at the bottom of the community is about an order of magnitude lower than that at the top.

Euphausiids, like other animals in this sonic-scattering layer must be able to distinguish and avoid ambient irradiance levels greater than 10^{-4} and less than $10^{-5} \, \mu\text{W} \, \text{cm}^{-2} \, \text{nm}^{-1}$ at 470 nm in order to remain within the community during the long midday periods when the incoming irradiance does not vary appreciably. Such adherence is presumably a part of the animals' survival pattern.

All of these sonic-scattering layer euphausiids are bioluminescent, and the luminescence of one of them, *Euphausia pacifica*, yielded an irradiance level of $2 \times 10^{-3} \, \mu\text{W} \, \text{cm}^{-2} \, \text{nm}^{-1}$ at 470 nm, at a distance of 1 cm, in the laboratory (Kampa and Boden, 1957). We have, then, a concrete value for the lowest level of light known to be perceptible to euphausiids, as well as an estimate of their luminescent potential.

To physiologists, light values are more meaningful when expressed as quanta, for it is on such particulate bundles of energy that visual molecular excitation is based. Therefore, from Planck's equation,

$$Q = h\nu,$$

where Q is the energy of a single quantum in ergs, h is Planck's constant $(6 \cdot 6 \times 10^{-27} \, \text{erg} \times \text{s})$, and ν is the frequency of the light,

$$1 \text{ quantum at } 470 \text{ nm} = 4 \cdot 25 \times 10^{-12} \text{ ergs}.$$

The observed irradiance at the bottom of the animal community, then is equal to $2 \cdot 35 \times 10^{7}$ quanta $\text{cm}^{-2} \, \text{s}^{-1}$ at 470 nm.

We have measured the corneal facets in fresh preparations from the eyes of *E. pacifica*; the diameter of the facet is about 60 μ, and, from this, the area is roughly $2 \cdot 8 \times 10^{-5} \, \text{cm}^2$. These facets are the outer surfaces of the visual units, the ommatidia, of the compound eye. An extension of the calculation shows that at the bottom of the community transmitted light from sun and sky provide individual ommatidia with approximately 658 quanta s^{-1} at 470 nm.

Similar calculations can be made for the possible effect of the luminescent glow of one *E. pacifica* upon the eye of another. If one assumes the optimal orientation of glowing and perceiving animals, the glow should deliver $2 \times 10^{-3}\ \mu W\ cm^{-2}$ at 470 nm if the luminescent emitter is 1 cm distant from the eye of the receiver. This is equivalent to $4 \cdot 7 \times 10^{9}$ quanta cm^{-2} s^{-1} at 470 nm or $1 \cdot 32 \times 10^{5}$ quanta/ommatidium s^{-1} at 470 nm.

The distance of 1 cm is absurd, of course, and at a more reasonable distance of 10 cm, the value would be reduced to 1320 quanta/ommatidium s^{-1} at 470 nm. At 50 cm, the level of the luminescent glow would be reduced to 528 quanta/ommatidium s^{-1} at 470 nm, a value of the same order of magnitude as that provided by transmitted sun and sky light.

It seems reasonable to conclude that euphausiids can and probably do utilize each others' luminescent signals up to distances approaching 50 cm. No distinction can be made between the wide-field irradiance source provided by transmitted sun and sky light and the narrow-field source provided by the glow of another animal until the visual capabilities of euphausiids have been more thoroughly examined. It is interesting to note, however, that even at a distance of 1 m, the glow of a single euphausiid delivers some 13 quanta/ommatidium s^{-1} at 470 nm—a value well within the range required to elicit a threshold response in other visual systems.

General considerations

Recommendations on the observations of bioluminescent phenomena are made by Turner (1966). He stresses the need of attention to the date, time, duration, position (geographical), course of vessel, wind, sea-state, weather, visibility, extent of luminescence as seen from ship, colour and intensity, quality of luminescence (defined in Table 1), movement or spread, possible stimuli, concurrent phenomena such as odours, change in sea-temperature or state.

Tarasov and Gitel'zon (1961) advocate more objective scrutiny by means of undersea instruments and laboratory experiments. They raise the question as to whether the ability to luminesce is an independent function or reliable on symbionts such as bacteria. This appears to have been resolved largely by the work of Hastings and his colleagues. Tarasov and Gitel'zon's commendation of future research in

20

this field deserves serious consideration. They stress the need for information on the spectra of luminescence of as many species of organisms of various taxonomic groups as possible—this to be obtained with the high degree of monochromatization achievable with modern recording spectrophotometers. A wide spectral band in the visible and contiguous infrared and ultraviolet ranges should be investigated with a view to making quantitative calculations of quantum results which may assist in elucidating the physical-chemical nature of the reaction. The bio-energetics of the system is of great interest to cellular biologists. They also think that an investigation should be undertaken of luminescent systems ranging from general bacterial luminescence up to complicated reflective and lenticular organs. Emphasis is placed on the need for recording the spectra of reflections from the pigments of the epidermis of animals such as fish, cephalopods, crustaceans, etc., and to compare them with the spectra of ambient radiation.

Much of this work has been embarked upon but in such a divergent eld it is obvious that much remains to be done. It is quite apparent that the co-operation they urge between biophysicists, biochemists, oceanographers, planktologists, ichthyologists, benthologists, physicists and all contingent disciplines is essential.

REFERENCES

Anctil, M. (1971). *Sci. Rep. of the Yokosuka City Mus.*, 18, 1–14.

Anctil, M. (1972). *Can. J. Zool.*, 50, (2), 233–237.

Bayer, M. and Meyer-Arendt, J. (1959). *Science, N.Y.*, 129, 644 only.

Beebe, W. (1934). "Half Mile Down". Harcourt Brace and Co., New York, 344 pp.

Bernard, F. (1955). *Bull. Inst. Oceanogr. Monaco*, 1063, 16 pp.

Bertelson, E. (1943). *Vidensk. Medd. Naturh. Foren. K.bh.*, 107, 185–206.

Bityukov, E. P. (1969). *Akad. Nauk SSSR, Okeanologiia*, 11, (1), 127–133.

Boden, B. P. (1969). *J. Mar. Biol. Ass. U.K.*, 49, 669–682.

Boden, B. P. and Kampa, E. M. (1957). *Pac. Sci.*, 11, 229–235.

Boden, B. P. and Kampa, E. M. (1958). *Vie et Milieu*, 9(1), 1–10.

Boden, B. P. and Kampa, E. M. (1964). *Oceanogr. Mar. Biol. Ann. Rev.*, 2, 341–371.

Boden, B. P. and Kampa, E. M. (1967). *Symp. Zool. Soc. Lond.*, 19, 15–26.

Boden, B. P., Kampa, E. M. and Snodgrass, J. (1965). *Nature (London)*, 208 (5015), 1078–1080.

Boyle, R. (1672). *Phil. Trans. Roy. Soc. London*, 7, 5108–5116.

Breslau, L. R. (1959). *Woods Hole Oceanogr. Instn. Techn. Rep.*, No. 59–28 (Unpublished manuscript), pp. 1–6.

Clarke, G. L. (1933). *Biol. Bull. Woods Hole*, 65, 402–436.

Clarke, G. L. and Backus, R. H. (1956). *Deep-Sea Res.*, 4, 1–14.

Clarke, G. L. and Breslau, L. R. (1960). *Bull. Inst. Oceanogr. Monaco*, 1171, 32 pp.

Clarke, G. L. and Denton, E. J. (1962). *In* "The Sea", (M. N. Hill, ed.), Vol. 1, pp. 456–458. Interscience, New York, London.

Clarke, G. L. and Hubbard, C. J. (1959). *Limnol. Oceanogr.*, **4**, 163–180.

Clarke, G. L. and Kelly, M. G. (1965). *Limnol. Oceanogr.*, **10**, 54–66.

Clarke, G. L. and Wertheim, G. K. (1956). *Deep-Sea Res.*, **3**, 189–205.

Clarke, G. L., Conover, R. J., David, C. N. and Nicol, J. A. C. (1962). *J. Mar. Biol. Ass. U.K.*, **42**, 541–564.

Cormier, M. J. and Strehler, B. L. (1953). *J. Am. Chem. Soc.*, **75**, 4864 only.

Cormier, M. J. and Totter, J. (1964). Bioluminescence. *Ann. Rev. Biochem.*, **33**, 431–458.

Crescitelli, F. (1958). *Ann. N.Y. Acad. Sci.*, **74**, Art. 2, 230–255.

Dahlgren, U. (1916). *J. Franklin Inst.*, **181**, 243–261.

Dartnall, H. J. A. (1964). *Ann. Roy. Coll. Surg. Engl.*, **35**, 131–150.

Dartnall, H. J. A. and Lythgoe, J. N. (1965). *J. Vis. Res.*, **5**, Nos. 3/4, 81–100.

Darwin, C. (1833). "Journal of researches into the natural history and geology of the countries visited during the voyage of H.M.S. *Beagle* round the world, under the command of Capt. Fitz Roy, R.N." New edition, 519 pp.

Davenport, D. and Nicol, J. A. C. (1955). *Proc. Roy. Soc. London, Ser. B*, **144**, 399–411.

David, C. N. and Conover, R. J. (1961). *Biol. Bull., Woods Hole*, **121**, 92–107.

Dennel, R. (1940). *Discovery Rept.*, **22**, 307–382.

Dennel, R. (1954). *J. Linn. Soc. (Zool.)*, **42**, 393–406.

De Sa, R., Hastings, J. W. and Vatter, A. E. (1963). *Science, N.Y.*, **141**, 1269–1270.

Dubois, R. (1887). *C.R. Soc. Biol.*, **39**, 564–566.

Eckert, R. (1966a). *Science, N.Y.*, **151**, 349–352.

Eckert, R. (1966b). *In* "Bioluminescence in Progress" (F. H. Johnson and Y. Haneda, eds.), pp. 269–300. Princeton University Press, Princeton, New Jersey.

Galloway, T. W. and Welch, P. S. (1911). *Trans. Amer. Microsc. Soc.*, **30**, 13–39.

Goldsmith, T. H. and Fernández, H. R. (1966). *In* "Proc. Int. Symp. on the Functional Organization of the Compound Eye", pp. 125–143. Pergamon Press, Oxford and New York.

Goldsmith, T. H. and Fernández, H. R. (1968). *Z. Vergl. Physiol.*, **60**, 156–175.

Gurney, R. (1942). "Larvae of Decapod Crustacea", pp. 306. Ray Soc., Bernard Quaritch Ltd.

Haneda, Y. (1950). *Pac. Sci.*, **4**, 214–227.

Haneda, Y. (1955). *In* "The Luminescence of Biological Systems" (F. H. Johnson, ed.), pp. 335–386. Amer. Ass. Advanc. Sci., Washington, D.C.

Hardy, A. C. (1956). "The Open Sea", pp. 335. Collins, London.

Hardy, A. C. and Kay, R. H. (1964). *J. Mar. Biol. Ass. U.K.*, **44**, 435–484.

Hardy, M. G. (1962). *Nature, London*, **196**, 790–791.

Harvey, E. N. (1952). "Bioluminescence", pp. 649. Academic Press, New York.

Harvey, E. N. (1956). *Quart. Rev. Biol.*, **31**, 270–287.

Harvey, E. N. (1957b). *In* "The Physiology of Fishes", Vol. 2, (M. E. Brown, ed.), pp. 345–366. Academic Press, New York.

Hastings, J. W. (1966). *Cur. Top. Bioenerg.*, **1**, 113–152.

Hastings, J. W. (1968). *Ann. Rev. Biochem.*, **37**, 597–630.

Hastings, J. W. (1971). *Science, N.Y.*, **173**, 1016–1017.

Hastings, J. W. and McElroy, W. D. (1955). *In* "The Luminescence of Biological Systems" (F. H. Johnson, ed.), pp. 257–264. Amer. Ass. Advanc. Sci., Washington, D.C.

Hastings, J. W. and Mitchell, G. W. (1971). *Biol. Bull.*, **141**, 261–268.
Hastings, J. W. and Sweeney, B. M. (1957a). *J. Cell. Comp. Physiol.*, **49**, 209–226.
Hastings, J. W. and Sweeney, B. M. (1957b). *Proc. Nat. Acad. Sci. U.S.A.*, **43**, 804–811.
Hastings, J. W. and Sweeney, B.M. (1958). *Biol. Bull.*, *Woods Hole*, **115**, 440–458.
Haxo, F. T. and Sweeney, B. M. (1955). *In* "The Luminescence of Biological Systems" (F. H. Johnson, ed.), pp. 415–420. Amer. Ass. Advanc. Sci., Washington, D.C.
Johnson, F. H. (1955). *In* "The Luminescence of Biological Systems" (F. H. Johnson, ed.), pp. 265–298. Amer. Ass. Advanc. Sci., Washington, D.C.
Johnson, F. H. (1966). *In* "Bioluminescence in Progress" (F. H. Johnson and Y. Haneda, eds.), pp. 3–21. Princeton University Press, Princeton, New Jersey.
Johnson, F. H. and Haneda, Y. (eds.) (1966). "Bioluminescence in Progress." Princeton University Press, Princeton, New Jersey, 650 pp.
Johnson, F. H., Zworykin, N. and Warren, G. (1943). *J. Bacteriol.*, **46**, 167–184.
Johnson, F. H., Eyring, H. and Polissar, M. J. (1954). "The Kinetic Basis of Molecular Biology", John Wiley, New York; Chapman and Hall, London, 874 pp.
Kampa, E. M. (1970). *J. Mar. Biol. Ass. U.K.*, **50**, 397–420.
Kampa, E. M. and Boden, B. P. (1954). *Nature (London)*, **174**, 869–873.
Kampa, E. M. and Boden, B. P. (1956). *Deep-Sea Res.*, **4**, 73–92.
Kay, R. H. (1965). *Proc. Roy. Soc., London*, *B*, **162**, 365–386.
Kelly, M. G. and Katona, S. (1966). *Biol. Bull.*, **131**, 115–126.
McElroy, W. D., Hastings, J. W., Sonnenfield, V. and Coulombre, J. (1953). *Science, N.Y.*, **118**, 385–386.
Marshall, N. B. (1954). "Aspects of Deep Sea Biology", 380 pp. Philosophical Library, New York.
Mauchline, J. (1960). *Proc. Roy. Soc. Edinburgh, Ser. B.*, **67**, 141–179.
Mauchline, J. and Fisher, L. R. (1969). "The Biology of Euphausiids", *Adv. in Mar. Biol.* (F. S. Russell and C. M. Yonge, eds.), 454 pp. Academic Press, London and New York.
Meighen, E. A., Nicoli, M. Z. and Hastings, J. W. (1971). *Fed. Proc.*, **30**, 1057.
Minnaert, M. (1954). "Light and Colour in the Open Air", pp. 347–349. Dover Publications, New York.
Miyoshi, D. S. (1972). *IEEE Trans. Nucl. Sci.*, (Pers. comm.).
Monod, T. (1954). *C.R. Acad. Sci., Paris*, **238**, 1951–1953.
Nealson, K. H. and Hastings, J. W. (1972). *J. Biol. Chem.*, **247** (3), 888–894.
Nicol, J. A. C. (1958a). *J. Mar. Biol. Ass. U.K.*, **37**, 535–549.
Nicol, J. A. C. (1958b). *J. Mar. Biol. Ass. U.K.*, **37**, 705–752.
Nicol, J. A. C. (1960). "The Biology of Marine Animals", 707 pp. Pitman and Co., London.
Nicol, J. A. C. (1962). *In* "Advances in Comparative Physiology and Biochemistry" (O. Lowenstein, ed.), Vol. 1, pp. 217–273. Academic Press, New York and London.
Nicol, J. A. C. (1963). *Endeavour*, **22**, 37–41.
Nicol, J. A. C. (1967). *Symp. Zool. Soc. London*, **19**, 27–55.
Okada, Y. K. and Baba, K. (1938). *Annot. Zool. Jap.*, **17**, 276–281.
Peko, (Balls, R.) (1954). *Wild. Fishg.*, **3**, 267–270.
Piccard, J. and Dietz, R. S. (1957). *Deep-Sea Res.*, **4**, 221–229.

Rechnitzer, A. B. (1962). *USNEL Res. Rep.*, No. 1095, 63 pp. (Unpublished manuscript.)

Rodewald, M. (1954). *Mar. Obs.*, **24**, 233–234.

Seliger, H. H. and McElroy, W. D. (1968). *J. Mar. Res.*, **26**, 245–255.

Shimomura, O. and Johnson, F. H. (1966). *In* "Bioluminescence in Progress". (F. H. Johnson and Y. Haneda, eds.), pp. 495–521. Princeton University Press, Princeton, New Jersey.

Shimomura, O., Johnson, F. H. and Saiga, Y. (1962). *J. Cell Comp. Physiol.*, **59**, 223–240.

Smith, H. T. (1931). *Mar. Abstr. London*, **8**, 230–234.

Strehler, B. L. (1953). *J. Am. Chem. Soc.*, **75**, 1264 only.

Strehler, B. L. (1955). *In* "The Luminescence of Biological Systems" (F. H. Johnson, ed.), pp. 209–255. Amer. Ass. Advanc. Sci., Washington, D.C.

Strehler, B. L. and Cormier, M. J. (1954). *J. Biol. Chem.*, **211**, 213–225.

Strehler, B. L., Harvey, E. N., Chang, J. J. and Cormier, M. J. (1954). *Proc. Nat. Acad. Sci. U.S.A.*, **40**, 10–12.

Sweeney, B. M. and Hastings, J. W. (1957). *J. Cell. Comp. Physiol.*, **49**, 115–128.

Tarasov, N. (1956). *Moscow Acad. Sci. SSSR*, 1–203.

Tarasov, N. and Gitel'zon, I. I. (1961). *Biul. Okeanografischeskoy Komissii SSSR*, **8**, 75–80.

Tett, P. B. (1969). Final Rept. No. AT/992/03/R.L. *Scot. Mar. Biol. Assoc.*, 1–76.

Turner, R. J. (1965). N.I.O. Internal. Rept. No. B4 (Unpublished ms.)

Turner, R. J. (1966). *Mar. Observ.*, January, 1966, 20–29.

Yentsch, C. S. and Laird, J. C. (1968). *J. Mar. Res.*, **26**, 127–133.

Wald, G. (1968). *J. Gen. Physiol.*, **51**, 2, 125–156.

Waterman, T. H. (1939). *Bull. Mus. Comp. Zool., Harvard Univ.*, **85**, 65–94.

Yunsalus-Miguel, A., Meighen, E. A., Nicoli, M. Z., Nealson, K. H. and Hastings, J. W. (1972). *J. Biol. Chem.*, **247**, 398–404.

Author Index

(The numbers in *italics* refer to pages in the References at the end of each Chapter.)

A

Abraham, G., 285, *288*
AGARD, 122, *133*
Allen, J. L., 424, *442*
Allen, M. B., 379, *387*
Allison, L., 302, *316*
Ambarzumian, V. A., 123, 125, *133*, 185, *217*
Anctil, M., 452, *466*
Armstrong, F. A. J., 22, *23*, 35, 36, *48*
Arrhenius, G. O. S., 36, *49*
Aruga, Y., 373, *387*
Arvesen, J. C., 96, *117*, 407, *412*
Aschoff, J., 429, *441*
Ashley, L. E., 42, *48*
Austin, R. W., 96, 97, 114, 116, *117*, *118*, 123, 129, *134*

B

Baba, K., 450, *468*
Backus, R. H., 428, *441*, 454, 457, 461, *466*
Bader, H., 38, *48*
Bainbridge, R., 348, *359*, 440, *441*
Baker, M., 35, *48*
Barber, N. F., 69, *75*
Barret, J., 18, *23*
Bauer, D., 30, 31, 33, *48*
Bayer, M., 451, *466*
Bayliss, L. E., 419, *441*
Beardsley, G. F., Jr., 28, 30, 31, 32, 37, 38, 39, 40, 41, 42, *48*, *49*, 126, 127, *133*, *134*, 228, 229, *235*
Beatty, D. D., 421, *441*
Beebe, W., 454, *466*
Beer, R. M., 231, *235*

Benoit, H., 8, *23*
Berg Olsen, N., 29, 30, 31, 33, 43, *49*
Bernard, F., 454, *466*
Bernard, G. D., 424, *442*
Berreman, D. W., 126, *133*
Berry, D. A., 430, *443*
Berry, S. S., 423, *441*
Bertelsen, E., 451, *466*
Betzer, P. R., 37, *48*
Birge, E. A., 19, 21, 22, *23*
Bityukov, E. P., 458, *466*
Blaxter, J. H. S., 428, *441*
Blinks, L. R., 352, 353, 355, 356, *359*, *360*
Blouin, F., 179, 185, *218*
Boden, B. P., 428, *441*, *442*, 447, 449, 450, 451, 457, 458, 459, 461, 462, 463, 464, *466*, *468*
Bogoroy, B. G., 428, *441*
Boileau, A. R., 96, 114, *117*, 331, *343*
Booth, A. L., 305, *316*
Boyle, R., 452, *466*
Boysen Jensen, P., 371, *387*
Breslau, L. R., 450, *466*
Brody, M., 356, *359*
Brody, S. S., 356, *359*
Bowden, K. F., 253, *255*
Braarud, T., 381, *387*
Brauer, A., 423, 424, 427, *441*
Bravo-Zhivotovskiy, D. M., 125, *133*
Brice, A., 10, *23*
Bridges, C. D. B., 419, 420, 421, *441*
Brown, C. M., 127, *133*
Brown, J. S., 352, *359*
Brown, O. B., 43, *48*, 130, *133*
Brown, W. P., 126, *133*
Brun-Cottan, J.-C., 38, 39, *48*
Bünning, E., 429, *441*
Burkholder, P. R., 371, *387*
Burt, W. V., 41, *48*, 80, 81, *94*

20*

Subject Index

(Numbers in bold type indicate the page on which a subject is treated most fully.)